VIROLOGY MONOGRAPHS

DIE VIRUSFORSCHUNG IN EINZELDARSTELLUNGEN

CONTINUING/FORTFÜHRUNG VON
HANDBOOK OF VIRUS RESEARCH
HANDBUCH DER VIRUSFORSCHUNG
FOUNDED BY/BEGRÜNDET VON
R. DOERR

EDITED BY/HERAUSGEGEBEN VON
S. GARD · C. HALLAUER · K. F. MEYER

8

1971
SPRINGER-VERLAG
WIEN   NEW YORK

SPONTANEOUS AND VIRUS INDUCED TRANSFORMATION IN CELL CULTURE

BY

J. PONTÉN

1971

SPRINGER-VERLAG

WIEN   NEW YORK

ISBN-13:978-3-7091-8260-4   e-ISBN-13:978-3-7091-8258-1
DOI: 10.1007/978-3-7091-8258-1

Spontaneous and Virus Induced Transformation in Cell Culture

By

J. Pontén

The Wallenberg Laboratory, University of Uppsala,
Uppsala, Sweden

With 35 Figures

Table of Contents

I. Introduction

A. Definitions of Transformation in vitro

When normal tissues or organs are explanted to conditions favoring the growth of cells as individual units ("cell culture"), the original cell population undergoes a large variety of modifications. Only a minority of the cells will thrive and multiply and within a rather short period of time, the complex composition of the original explant is replaced by a much simplified one of only a few recognizably different cell types. With most organs fibroblast-like cells survive longest and outgrow other types. This is then a stable state of affairs for many generations. This treatise will not discuss whether this simplification and stabilization represents selection of certain pre-existing cell types or a modification of cells into only a few recognizably different categories; for an excellent review see HARRIS (1964).

Table 1. *Terminology Employed to Describe Transformations in vitro*

Type of transformation	Essential features
Irregular growth	Lack of contact inhibition of cell membrane movement ("ruffled membranes") between juxtaposed cells
Unrestrained growth	Deficient inhibition of the cell cycle (mitosis) in a crowded culture
Infinite growth	Capacity of cells to undergo an infinite number of divisions (formation of established cell lines)

Cells may depart from this typical behavior in numerous ways involving for instance cellular morphology, immunology, chromosomes or metabolism. Such changes have, sometimes rather vaguely, been called "transformations". This is unprecise and the term "transformation" will here be used exclusively to indicate disturbances in cell growth related to neoplasia. Reversible phenotypical alterations, for instance these dependent on a modified medium composition, will not be considered.

Table 1 lists the three different types of disturbances in the regulation of cell growth *in vitro* which have been recognized and defined as transformations in this review. Although they may exist independently they are most often combined.

The Tissue Culture Association has issued a Proposed Usage of Animal Tissue Culture Terms (FEDOROFF, 1967). The recommended terminology has been followed with one important exception. The T.C.A., proposed that 'the term "cell transformation" should be reserved to mean changes induced in the cells by the introduction of new genetic material and that the new genetic material should be specified'. They proposed that persistent changes in 'morphology, chromosome constitution, virus susceptibility, nutritional requirements, proliferative capacity, malignant character, etc.' should be termed culture alteration.

This definition of transformation is obviously inspired by bacterial genetics. Somatic cell genetics is, however, not yet sufficiently developed to meet the requirements called for by an analogy with bacterial transformation. The definition of cell transformation by the T.C.A., read literally, would include the effects of any lytic or non-lytic virus infection, since new genetic material would be introduced into the cells. An artificial division would be created between for instance SV 40 infected cells where there is definite evidence for the permanent presence of newly introduced DNA which is transcribed and translated into "T-antigen", and spontaneously altered mouse cells where there is no such evidence. Both cells have acquired the permanent changes listed in Table 1 and are also tumorigenic after animal implantation. This review accepts that transformation has a different meaning in bacterial genetics and in the study of animal cells, which is in accord with the most common usage of the word.

1. Irregular Growth Transformation
The Lack of Contact Inhibition of Cell Membrane Movement Makes Cells Capable of Moving over each Other to Form Randomly Arranged Multilayers

The basis for this concept is an important observation by ABERCROMBIE and HEAYSMAN (1953 and 1954) on the interaction of normal fibroblasts when they come in close contact *in vitro*. Using time-lapse cinematography and direct observation, these authors found that isolated chick heart fibroblasts showed irregular movements of their cell membrane. When cells approached each other, membrane movement was paralyzed along the line of contact and adhesions were formed. In this way the cells ceased to be randomly oriented and became arranged as regular parallel units. The phenomenon was called contact inhibition of locomotion. A detailed and very informative interference microscope study of living fibroblasts showed the undulating movement (ruffled membrane) to be confined to the lateral cell borders (ABERCROMBIE and AMBROSE, 1958). Ruffling did not seem to be solely responsible for net directional cell movement but was intimately concerned with pinocytosis. Contact inhibition of ruffled membranes explained why cells move centrifugally from an explant and do not arrange themselves as a pile of irregularly overlapping units, as would be expected if there were no restrictions in cell movements (ABERCROMBIE and HEAYSMAN, 1954; CURTIS, 1961). Even if contact inhibition seems to be the most important factor controlling cell movement *in vitro* other phenomena such as contact guidance (HARRISON, 1914; P. WEISS, 1934, 1945, 1958) and possibly also negative chemotaxis play a role. However, ABERCROMBIE and GITLIN (1965) recorded the movement of small groups of chick embryo heart fibroblasts and concluded that contact inhibition rather than chemotactic gradients is the decisive factor in the determination of cell locomotion.

In striking contrast to primary fibroblasts, sarcoma cells were found to have lost their contact inhibition (ABERCROMBIE, HEAYSMAN, and KARTHAUSER, 1957). When they approached each other or normal fibroblasts the ameboid movement of the malignant cell's membrane continued and they moved freely on top of the other cells. This produced a disorderly three-dimensional arrangement which was easily distinguished from the regular two-dimensional pattern of normal fibroblasts.

Contact inhibition and its absence have only been extensively studied in populations of fibroblast-like cells (ELSDALE, 1968). Recent evidence (VAUGHAN and TRINKAUS, 1966) suggests, however, that normal epithelium in culture is also subject to contact inhibition. Whether carcinoma cells *in vitro* display the same lack of contact inhibition as sarcoma cells is unknown.

PONTÉN, WESTERMARK, and HUGOSSON (1969) studied astrocyte-like cells SHEIN, 1965) which form the predominant outgrowth from human brain. Cinematographic recordings and light microscopic observations showed essentially the same features as in fibroblast-like cells. Astrocyte-like cells moved by way of ruffled membranes and were strongly inhibited after mutual contacts had been established. Glioma cells differed by their deficient contact inhibition of cell membrane movements.

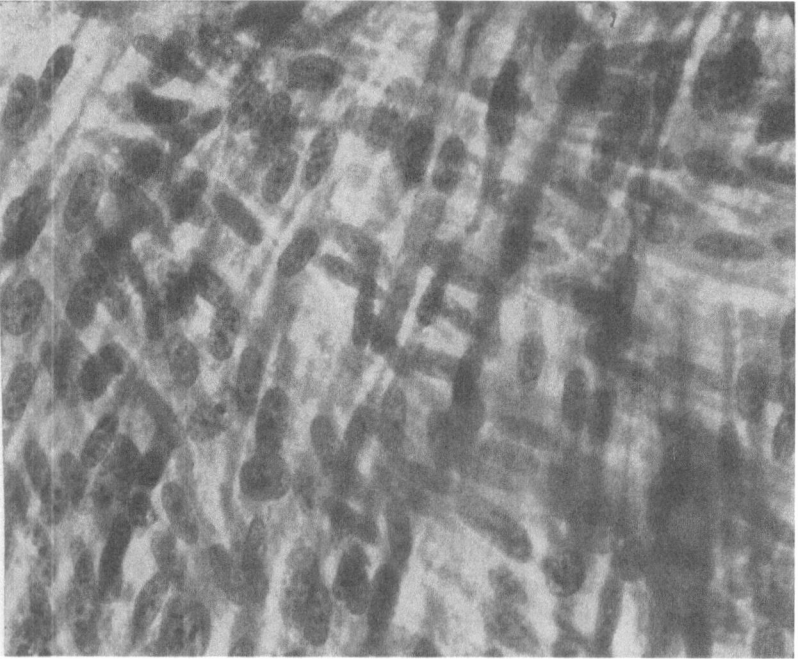

Fig. 1. Normal bovine fibroblasts in a culture subjected to daily medium changes. The cells are arranged in at least two clearly identified layers. Within each the nuclei lie in parallel array. The kind of ordered nuclear overlapping illustrated should not be considered an indication of irregular growth transformation (lack of contact inhibition). May-Grünwald-Giemsa. Appr. magn. × 800

Macrophages (amoebocytes) (CHANG, 1964; VIROLAINEN and DEFENDI, 1967) may also survive long in culture but have not been studied with respect to cell-cell interactions.

An ideal method for measuring the contact inhibition of cell locomotion and membrane movement has not been found. The original and very reliable method of direct observation supplemented by time-lapse cinematography (ABERCROMBIE and HEAYSMAN, 1966) is not well suited for experiments on a large scale. A method has been proposed whereby the growth pattern of large numbers of cells can be evaluated in fixed preparations (ABERCROMBIE and HEAYSMAN, 1954). The number of overlapping nuclei is compared with that expected for a random arrangement of nuclei in the sample area. Later a modification of the original method was introduced where the nuclei were idealized as ellipses rather than circles (CURTIS,

1961). The method avoids the time-consuming and subjective direct observations of living cells under phase contrast but is not satisfactory because it fails to differentiate between a completely disordered arrangement and cell sheets on top of each other but with a normal array within each layer. The latter arrangement is observed among normal fibroblasts, particularly after prolonged cultivation (cf. ABERCROMBIE, 1961), and scores significant nuclear overlapping, but it probably does not reflect any disturbance of the normal contact inhibition (ELSDALE and FOLEY, 1969). Figures 1 and 2 clarify the difference between the two types of multilayering.

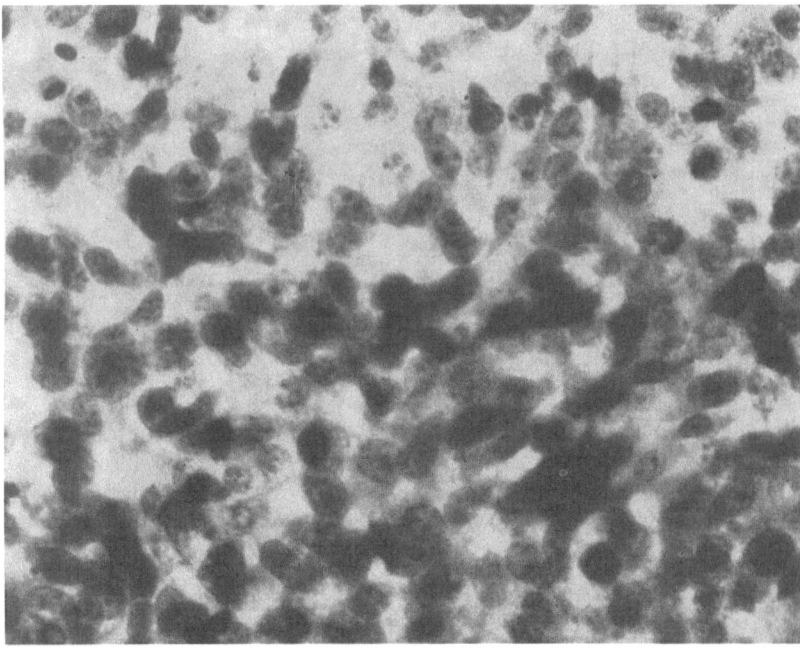

Fig. 2. Bovine fibroblasts after transformation by polyoma virus. The total cell density is about the same as in Fig. 1. The essential difference is the complete randomness by which the transformed cells are distributed in all three dimensions. This appearance is typical of irregular growth transformation (= lack of contact inhibition). May-Grünwald-Giemsa. Appr. magn. × 800

Cells may have such a strong adhesion to their plastic or glass support that they remain anchored there in spite of undiminished membrane activity or they may detach as soon as they reach the upper surface of a neighbouring cell. Both these instances will show no nuclear overlapping in spite of a lost contact inhibition of ruffled membrane movements. In this review, a requirement for the acceptance of an irregular growth transformation will be a documentation not only of an abnormally high cell density and growth in multilayers but also a random arrangement of the cells in all three dimensions of the culture. Loss of contact inhibition of membrane movement in a monolayer is only accepted on cinematographic evidence.

The mechanism by which contacts induce membrane paralysis is unknown. It has been suggested that the surface charge is important (ABERCROMBIE and

AMBROSE, 1962; FORRESTER, AMBROSE, and MACPHERSON, 1962); it could be altered by uncovering the acid groups of mucopolysaccharides (DEFENDI and GASIC, 1963). Attempts to distinguish cells with and without contact inhibition with electrophoresis have only met with partial success (FORRESTER, AMBROSE, and MACPHERSON, 1962; DEFENDI, 1966), but more refined methods (L. WEISS, 1968; SWIFT and TODARO, 1968) may lead to a clearer distinction. An excellent account of the problem is found in a survey article by ABERCROMBIE and AMBROSE (1962), where the suggestion is also made that invasiveness *in vivo* is a reflection of lack of contact inhibition combined with "mobilization". By the latter is meant an abnormal decrease in intercellular adhesiveness which frees cells to escape from their surroundings. A comprehensive review stressing biophysical aspects of the cell periphery has been written by L. WEISS (1967).

2. Unrestrained Growth Transformation

The Loss of Density Dependent Proliferation Restraint Permits the Cell Cycle to Continue at a Cell Density where it is Normally Inhibited

A normal cell growth curve can be divided into three phases:

a) a lag during which there is no increase in number;

b) an exponential period and

c) a post-logarithmic resting phase in which the cell number is constant or only increases slowly.

The last phase is believed to reflect a growth control shared by all normal cells in culture. It is entered at a certain "terminal cell density" expressed as number of cells per unit area of solid substrate supporting the cells. The terminal (saturation) cell density is independent of the size of the inoculum used to start the culture, but differs between different types of cells and is also influenced by the composition of the medium.

Unrestrained growth transformation may be defined as the absence of a post-logarithmic resting phase with lack of cell cycle inhibition. It would theoretically be expected to show up as uninterrupted exponential growth. In practise this is never achieved because all cell populations will eventually decrease their rate of multiplication probably due to the difficulty in ensuring adequate nutrition of the cells which become deeply buried at the bottom of a growing cell mass. In these cases the growth curve gradually flattens out but the lacking inhibition of the cell cycle manifests itself as continuation of DNA synthesis in the post-logarithmic phase.

A flat post-exponential growth curve does not always denote absence of unrestrained growth transformation. It may reflect a balance between cell multiplication and detachment rather than a blocked cell cycle (HAHN, STEWART, YANG, and PARKER, 1968) and is then revealed by the absence of a decrease in the mitotic index or rate of DNA synthesis (MACIEIRA-COELHO, 1967 b, c; NILSSON and PHILIPSON, 1968) (see Fig. 3).

Since all multiplying cell populations eventually will show at least some suppression of cell division, unrestrained growth transformation has to be assessed by careful comparison with normal control cells, which under identical well defined conditions display a definite post-logarithmic growth arrest. The literature on cell cycle inhibition is difficult to interpret partly because these factors have sometimes been neglected. It is not definitely known if unrestrained growth transformation reflects a quantitative or qualitative defect in the cell's growth regulating mechanism.

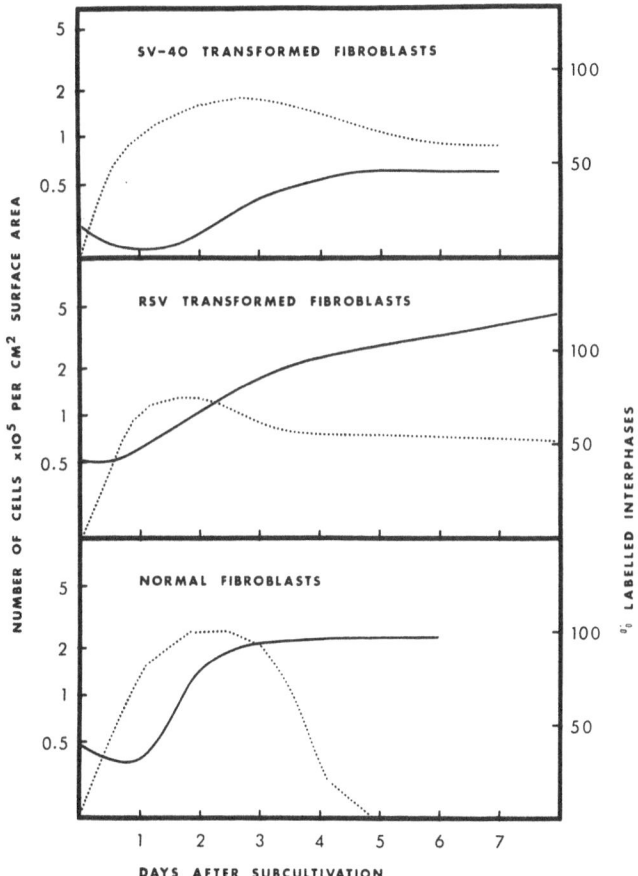

Fig. 3. Examples of unrestrained growth transformation (deficient inhibition of the cell cycle). Cell proliferation is indicated by uninterrupted lines, the dotted curves represent DNA synthesis measured by the fraction of cells capable of incorporating tritiated thymidine during an exposure time of 24 hours. Normal fibroblasts show a density-dependent inhibition of DNA synthesis. The RSV transformed fibroblasts exemplify unrestrained growth transformation with continuing DNA synthesis and progressive increase in cell number. The SV40 transformed fibroblasts exemplify unrestrained growth transformation with continuing DNA synthesis and a flat post exponential growth phase which reflects an equilibrium between rate of detachment and rate of proliferation. Modified from MACIEIRA-COELHO, PONTÉN, and PHILIPSON (1966b) and MACIEIRA-COELHO (1967c). The experiments with virus transformed cells were carried out with embryonic bovine lung fibroblasts

A post-logarithmic cell cycle inhibition was first observed among normal fibroblast-like cells from a variety of species (SIMINOVITCH and AXELRAD, 1963; TODARO, WOLMAN, and GREEN, 1963; STOKER, 1964; LEVINE, BECKER, BOONE, and EAGLE, 1965; MACIEIRA-COELHO, PONTÉN, and PHILIPSON, 1966b; STOKER, SHEARER, and O'NEILL, 1966; BOREK and SACHS, 1966a; MACIEIRA-COELHO, 1967c; EAGLE and LEVINE, 1967). In fibroblast-like cells the rate of DNA synthesis is suppressed to a level of about 10% of that of the exponential phase (LEVINE, BECKER, BOONE, and EAGLE, 1965; NILAUSEN and GREEN, 1965; MACIEIRA-COELHO, PONTÉN, and PHILIPSON, 1966b). Normal adult human glia cells show almost perfect cell cycle inhibition since less than 1% of the population enters DNA synthesis during a 24 hour interval (PONTÉN, WESTERMARK, and HUGOSSON, 1969). Mouse kidney epithelium shows a more pronounced post-logarithmic arrest than the corresponding fibroblasts (WEIL, PÉTURSSON, KÁRA, and DIGGELMANN, 1967). Early passage cells have a higher saturation density than the corresponding phase III (page 15) cells (MACIEIRA-COELHO, PONTÉN, and PHILIPSON, 1966a) and embryonic cells will reach higher densities than those of adult origin (MACIEIRA-COELHO and PONTÉN, 1969). Early passage embryonic fibroblasts are consequently least suitable for contact inhibition studies. This can to some extent be circumvented by the use of low serum concentration in the medium (BÜRK, 1966; FRIED and PITTS, 1968).

Cell cycle inhibited diploid fibroblasts show an overall reduction in the rate of RNA and protein synthesis (LEVINE, BECKER, BOONE, and EAGLE, 1965). 5s RNA, tRNA and rRNA (ribosomal RNA) were depressed to 5% of the maximal rate during exponential growth, whereas the fourth major RNA fraction believed to contain most of the mRNA only was reduced to about 25% (RHODE and ELLEM, 1968).

Most established cell lines (page 20) show unrestrained growth transformation. An exceptional line—3T3—has been isolated by TODARO and GREEN (1963) by serial subcultivations starting at low cell densities. This widely employed, heteroploid line is composed of polygonal "epithelial-like" cells which stop growing once a monolayer has formed at a density of about 50,000 cells per cm² solid substrate. This ideal situation is only obtained under well defined medium conditions. Variants with unrestrained growth transformation will always tend to become predominant and repeated cloning is necessary to maintain the desired property.

For the 3T3 cells the post-exponential cell cycle arrest occurred in the G1 period (NILAUSEN and GREEN, 1965) while the human lung fibroblasts were delayed in G1 or G2[1] (MACIEIRA-COELHO, PONTÉN, and PHILIPSON, 1966b; MACIEIRA-COELHO, 1967c). The proportion of cell delayed in G2 was variable and under most conditions small. The block in the cell cycle is reversible as demonstrated by subcultivation or creation of a defect in the layer of cells (TODARO, LAZAR, and GREEN, 1965; MACIEIRA-COELHO, 1967c).

Several explanations have been offered for the post-logarithmic growth retardation.

a) Non-specific Medium Depletion

With normal fibroblasts and glia cells this mechanism seems to have been excluded because fluid from stationary cultures is fully capable of supporting growth if the cell layer is broken up (PONTÉN, WESTERMARK, and HUGOSSON, 1969). Conversely, even daily medium changes fail to support continuous rapid growth (STOKER, SHEARER, and O'NEILL, 1966; MACINTYRE and PONTÉN, 1967).

---

[1] The cycle is divided in four periods (M = mitotic time; G1 = post-mitotic time without any DNA synthesis; S = DNA synthesis and G2 = pre-mitotic time without any DNA synthesis.

Results with the established 3T3 line have indicated that medium from stationary cultures is fully capable of supporting exponential growth in conformity with the above data from normal cells (TODARO, LAZAR, and GREEN, 1965). This was most convincingly demonstrated by scraping off a strip of cells from stationary monolayers with a Teflon policeman. A wawe of mitosis was seen in the vicinity of the "wound" even in the absence of a medium change indicating that non-specific depletion of the medium could not have been solely responsible for the stationary character of an undamaged cell sheet (TODARO, MATSUYA, BLOOM, ROBBINS, and GREEN, 1967).

HOLLEY and KIERNAN (1968) obtained results at variance with those of TODARO et al. In the hands of the former, depleted serum from stationary 3T3 cells did not support exponential growth and it was concluded that the post exponential resting phase of 3T3 cells was caused by consumption of a non-specific growth factor. The discrepancy has not been explained. 3T3 cells are, however, heteroploid and require repeated cloning to maintain their capacity to enter a distinct resting phase. Variant populations with a different capacity to respond to metabolites contained in the medium may have been selected in the different laboratories.

A very elegant demonstration of density dependent inhibition of proliferation in epithelial cells was provided by ZETTERBERG and AUER (1970). Using primary mouse kidney they took advantage of the fact that islets of epithelial cells are formed which differ considerably in density. In the same monolayer they related the local cell concentration to the fraction of cells engaged in the cell cycle. DNA synthesis decreased sharply in the interval between 10,000 to 50,000 cells/cm². At 120,000 it was virtually zero and the inhibited cells were retained in G1. These results are incompatible with non-specific medium depletion.

b) Cell Cycle Inhibition is Caused by Cellular Release of Factors Inhibiting Cell Division

Stable long range substances diffusely distributed through the entire medium seem unlikely mainly because of the failure of most experimenters to demonstrate an inhibitory effect on multiplying cells of fluid removed from stationary cultures (RUBIN, 1966a; RUBIN and REIN, 1967 and others cf. under a)). Recently YEH and FISHER (1969) observed that 3T3 cells in mutual contact released a dialyzable factor capable of counteracting the stimulation by fresh serum of uridine incorporation into stationary 3T3 cultures (page 13). It is not yet known if this is related to density-dependent proliferation inhibition.

A more likely alternative assumes the existence of short range inhibitory factors which are only effective in the immediate vicinity of a particular cell. They cannot be detected in bulk medium either because of excessive dilution, chemical instability or because they are non-diffusible and perhaps attached to cell membranes or the solid surface used to support the cells.

BÜRK (1966, 1967) has claimed the existence of a cell-associated growth inhibitor, 'anomin', capable of affecting normal cells (and cells transformed by polyoma virus). This substance can hardly be implicated in normal cell cycle inhibition because of its presence in BHK-21 cells, which lack such inhibition

(STOKER, 1964; HOUSE and STOKER, 1966) and its unexplained failure to effect DNA synthesis in spite of the alleged growth inhibition (BÜRK, 1967).

RUBIN and REIN (1967) observed that X-irradiated chicken fibroblasts inhibited proliferation of normal cells if the two kinds of cells were brought within 1 mm from each other but that direct physical contact was not required. These data are compatible with the secretion of inhibitory short range molecules, but the experimental design with cell layers growing on two closely apposed cover slips did not exclude such trivial factors as local medium depletion or drastic changes in the pH.

There is thus no proof of an extracellular substance specifically controlling the proximity dependent growth inhibition in cultures during the post-exponential phase. A recent direct attempt to detect an inhibitor in lyophilized cell cycle inhibited 3T3 cells gave a negative result (SCHUTZ and MORA, 1968b).

c) Cell Cycle Inhibition is Triggered by Close Proximity (or Physical Contact) between Cells without the Transmission of any Extracellular Material

The influence of local cell density required by this hypothesis was observed by LOEB and FLEISHER (1919), FISCHER and PARKER (1929) and FISHER and YEH (1967) in explants or colonies of cells sensitive to cell cycle inhibition. In the dense center division and DNA synthesis were considerably reduced in comparison with the sparse periphery where the cell cycle proceeded without suppression. This is in contrast to cells with unrestrained growth transformation where mitosis is diffusely distributed independently of the local cell density (FISCHER and PARKER, 1929; FISHER and YEH, 1967).

Normal fibroblasts, glia cells and 3T3 cells seeded on top of stationary isogeneic, allogeneic or xenogeneic cells are strongly inhibited (BOREK and SACHS, 1966a; MACINTYRE and PONTÉN, 1967; EAGLE, LEVINE, and KOPROWSKI, 1968; WESTERMARK, 1970). With 3T3 cells this effect was abolished by the insertion of a Millipore filter (pore size 0.45 μ, thickness 25 ± 5 μ) between the two cell sheets strongly suggesting that physical interaction was necessary to trigger cell cycle inhibition (SCHUTZ and MORA, 1968a). More indirect evidence for the importance of direct contact has been provided by STOKER (1964) and LEVINE, BECKER, BOONE, and EAGLE (1965).

Electrophysiological data have shown an intercellular communication among normal cells facilitating flow of small ions and permitting passage of molecules up to a size corresponding to a molecular weight of $10^3 - 10^4$. This communication is interrupted in malignant tumors (cf. LOEWENSTEIN, 1968b). Use of chemical markers has established that enzymatic activity can be interchanged between adjacent cells (SUBAK-SHARPE, BÜRK, and PITTS, 1966; FRIEDMANN, SEEGMILLER, and SUBAK-SHARPE, 1968).

The speculation (STOKER, 1967) that the intercellular transfer of molecules is responsible for cell cycle inhibition has, however, not been experimentally verified.

Viable cells are necessary to trigger cell cycle inhibition. Cells irradiated with non-lethal X-ray doses are still effective whereas lyophilized cells are powerless (STOKER, 1964; SCHUTZ and MORA, 1968b). The results with X-irradiated cells pose a paradox, because such cells have been extensively used as "feeder layers" in other systems where they have enhanced rather than inhibited normal cell growth (PUCK, CIECIURA, and FISHER, 1957).

The proximity related cell cycle inhibition can temporarily be neutralized by a component of fresh serum. If stationary 3T3 cells are exposed to fresh serum, DNA synthesis and division is stimulated in a proportion of the cells which is directly correlated with the serum concentration (TODARO, LAZAR, and GREEN, 1965; BLOOM, TODARO, and GREEN, 1966; TODARO, MATSUYA, BLOOM, ROBBINS, and GREEN, 1967; HOLLEY and KIERNAN, 1968). In this reaction the serum factor is consumed with an estimated half life of 4 hours (TODARO, MATSUYA, BLOOM, ROBBINS, and GREEN, 1967).

A similar temporary stimulation of growth may be observed in stationary normal fibroblasts or glia cells (MACINTYRE and PONTÉN, 1967; MACIEIRA-COELHO, 1967c; PONTÉN, WESTERMARK, and HUGOSSON, 1969). Primary chick embryo fibroblasts have, on the contrary, been reported unresponsive to stimulation by fresh medium (BECKER, 1967) but to consume a serum factor resembling that used up by 3T3 cells at confluency (TEMIN, 1966).

Rabbit lens epithelium kept in organ culture responds to fresh serum with proliferation and cell migration apparently in the same manner as dispersed cells (HARDING, WILSON W., WILSON J., REDDAN, and REDDY, 1968).

Fresh medium perfused at the rate of 59 ml per day per 60 m² culture vessel bottom surface strongly counteracted the proximity dependent cell cycle inhibition of normal embryonic human lung fibroblasts (KRUSE and MIEDEMA, 1965). The population doubling time increased from 1.3 to 5.9 days indicating that even this rapid medium renewal could not completely abolish a retarding influence on cell multiplication in dense cultures.

The serum factor stimulating division in inhibited 3T3 cells coprecipitates with gamma globulin but is apparently not a gamma globulin since it is present in naturally agammaglobulinemic serum (JAINCHILL and TODARO, 1970). Its chemical nature has not been determined.

Initiation of cell division is preceded by 10—20 fold increase in uridine incorporation into RNA starting almost immediately after addition of fresh serum. If this reflects an increased RNA synthesis or only an increased uptake of exogenous uridine was not established. Somewhat later protein synthesis started to rise reaching a peak at 4 hours. DNA synthesis began at 14 hours and was maximal at 20 hours. The mitotic wave reached its peak 30 hours after addition of the stimulatory factor (TODARO, LAZAR, and GREEN, 1965; BLOOM, TODARO, and GREEN, 1966; TODARO, MATSUYA, BLOOM, ROBBINS, and GREEN, 1967). The observations with 3T3 cells have been confirmed in another established mouse line—C3H2K—by YOSHIKURA and HIROKAWA (1968).

Most authors have concluded that fresh serum cannot completely prevent the restraining effects of cell proximity. KRUSE, WHITTLE, and MIEDEMA (1969), on the other hand, suggested that cell-to-cell contacts are unimportant. They were able to overcome density dependent inhibition of mitosis almost completely by intense perfusion of fibroblasts. It was, however, not excluded that the apparently unrestrained growth was not due to the synthesis of an intercellular matrix which prevented effective intercellular membrane contact.

Transformed as well as normal cells require serum for optimal multiplication

(EAGLE, 1955; SANFORD, WESTFALL, FIORAMONTI, MCQUILIKEN, BRYANT, PEPPERS, EVANS, and EARLE, 1956). In most instances the serum requirement seems to be considerably less among transformed than normal cells (STANNERS, TILL, and SIMINOVITCH, 1963; TEMIN, 1966, 1967 c; BÜRK, 1967; JAINCHILL and TODARO, 1970).

In most systems inhibition of mitosis and cell movement coexist and it has often been assumed that contact inhibition of cell membrane movement is also responsible for cessation of mitosis (GOLDÉ, 1962; TODARO, LAZAR, and GREEN, 1965). However, a few recent observations suggest a possible dissociation between the two. MACIEIRA-COELHO (1967 a) found that human fibroblast cells transformed by Schmidt-Ruppin strain of Rous sarcoma virus could attain twice the cell density of controls. When the terminal stage was reached — contrary to expectation — contact inhibition (measured as degree of nuclear non-overlapping) was enhanced among the transformed cells. The interpretation was that Rous virus released the cells temporarily from the normal inhibition of mitosis without effecting a corresponding release from inhibition of cell movement.

A dissociation between inhibition of cell movement and mitosis has been claimed in the established BHK-21 line of embryonic hamster cells. Even under crowded conditions the fibroblast-like cells are, despite high mitotic activity, regularly arranged with little or no overlapping (STOKER, 1964; HOUSE and STOKER, 1966; STOKER and RUBIN, 1967). None of these reports should be regarded as proofs that a dissociation may indeed occur, since they were not supplemented by cinematographic records of cell membrane movements. Even if the precise relationship between unrestrained and irregular growth transformation as defined in Table 1 is still unclear, the two types of growth disturbances can be separated operationally. Irregular growth is reflected in the morphology of cell interrelations, while unrestrained growth is reflected by unabated DNA synthesis and cell proliferation at high cell density under specified environmental condition (Fig. 3).

Although much experimental evidence favors hypothesis c), data are still too contradictory and uncertain to permit any firm conclusions. The situation may also prove to be more complex than coutlined above and the different alternatives may very well be valid in part. CASTOR (1968, 1970) attempted to correlate inhibition of net cell movement, cell surface area and serum concentration on one hand with growth rate on the other. From time-lapse recordings of an established epithelial-like mouse line with density dependent cell cycle inhibition he calculated the average bottom surface of the cells and found a direct correlation to the mitotic index per square cm. The interesting implication from these findings is the possibility that proliferation can be regulated solely by two factors: the serum concentration and the size of the contact area between the cell and its solid substrate. The lower surface of the cell in contact with a solid substrate is assumed to be the active part of the membrane in the uptake of growth stimulatory serum constituents. At any given serum concentration cells will proliferate as long as space permits them to expand sufficiently to reach a critical size. As cells make contact less solid support becomes available and eventually a stage is reached where only few or no cells can stretch sufficiently to reach the necessary size of the bottom area to permit adequate ingestion of growth promoting medium components. This theory then only assigns an indirect role to lateral cell contacts and ruffled membranes. Unrestrained growth transformation is characterized by an absence of correlation between cell surface area and growth rate (CASTOR, 1970).

Hemic cells are normally ameboid and show no contact inhibition of membrane movement. Their multiplication is not specifically repressed by increased cell density. It follows that unrestrained and irregular growth transformation as defined in Table 1 cannot be applied to normal or neoplastic hemic cells in culture. The same considerations apply to suspension cultures.

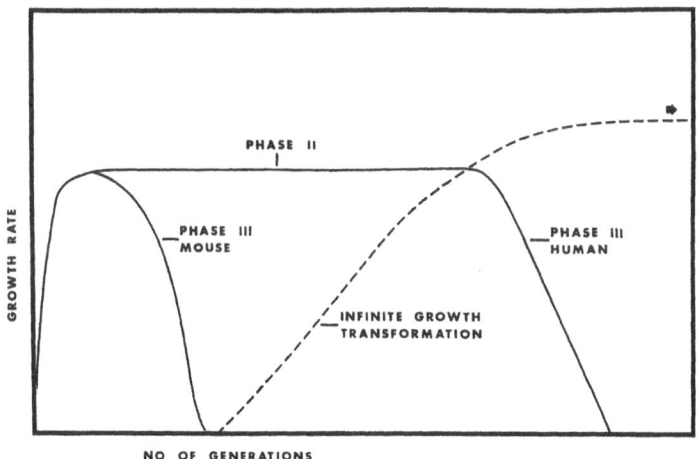

Fig. 4. Infinite growth transformation (formation of established cell lines). Human as well as mouse cells enter a degenerative phase III after a longer or shorter period of time in culture. In the human system this stage is irreversible and no infinite growth transformation will take place. In the murine system a large proportion of cultures will resume rapid proliferation at the end of phase III. This process is irreversible and permanent cell lines capable of infinite multiplication will be obtained. Modified from ROTHFELS, KUPELWIESER, and PARKER (1963) and HAYFLICK (1965)

3. Infinite Growth Transformation

The Formation of Established Cell Lines, Loss of Aging Properties Permits Cells to Undergo Unlimited Division in Culture

Normal fibroblasts — unlike bacteria or certain neoplastic cells—are not thought to be capable of unlimited division; a conclusion mainly from studies with human fibroblast-like cells. SWIM and PARKER (1957) were the first to realize that normal human fibroblasts fail to survive indefinitely *in vitro* in spite of initially excellent growth and a continued supply of adequate medium. This has later been confirmed and expanded by HAYFLICK and MOORHEAD (1961). It was shown that the *in vitro* life history of human fibroblasts (Fig. 4) could be divided into three phases: I = the original outgrowth and migration of explanted cells; II = luxuriant growth; III = declining growth, degenerative cell changes and aneuploidy culminating in death of the cultures. Growth of phase III cells could not be restored by the addition of phase II cells or by frequent changes of medium. It has been suggested, that the phase III phenomenon represents senescence on a cellular level (HAYFLICK, 1966). HAYFLICK (1965) calculated that each clonable embryonic lung cell was endowed with a potential of 50 ± 10

doublings. Cell lines may be preserved in the frozen state, which suggests that the total number of cell doublings rather than' chronological age is the proper measure in the determination of the life span (HAYFLICK, 1965). Normal human fibroblasts, regardless of tissue of originᵛand age of the donor, always seem to enter phase III *albeit* at different passage levels (HAYFLICK, 1965).

Subsequent studies employing normal mouse (TODARO and GREEN, 1963; ROTHFELS, KUPELWIESER, and PARKER, 1963), hamster (TODARO, NILAUSEN, and GREEN, 1963), chicken (PARKER, 1961; PONTÉN, 1970) and bovine (STENKVIST, 1966a; LITHNER and PONTÉN, 1966) fibroblasts have substantiated the findings by HAYFLICK and MOORHEAD (1961); each of these seem to have a finite potential for multiplication *in vitro*. For an assessment of their tendency to spontaneous transformation into established cell lines capable of infinite multiplication and further discussion see part II. In the experiments with bovine fibroblasts, homologous serum was used in the medium, whereas the other systems were grown in heterologous serum. Evidently the presence of foreign serum cannot explain the limited life span *in vitro*.

Infinite growth transformation is, by its very nature, difficult to authenticate. The limit after which a particular cell line can safely be considered to be established is not fixed although seventy total population doublings have been suggested as based on experience with human fibroblasts (FEDOROFF, 1967). This seems to be low in view of the reports of a late 'crisis' with degeneration in long-term SV40 transformed human lines (MOYER, WALLACE, and COX, 1964; SHEIN, ENDERS, PALMER, and GROGAN, 1964; GIRARDI, JENSEN, and KOPROWSKI, 1965), one exceptionally long-lived fibroblast line from normal human skin (PONTÉN and SAKSELA, 1967) and the long survival of bovine fibroblasts in culture (STENKVIST, 1966a). Present data would rather place a safe limit at 150 passages or more than 2 years of continuous serial cultivation. This is, however, based on the behavior of the longest lived normal cells so far studied and an infinite growth transformation can be recognized much earlier if the parent cells normally have a short *in vitro* life span.

As discussed below infinite growth transformation (the formation of established cell lines) is an important result of infection by oncogenic viruses. It can also arise as a spontaneous event.

Certain observations suggest that somatic cells from higher animals also have a limited survival potential *in vivo*. Serial transplantation of normal skin, mammary glands and hemopoietic cells seems impossible for more than a finite number of generations in contrast to results with tumors which can often be transplanted "indefinitely" (for a review see HAYFLICK, 1966).

B. Relationship between Transformation in vitro and Neoplasia in vivo Use of Implantation Tests

Neoplasia in the intact animal can only be described as a relationship between a cell population and the host. An assemblage of cells is regarded as neoplastic if it increases continuously in number within an organism whose physiologic growth control is unimpaired. At the present it is not possible to adequately simulate *in vivo* conditions *in vitro* and conclusions about the neoplastic or "malignant" potential of cells cannot be deduced without implanting cells into animals. At first sight such a procedure would seem to provide a clear-cut answer; however further analysis shows that definite conclusions can be made only with caution.

1. Genetic Incompatibility

Outbred animals contain segregating histocompatibility antigens and an absence of progressive growth after implantation might depend either on histo-incompatibility or inherent lack of neoplastic potential. This limits the usefulness of implantation tests to inbred syngeneic animal strains, which are so far generally available only among certain rodents, particularly mice. Even if such animals are used, the possibility exists that the cells grown *in vitro* will mutate at a histocompatibility locus which may result in a false negative result because of rejection of neoplastic but genetically altered cells (for an excellent review see G. KLEIN, 1963).

2. Tumor-cell Antigens

Cells transformed by viruses (for a review see SJÖGREN, 1965) or spontaneously (SANFORD, MERWIN, HOBBS, FIORAMONTI, and EARLE, 1958) may have more or less easily identified transplantation antigens whose presence may result in a false negative result because of a rejection response.

3. Size of the Implanted Dose of Cells

When cultivated tumor producing cells are implanted into animals there exists a threshold dose below which no takes are obtained. In many cases this threshold is in the range of 10^4-10^6 cells. For practical reasons higher doses than 10^7 can rarely be inoculated and a negative result may simply be due to the use of too few cells. Conversely, when many cells are needed only a small minority may be truely tumor producing whereas the majority may not have this capacity; the positive result does then not reflect the state of the bulk of the cell population. Evidence for considerable heterogeneity in transformed cell populations has been obtained in cloning experiments (SANFORD, 1958; GOTLIEB-STEMATSKY and SHILO, 1964), involving alternate passages of tumor cells *in vitro* and *in vivo* (SANFORD, LIKELY, and EARLE, 1954), direct measurements of DNA content of individual cells (reviewed by CASPERSSON, 1950; KILLANDER, 1966) and chromosome analysis (HAUSCHKA and LEVAN, 1951; MAKINO, 1952). For a general discussion of the importance of heterogeneity in neoplastic cell populations see FOULDS (1954). Of relevance in this respect is an old suggestion that all normal tissues and by inference all populations of cells contain a few neoplastic cells which could develop into tumors given the right set of environmental conditions.

4. Site of Inoculation and Nature of the Tumors Formed

It is well known that embryonic cells after implantation may give rise to nodules, "embryomas", which cannot grossly or sometimes even microscopically be differentiated from true neoplasia. Particularly if such shielded sites as the hamster cheek pouch (BILLINGHAM, FERRIGAN, and SILVERS, 1960) or the anterior chamber of the eye are chosen (GREENE, 1955; ANDRESEN, EVANS, PRICE, and DUNN, 1966), it becomes a difficult task to ascertain whether a given growth is neoplastic or merely a proliferation analogous to that obtained after subcultivation *in vitro*. Transfer and growth of cells in such shielded sites cannot indiscriminately be accepted as proofs of neoplastic properties.

5. Selective Pressure in vitro

Survival and growth *in vitro* usually demands a capacity to attach to and grow on glass or plastic, multiplication at low densities, a capacity to utilize simple media lacking many of the constituents of the *in vivo* environment, a capacity to thrive on heterologous serum etc. Cells particularly fit for these artificial requirements will have a selective advantage, however there is no reason to believe that all tumor cells will have these capacities. A highly neoplastic population of cells may then lose much of its neoplastic potential *in vitro* (EARLE, SHELTON, and SCHILLING, 1950; DAWE, POTTER, and LEIGHTON, 1958). Striking examples of this dissociation have been published by YERGANIAN (1966). One can therefore conceive of a situation where the same oncogenic agent gives rise to malignant tumors *in vivo*, but *in vitro* to a population of cells which is transformed, but only weakly or not at all tumorigenic *in vivo*.

6. Induction of Neoplasia in the Host

If cells transformed by virus *in vitro* give rise to tumors *in vivo* it is possible that primary tumors are induced by released complete virus or by subviral transfer of the information necessary for neoplasia (SVOBODA, 1962). However, a short latent period, particularly in adult immunocompetent recipients, provides strong but not conclusive evidence for transplantation rather than induction. Proof of transplantation will only be obtained by the use of a reliable marker such as a sex difference between inoculum and recipient (PONTÉN, 1962c).

C. The Nature of Transformations Involving the Control of Cell Growth

The irregular growth transformation has been suggested as a necessary condition for invasiveness *in vivo* (ABERCROMBIE and AMBROSE, 1962). In this context it should be remembered that many embryonic cells, trophoblasts, macrophages, blood cells, etc. are normally capable of migrating through the body and in the case of trophoblasts and leukocytes of even passing through blood vessel walls (see WILLIS, 1967). Therefore a perfect correlation between pure irregular growth transformation and malignancy cannot be expected. In many systems as *e.g.* those employed by ABERCROMBIE and HEAYSMAN (1954) in their original experiments, cells which showed a disordered growth pattern were also tumorigenic after implantation *in vivo*. The extensive data of STOKER, MACPHERSON, and collaborators (see part III), where irregular growth is used as a criterion to distinguish cells transformed by polyoma virus from normal cells also point in the same direction. Clones of cells arranged in regular, parallel array did not give rise to tumors after implantation of 10^6 cells, whereas clones of cells growing in a disorderly "random" fashion regularly produced tumors under the same circumstances (STOKER, 1962). Clones of uninfected mouse cells which underwent spontaneous irregular growth transformation were tumorigenic in contrast to those with parallel monolayer arrangement (WEISBERG, 1964). There is no definite exception to the rule that cells showing an irregular growth transformation are tumorigenic. It was, however, observed that cells from chicken sarcomas induced by RSV, which show a pattern indicative of irregular growth transformation after explantation *in vitro*, failed to breed true and establish a transplantable tumor cell line after transfer to histocompatible recipients *in vivo* (PONTÉN, 1964). Of significance in this respect is the observation by DEFENDI, LEHMAN, and KRAEMER (1963) that the inoculation of the established embryonic BHK-21 cell line caused a high incidence of tumors in spite of its regular fibroblastic arrangement *in vitro*.

DEFENDI and LEHMAN (1965) exposed a number of hamster embryo cultures to polyoma virus and obtained many lines which showed irregular growth transformation and which also formed tumors after implantation in test animals. One line, P84, reverted from the irregular criss-cross growth pattern to a regular pattern suggestive of contact inhibition of cell movement. Nevertheless its tumorigenicity remained.

The unrestrained growth transformation has only recently been separated from other related phenomena and no data exist on the neoplastic behavior of cell populations showing this type of transformation in a pure form. It bears a striking resemblance to neoplastic growth *in vivo*, because a permanent failure to respond to physiologic growth control seems to exist in both cases. Established mouse lines with different degrees of unrestrained growth showed a direct correlation between tumorigenicity and loss of density dependent proliferation restraint (AARONSON and TODARO, 1968b).

Established cell lines, which by definition have undergone an *infinite growth transformation*, have been tested during more than two decades for their neoplastic potential in a remarkably extensive and complete study by EARLE, SANFORD, EVANS and collaborators (for a review see SANFORD, 1965). All adequately tested established murine cell lines were tumorigenic (SANFORD, 1965). But sometimes more than a year elapsed before tumors developed and with some lines at least 100 animals whose immunologic defense were depressed by X-irradiation had to be employed. In view of this, the investigator who claims the existence of a non-neoplastic established mouse line will have to bear the burden of proof. A few such claims have been made (*e.g.* J. C. KLEIN, 1966; AARONSON and TODARO, 1968b); however in none of these instances have the tests been sufficiently extensive. JARRETT and MACPHERSON (1968) have calculated that only about 0.001% of the cells of the established BHK-21 hamster line are highly tumorigenic variants.

FOLEY and HANDLER (1957, 1958) tested a number of established cell lines after heterologous inoculation into the hamster cheek pouch. Although the threshold dose varied, all lines were tumorigenic when a minimal quantity of 10^6 cells were administered. These authors attempted a separation of their lines into "benign" and "malignant" on the basis of the size of the threshold dose. Since, however, the differences were quantitative, there seems to be no compelling reason to accept the existence of the qualitative difference between non-neoplastic and neoplastic.

Results of implantation tests with rat and hamster cells (see page 16) also indicate that established lines are capable of forming neoplasms in syngeneic recipients.

*

The word "transformation" without further specification will imply that one or more of the features defined in Table 1 are present. Cultures with all three derangements of growth control seem without exceptions to be tumorigenic after implantation in adequate recipients. This correlation forms a firm basis for the assumption that transformation *in vitro* and induction of tumor formation *in vivo* are phenomena which are fundamentally related to each other.

Much evidence accumulated during recent years indicates that the induction of "new" cellular antigens is a regular sequel of infection and transformation by oncogenic viruses. They persist long after transformation and it has even been suggested that the polyoma virus induced transplantation antigen in mice is indispensable for the maintenance of the neoplastic character (Sjögren, 1964d). The fact that the information for the production of these antigens is transmitted to daughter cells through very many generations has provided a strong impetus for the idea that transformation is genetic. Formal proof for this hypothesis has, however, not been delivered because it is not possible to make a genetic analysis of somatic cells (G. Klein, 1963). The question must still be regarded as open. The present facts can be explained by epigenetic phenomena and the final solution of this important problem will have to await techniques capable of very fine resolution. A promising avenue has been opened by the application of homology tests between messenger RNA in virus transformed cells and viral DNA (Benjamin, 1966a, b; Fujinaga and Green, 1966).

In this review the various types of transformation are defined in operational terms and it is not implied that they are necessarily genetic. As already stressed, neoplastic behavior of cells or transformation in tissue culture are in all likelihood not dependent on any single alteration identical in every case but rather result from the interplay of a host of disturbances some of which may be genetic, while others may be epigenetic and involve abnormal repression or derepression of certain otherwise normal genes.

II. Spontaneous Transformation of Non-haematopoietic Cells

A spontaneous transformation is defined as any of the growth control disturbances of Table 1 occurring *in vitro* in the absence of viruses, hydrocarbons or other known carcinogens.

Spontaneous transformation is only recognized long after explantation and has only been studied in mass populations. Of the three possible types, only the infinite growth transformation has been studied as an independent phenomenon; it strongly correlates with neoplasia *in vivo*.

The important discovery that cells from normal animals will—after a variably long sojourn *in vitro*—produce established cell lines with tumorigenic properties was made independently in two laboratories.

In October, 1938, Gey's group (Firor and Gey, 1945; Gey, Gey, Firor, and Self, 1949; Gey, 1955) started to culture adult rat fibroblasts in plasma clots. Four months to 2.5 years after explantation, 3 individually maintained sublines altered to atypical polygonal or rounded cells with increased frequency of mitosis, many abnormal division figures and decreased cohesiveness. Such altered cultures produced sarcomas when implanted in young non-inbred rats (GEY, GEY, FIROR, and SELF, 1949).

EARLE (1943), EARLE and NETTLESHIP (1943) and NETTLESHIP and EARLE (1943) cultivated cells from subcutaneous adipose tissue of a 100 day old C3H mouse for long periods. They hoped to achieve an *in vitro* malignant transformation with various concentrations of methylcholanthrene but found that both treated and control sublines easily formed established cell lines with the morphologic characteristics of the rat lines of GEY above. These often gave rise to tumors when injected into the mouse strain of origin (NETTLESHIP and EARLE, 1943). A valuable by-product of these studies is the well known L strain of mouse fibroblasts from a subline exposed to methylcholanthrene for 111 days about a year after explantation in October, 1940 (EARLE, 1943).

Both groups made rigorous attempts to exclude accidental contamination by chemical carcinogens (EARLE, 1943; SANFORD, EARLE, SHELTON, SCHILLING, DUCHESNE, LIKELY, and BECKER, 1950) or gamma irradiation (GEY, GEY, FIROR, and SELF, 1949) and could conclude that no known carcinogen including the mammary tumor agent was responsible for the changes which included a capacity to produce transplantable tumors. It is not influenced by the oxygen supply (SANFORD and PARSHAD, 1968).

A considerable number of studies with a large variety of different cells and experimental conditions have confirmed and expanded GEY's and EARLE's early observations. Most of these are of a descriptive nature giving little insight into the mechanisms responsible for spontaneous transformation *in vitro*. One significant, but still unexplained fact is the wide variation in the frequency with which tissues of different animal species tend to transform spontaneously into established cell lines.

A. Frequency of Spontaneous Infinite Growth Transformation in Different Species
Mouse Cells

Pioneering experiments employed cells from a single animal and it could not be excluded that it had been especially prone to spontaneous transformation. This was tested in a large scale experiment by SANFORD, EARLE, SHELTON, SCHILLING, DUCHESNE, LIKELY, and BECKER (1950) where 12 different lines were started from the striated muscle or subcutaneous tissue of five C3H mice. All underwent the infinite growth transformation and, with the exception of one which was lost accidentally after 41 months, produced sarcomas after isologous implantation. Mass cultures of minced mouse embryo tissue (EVANS, PARKER, and DUNN, 1964; WEISBERG, 1964), kidneys from 3 day old C3H mice (ANDRESEN, EVANS, PRICE, and DUNN, 1966), mouse epidermis (SANFORD, LIKELY, and EARLE, 1954) and murine liver (EVANS, HAWKINS, WESTFALL, and EARLE, 1958) were all found to give rise to established tumorigenic cell lines. BARSKI and his group (BARSKI and CASSIGENA, 1963; BARSKI and WOLFF, 1965 and BARSKI, BILLARDON, JULLIEN, and CARSWELL, 1966) reported that C57B1 lung tissue and C3H subcutaneous tissue always achieved infinite growth transformation.

TODARO and GREEN (1963) analysed early events during infinite growth transformation systematically. Minced Swiss mouse embryos were serially cultivated under well defined conditions with transfers of standard sized inocula at 3 or 6 day intervals. The growth rate declined progressively to reach a minimum during the first 15—30 cell generations. In 9/11 tested cultures this process then reversed itself and the growth rate increased until it reached or exceeded the original. The lines could then be propagated infinitely. All but one of the established lines lost at least some contact inhibition of cell movement. The exceptional line—3T3—had no tendency to multilayering and showed a pronounced inhibition of the cell cycle, (*i.e.* no unrestrained growth transformation) at the low density of 5×10^4 cells per cm². The 3T3 line thus provided the first clear example of a dissociation between "infinite growth transformation" on one hand and "irregular" or "unrestrained growth transformation" (Table 1) on the other.

ROTHFELS, KUPELWIESER, and PARKER (1963) obtained independent results similar to those of TODARO and GREEN (1963); cell lines started from skin, lung or kidney of C3H mice often underwent infinite growth transformation. Twelve of fifteen were found to be tumorigenic but the other three were only tested in a few animals so that the authors concluded that all established mouse lines would probably eventually become tumorigenic. The establishment of a few cell lines was observed in detail and these followed the same pattern of declining growth followed by rapid proliferation described by TODARO and GREEN (1963).

Hamster Cells

Whereas murine cells unequivocally have a very high frequency of spontaneous infinite growth transformation, hamster cells have given diverse results.

TODARO, NILAUSEN, and GREEN (1963) studied 7 strains originating from embryonic Syrian hamster by the techniques used in the mouse cell experiment described above (TODARO and GREEN, 1963). The hamster cultures followed the same initial course as the mouse; but then failed to undergo an infinite growth transformation.

From earlier data, however, it is clear that established hamster cell lines may be obtained spontaneously. One example is the widely studied BHK-21 line and its clonal derivative C13 derived from pooled baby Syrian hamster kidney (MACPHERSON and STOKER, 1962; STOKER and MACPHERSON, 1964). Other established Syrian hamster cell lines with oncogenic properties have been described by COOPER and BLACK (1963); GOTLIEB-STEMATSKY and SHILO (1964); DEFENDI and LEHMAN (1965); TSUDA (1965); YAMANE and TSUDA (1966); GOTLIEB-STEMATSKY, YANIV, and GAZITH (1966); VESELÝ, SVOBODA, and DONNER (1966) and by DIAMOND (1967a). The latter describes a two-year study of 23 primary embryo cultures. Two of these underwent spontaneous infinite growth transformation and also became tumorigenic.

DIAMANDOPOULOS and ENDERS (1966) maintained a number of cultures from pooled Syrian hamster embryos, apparently indefinitely. These lines, after a phase of slow growth, resumed rapid proliferation necessitating subcultivation at least once a week. Further evidence for a high rate of spontaneous transformation was obtained by SANFORD and HOEMANN (1967) who found that 8/8 Syrian embryonic hamster lines underwent spontaneous infinite growth transformation and also became neoplastic. YERGANIAN, LEONARD, and GAGNON (1961) have produced a number of established lines from normal tissues and YERGANIAN

(1966) has concluded that the majority of Chinese hamster cells develop into established lines.

One reason that hamster cell lines sometimes fail to become established may be that medium is deficient. The survival and proliferative capacity of hamster cells can be profoundly enhanced by bovine serum albumin. In this way TODARO and GREEN (1964c) prolonged the *in vitro* life span from about 10 to 100 passages. Although they lacked clear documentation, they stated that there had been no "infinite growth transformation" even after this high number of passages. MATSUYA and YAMANE (1968) found 100% "infinite growth transformation" and tumorigenicity in hamster lines carried in bovine serum albumin fortified medium.

In summary, embryonic hamster cells have a definite, and in certain hands, high tendency to undergo a spontaneous infinite growth transformation. However, with standard media, which probably are inadequate for hamster cells, the tendency may be lower than for murine cells.

Rat, Monkey, Rabbit, Goat, Pig, and Marsupial Mice Cells

A fairly large number of spontaneously formed established cell lines have been reported from rat and monkey tissue (GOLDBLATT and CAMERON, 1953; GEY, 1955; LEVAN and BIESELE, 1958; HOPPS, BERNHEIM, NISALAK, TJIO, and SMADEL, 1963; CHAPIN and DUBES, 1964; KROOTH, SHAW, and CAMPBELL, 1964; FERGUSON and TOMKINS, 1964; VESELÝ, DONNER, and KOCEROVÁ, 1968; SATO, NAMBA, USUI, and NAGANO, 1968; SHARON and POLLARD, 1969). A few instances of spontaneous transformation have also been noted in rabbit, goat and pig cells (HAFF and SWIM, 1956; MADIN and DARBY, 1958; RUDDLE, 1961; HARRIS, 1961; VARON, RAIBORN, SETO, and POMERAT, 1963; VALENTI and FRIEDMAN, 1968). MOORE and UREN (1966) have described established lines from marsupial mice. A recent report of four lines out of 18 attempts from rabbit kidney indicates a semistable character of this species (CHRISTOFINIS and BEALE, 1968).

Chicken Cells

Embryonic chicken cells have also been studied rather extensively (HARRIS, 1957; PARKER, 1961) and found to degenerate irreversibly without forming established lines. This was confirmed with techniques similar to those of TODARO and GREEN (1963) by PONTÉN (1970). Ten individual 9 day old embryos were studied. 10^6 cells were seeded for each passage into 50 mm Falcon Petri dishes and subcultivations were made every 3rd day. The growth rate declined after 16—20 transfers and no lines survived beyond passage 25—30. The cultures were routinely maintained on Eagle's minimal essential medium with 10% calf serum (EAGLE, 1959) and addition of bovine serum albumin failed to increase their life span.

An apparent exception to the rule that chicken fibroblasts do not form established cell lines *in vitro* is the famous heart fibroblast line of Carrel which reportedly was kept in continuous culture for 34 years before it was intentionally discarded (CARREL, 1912, 1914; EBELING, 1913, 1922; PARKER, 1961). HAYFLICK (1965) scrutinized the original protocol of Carrel and observed that unfiltered chicken extract was added at frequent intervals. He considered it likely that viable cells were included in this extract and that Carrel's observation of a spontaneous established chicken line is not acceptable. The entirely negative results of the more systematic investigations show that avian cells do not spontaneously form established lines *in vitro*.

Bovine Cells

Fibroblasts from fetal or newborn bovine lung have been studied rather extensively in our laboratory. These are carried in homologous serum. Thirteen different lines have been systematically transferred by 1:2 splits. Although the lines showed unusually long survival with maximum of 80—90 transfers, they all eventually degenerated and were lost (STENKVIST and PONTÉN, 1964; STENKVIST, 1966a). At least one definite instance where an established line developed spontaneously from normal adult bovine kidney is, however, on record (MADIN and DARBY, 1958) with appropriate chromosome verification (NELSON-REES, KNIAZEFF, and DARBY, 1964).

Human Cells

Next to mouse cells, human cells have been the most intensively studied. Fibroblast-like cells from the most varied sources including embryos, newborns, young and very old adults (SWIM and PARKER, 1957; HAYFLICK and MOORHEAD, 1961; FERGUSON and WANSBROUGH, 1962; TODARO and GREEN, 1964c; HAYFLICK, 1965; PONTÉN and SAKSELA, 1967) have invariably failed to form established cell lines, in spite of the varied techniques employed by the different investigators. Certain isolated instances of spontaneous transformations (CHANG, 1954; BERMAN and STULBERG, 1956; BERMAN, STULBERG and RUDDLE, 1957; WESTWOOD, MACPHERSON, and TITMUSS, 1957) have been criticized because proper identification techniques were not used (ROTHFELS, AXELRAD, SIMINOVITCH, McCULLOCK, and PARKER, 1959; DEFENDI, BILLINGHAM, SILVERS, and MOORHEAD, 1960; BRAND and SYVERTON, 1960, 1962). Claims that certain organs give rise to established cell lines could not be verified, see the review of HARRIS (1964). It is by now clear that contamination by HeLa, L cells or other similar established laboratory lines have been responsible for many of these reports and that rigorous precautions must be taken to avoid the accidental introduction of established cell lines into experimental cultures (GARTLER, 1968).

Our own observations (PONTÉN, unpublished) include 35 biopsies derived from normal uninvolved skin of children and adults bearing benign or malignant mesenchymal tumors. All specimens gave rise to fibroblastic cell strains, which could be serially cultivated during 20—80 passages. None developed into an established line. In addition, 6 lines derived from kidney, lung or synovial tissue failed to achieve the infinite growth transformation. Ten lines from normal glia have not undergone infinite growth transformation (PONTÉN and MACINTYRE, 1968).

It is possible that human amnion is prone to infinite growth transformation. FOGH and LUND (1957) and HAYFLICK (1961) have described the apparently spontaneous appearance of rapidly growing epithelial-like cells capable of indefinite propagation in cultures of full term amnion. In the latter case, at least, the transformed cells appeared originally as small foci and were heteroploid and the possibility exists that they may have resulted from cell contamination, a

suspicion strengthened by the presence of type A glucose-6-phosphate dehydro-
genase. This form of the enzyme has only been found in Negroes and almost
certainly indicates contamination by HeLa cells (which originally were derived
from a Negro woman) in Hayflick's line as well as a large number of established
lines masquerading under different designations (GARTLER, 1968). Larger series
of well controlled experiments seem necessary to assess the true potential of
human amnion cells for infinite growth transformation.

If only well controlled experiments are considered it seems safe to conclude
that cultures from human solid tissues (lymphoid organs and peripheral blood
are discussed in Part III) have very little, if any tendency to undergo spontaneous
infinite growth transformation.

B. Biological Significance of Inter-species Differences

Fibroblastic cells seem to fall into two major classes with respect to their
tendency to spontaneous infinite growth transformation. A. Stable species: man,
chicken, cow and B. Unstable species: mouse, rat and hamster. These differences
are correlated with the response to carcinogens *in vitro*. For instance, rodent
cells exposed to Rous sarcoma virus readily form established lines, whereas
human, bovine and chicken cells, in spite of showing unrestrained growth transformation, fail to become infinitely propagable (PONTÉN and LITHNER, 1966).
Certain facts suggest that the inherent stability also influences the response
in vivo. Mice are curiously prone to develop neoplasms even after such innocuous
procedures as the implantation of plastic films made of a chemically non-carci-
nogenic material (OPPENHEIMER B. S., OPPENHEIMER E. T., and STOUT, 1948;
OPPENHEIMER B. S., OPPENHEIMER E. T., and STOUT, 1952). In this case the mere
introduction of a sufficiently large foreign body seems enough to produce a
local sarcoma. Human patients have been subjected to apparently analogous
procedures in modern surgery where nails, plastic vessels, artificial heart valves
etc. have been implanted without untoward effects. This difference in response
may well be related to the different inherent stability of the mesenchymal cells.

Man, fowl and cattle have a considerably longer life span and are bigger
than rodents. This means that the average number of cell divisions per individual person or animal probably is larger than in a mouse or a hamster. One may
therefore speculate that the development of stability is an important step in the
evolution of large body size and longevity.

C. Chromosome Analysis of Established Cell Lines

There have been several reviews of chromosome studies of different established
cell lines (LEVAN, 1958; HSU, 1961; HARRIS, 1964).

A fundamental question is whether or not established lines of mammalian
cells can maintain a normal diploid chromosome complement.

Early observers noted that established cell lines derived from normal tissues
of mouse (HSU and KLATT, 1958; ROTHFELS and PARKER, 1959), rat (LEVAN and
BIESELE, 1958) or hamster (FORD and YERGANIAN, 1958; TJIO and PUCK, 1958;
FORD, 1959) were heteroploid and it was postulated that heteroploidy was necessary for establishment. Strong indirect support for this view was provided by

HAYFLICK and MOORHEAD (1961), who claimed that human strains, which regularly failed to undergo infinite growth transformation, remained diploid until the end of their finite life time *in vitro*. The persistence of diploidy was later shown to be only relative by SAKSELA and MOORHEAD (1963) and YOSHIDA and MAKINO (1963) who observed chromosome breaks and rearrangements during the last few passages before their cultures ceased to divide.

It is by now quite certain that nearly all lines which have undergone spontaneous infinite growth transformation are karyotypically abnormal. The possible existence of truly diploid but indefinitely propagable cell lines has been discussed by KROOTH, SHAW, and CAMPBELL (1964) in connection with the description of a rat cell strain of their own which was over 85% diploid. The authors concluded that among the numerous lines which had been established there were only 5, including their own, which seemed persistently and predominantly euploid (YERGANIAN and LEONARD, 1961; MACPHERSON, 1963; BASRUR and GILMAN, 1963; PÉTURSSON, COUGHLIN, and MEYLAN, 1964). However, any apparently diploid established line can only be provisionally accepted as such because it is not possible to rule out the presence of chromosome changes not resolved in the light microscope. Established diploid lines such as BHK-21 also have a high tendency to throw off heteroploid sublines (DEFENDI, LEHMAN, and KRAEMER, 1963).

It has not been possible to determine if there is a causal relationship between aneuploidy and spontaneous formation of established lines. In the two studies which bear most directly on this question the emergence of the infinite growth transformation was followed by the criterion of increased growth rate. TODARO and GREEN (1963) found that the growth rate in their mouse lines began to increase before the chromosome number became aneuploid and they concluded that the drastic rearrangements seen in old established lines are not essential for the initial establishment. However, they found a gross shift from normal to mixoploidy only a few passages after the upturn of the growth rate. On the other hand, ROTHFELS, KUPELWIESER, and PARKER (1963) found that the increased growth rate signifying establishment of their C3H mouse lines coincided exactly with the appearance of aneuploid metaphases. Fetal calf serum compared to horse (fetal and postfetal) or calf serum delayed the appearance of chromosome alterations and also tumorigenicity (EVANS, JACKSON, ANDRESEN, and MITCHELL, 1967; MITCHELL, ANDRESEN, and EVANS, 1969). There does then seem to be a close temporal association between visible aneuploidy and increased growth in murine cells. Whether this close association is peculiar to mouse cells remains to be determined.

D. The Significance of Genetic Heterogeneity in Established Cell Lines

A shift to heteroploidy creates a genetically heterogenous population. The heterogeneity is dynamic and selfperpetuating. KILLANDER (1965, 1967) identified individual cells of an established mouse line (CCRF-1210) of neoplastic origin by time lapse cinematography. The post-mitotic DNA determined by microspectrophotometry differed between the two sister cells of a pair indicating an unequal distribution of DNA at cell division. This explains the wide spread in chromosome number (HAUSCHKA and LEVAN, 1951, 1953; LEVAN and HAUSCHKA, 1952; MAKINO, 1952; LEVAN, 1956; HSU and KLATT, 1958) and DNA content

(CASPERSSON and SANTESSON, 1942; CASPERSSON, 1950; LEUCHTENBERGER C.,
LEUCHTENBERGER R., VENDRELY C., and VENDRELY R., 1951; MOBERGER, 1954;
CASPERSSON, FOLEY, KILLANDER, and LOMAKKA, 1963; KILLANDER, 1966) between
individual cells of a tumor *in vivo* or an established cell line *in vitro*.

The uneven distribution of DNA probably causes a constant reshuffling of
genes. The existence of two genetically identical cells in an established line or a
heteroploid tumor may be as unlikely as the probability of finding two identical
individuals within a species where constant genetic reassortment is achieved by
sexual reproduction.

Even if clones are started from individual cells, heterogeneity is rapidly
recreated. Clones of CCRF-1210 display the same spread in dry mass and DNA
content as the original mass population (CASPERSSON, FOLEY, KILLANDER, LAZA-
RUS, SAITO, and ZETTERBERG, unpublished). Early after cloning of BHK-21
hamster cells, the clones will differ with respect to sensitivity to infection by foot-
and-mouth disease virus. However, after only a few passages they all converged
towards the level of sensitivity of the original population (DIDERHOLM, unpublished). More indirect evidence for heterogeneity is provided by clone C13 of the
BHK-21 line, which shows widely different chromosome numbers and oncogenicity
depending on the selective pressure (DEFENDI, LEHMAN, and KRAEMER, 1963). The
antigenic heterogeneity of established aneuploid mammalian lines was discussed
by FRANKS (1966). He observed rapid and frequent changes from antigen-positive
to antigen-negative and *vice versa* (blood group antigens were studied) in cloned
populations.

Homogenous clones can only be produced if genetic information is distributed
equally among all progeny, a condition expected only in stable euploid cell lines
with a finite life span. Clones of established lines only have a limited usefulness.

A consequence of genetic heterogeneity is the capacity to withstand an adverse
environment by selection; new variants able to cope with extreme changes are
constantly created. This may be why heteroploidy and indefinite population
survival are correlated.

Non-established cell lines are preferred for studies of virus induced transformation and growth control because of their genetic homogenity.

E. Cytology of Spontaneously Established Cell Lines

The lack of generally accepted criteria for the detailed appearance of normal
cells in culture, apart from such crude descriptions as "fibroblast-like" or "epithelial-like", makes it virtually impossible to correlate morphology with infinite
growth transformation. This difficulty is further enlarged by the plastic response
cells show to environmental alterations.

Nevertheless, cell lines established from mice display abnormal mitoses,
nuclear blebs and pleomorphism. In addition fibroblastic cells often lose their
normal elongated shape and assume a compact polygonal "epithelial-like" form
(SANFORD, EARLE, SHELTON, SCHILLING, DUCHESNE, LIKELY, and BECKER,
1950; GEY, 1955; TODARO and GREEN, 1963). These changes are usually gradual
and do not appear in defined foci—a feature often reported in conjunction with
suspected or proven accidental cell contamination (PARKER, 1955; BERMAN and

STULBERG, 1956; HENLE and DEINHARDT, 1957; McCULLOCH and PARKER, 1957; BERMAN, STULBERG, and RUDDLE, 1957; PARKER, CASTOR, and McCULLOCH, 1957; SALK and WARD, 1957; WESTWOOD, MACPHERSON, and TITMUSS, 1957; DREW, 1957; PARKER, 1958).

Pleomorphism, abnormal mitosis, increased nucleocytoplasmatic volume ratio and other cytological alterations have never been systematically studied during the infinite growth transformation. Such a study might reveal a surprisingly good correlation, because cellular "atypia" is a common denominator of most malignant neoplasias.

F. Irregular Growth Transformation in Established Cell Lines

In a classical study ABERCROMBIE, HEAYSMAN, and KARTHAUSER (1957) showed that the murine sarcomas, S-37 and S-180 lacked contact inhibition of cell movement. Later 4 additional mouse tumors (ABERCROMBIE and KARTHAUSER cited by ABERCROMBIE and AMBROSE, 1962), were shown to conform with the same abnormal pattern, however, the locomotion pattern of spontaneously established cell lines has only been examined to a limited extent without measurements of membrane movement, cell overlapping etc.

The majority of the descriptions and illustrations of authentic established cell lines suggest at least a partial lack of normal contact inhibition. GEY (1955) described sublines of his rat fibroblast line and stressed their decreased intercellular cohesiveness and SANFORD, EARLE, SHELTON, SCHILLING, DUCHESNE, LIKELY, and BECKER (1950) and SANFORD, LIKELY, and EARLE (1954) remarked on the presence of cell "overlaps" in their established mouse lines.

The first clear demonstration that irregular growth is not an obligate feature of established lines was provided by the 3T3 mouse line of TODARO and GREEN (see page 21). DEFENDI, LEHMAN, and KRAEMER (1963) have also found that clone C13 of the established hamster BHK-21 line grows as regular parallel units without significant overlapping.

G. Unrestrained Growth Transformation in Established Cell Lines

Much of what has been said about the irregular growth transformation also applies to the unrestrained growth transformation. Thus, descriptions of established lines have often included high mitotic rates under crowded conditions. The only clear example of a lack of unrestrained growth transformation, in spite of infinite growth transformation, is provided by the 3T3 line in which DNA synthesis is curtailed at a rather low cell density (NILAUSEN and GREEN, 1965).

H. Nature of Spontaneous Infinite Growth Transformation Induction or Selection

Two hypotheses naturally present themselves for the mechanism of the spontaneous transformation *in vitro*. The *inductive hypothesis* assumes that all cells put into culture are normal; unknown factors *in vitro* transform certain cells directly. Accordingly, it should be possible to devise culture methods which will not lead to any transformation (PUCK, 1959). The *selective hypothesis* assumes that transformed cells already exist *in vivo* but are kept in check by controlling mechanisms. *In vitro* they are released from this control and become dominant due to the selective advantage of their capacity for infinite multiplication.

The long latent periods that precede visible morphologic alterations and attainment of malignant properties in rodent cultures has suggested that the basis for the transformation is induction rather than selection (see SANFORD, 1965, for a review).

TODARO and GREEN (1963) made a critical attempt to assess the significance of the latent period. They eliminated the possibility that any cells of the rapid rate of multiplication characteristic of the established lines, had been present in the original explant and also showed that the multiplication of the transformed cells was not inhibited by an excess of untransformed cells. This supports the inductive rather than the selective hypothesis. On the other hand, GOTLIEB-STEMATSKY, YANIV, and GAZITH (1966) cultured pooled hamster embryo cells diluted to a point where distinct colonies, presumably derived from single cells, could be isolated. Four of 65 colonies could be carried in serial passage and three of these produced tumors when inoculated into hamsters (the fourth was accidentally lost after only limited testing). At the same time, cells carried as mass cultures failed to become tumorigenic. This suggests that hamster embryos contain a certain proportion of cells which have the capacity of infinite growth *in vitro* and are also potentially tumorigenic and that when these are in close contact with normal cells either *in vitro* or *in vivo* they may not be capable of expressing this potential. Such a factor may help to explain the diverse opinions on the frequency of spontaneous transformation among hamster cells (page 22).

Indirect but strong evidence that latent tumor cells exist in normal mice is provided by experiments with intraperitoneally placed Millipore filter chambers, which exclude passage of cells but allow a free exchange of extracellular fluid. If normal connective tissue is placed in these, it will eventually be able to produce sarcomas after syngeneic transplantation (SHELTON, EVANS, and PARKER, 1963). The most likely interpretation is that preexisting latent tumor cells express their neoplastic potential when partially removed from direct contact with neighboring cells.

I. Antigenic Changes in Established Cell Lines from Normal Tissues

The antigenic changes associated with the spontaneous formation of established cell lines has been little studied in comparison to those of virus-induced transformation. Tests have only been made for transplantation antigens.

SANFORD and her collaborators tested established murine lines (SANFORD, MERWIN, HOBBS, and EARLE, 1959; SANFORD, MERWIN, HOBBS, YOUNG, and EARLE, 1959) in their C3H strain of origin and found in each a more or less pronounced rejection response. Therefore the spontaneous infinite growth transformation *in vitro* seems to be accompanied by new transplantation antigens (McCARTHY, 1968), but because of the long elapsed period between the origination of the tumorigenic line and the test, alternative explanations should be considered: Strains of mice and of cells can mutate at histocompatibility loci (see review by G. KLEIN, 1963), and the observed antigenic differences may not be related to the transformation phenomenon *per se*.

DIAMOND (1967a) immunized adult hamsters with cells from 2 spontaneously formed established hamster lines designated Nil-1 and Nil-2. She gave 3 courses of $8-10 \times 10^6$ irradiated (4,000 r) cells intraperitoneally. Subsequent subcutaneous challenge with Nil-1 cells failed to reveal any evidence for a transplantation antigen. BARSKI, MANDAL, BELEHRADEK, and BARBIERI (1967) failed to demonstrate transplant rejection after immunization with C57Bl mouse lung lines which had undergone spontaneous infinite growth transformation and were more than 3 years old.

In summary, infinite growth transformation is not a normal or inevitable tissue culture event. Chicken, bovine and human cells almost always show a finite life span *in vitro* in spite of a long preliminary phase of luxuriant growth and deliberate attempts to create ideal culture conditions. Rodent cells and particularly those of murine origin show a high frequency of spontaneous infinite growth transformation. Established lines invariably seem to be neoplastic and most if not all are heteroploid and genetically unstable. Infinite growth transformation is usually but not always combined with unrestrained and irregular growth transformation.

The mechanism behind the spontaneous infinite growth transformation is unknown. Certain data suggest that it may be caused by the existence of latent tumor cells *in vivo*, which after explantation *in vitro* are released from restraining influences.

III. Spontaneous Lymphoblastoid Transformation of Human Lymphoid Tissue or Peripheral Blood

In the many types of normal human tissues studied, the cells have almost always been reported to cease to divide and degenerate after a variably long period of active growth (see page 24).

Experience during the last few years, with human lymphoid cells from normal individuals indicates that these are exceptional in readily forming indefinitely perpetuable lines *in vitro*. The phenomenon will be referred to as lymphoblastoid transformation. It is included in this review as a form of infinite growth transformation (Table 1) and because of its possible relation to a hypothetical human leukovirus.

The conditions under which lymphoblastoid transformations have occurred have differed, mainly depending on whether peripheral blood or hematopoietic tissue was used. Since persistent growth of lymphoid blast cells has been a common denominator, the subject will be treated as a unified concept. It should not be confused with the induction of a blast-like appearance and mitosis in peripheral blood lymphocytes after stimulation by phytohemagglutinin as first described by NOWELL (1960). There is no evidence that Nowell's phenomenon involves a transformation in the sense of this review.

Lymphoblastoid transformation was originally observed in a small proportion of bone marrow cultures from children mainly suffering from acute leukemia (BENYESH-MELNICK, FERNBACH, and LEWIS, 1963) and in cultures of tumor tissue from a patient with Burkitt's lymphoma (EPSTEIN and BARR, 1965). Lymphoblastoid transformation was almost without exception absent in material from patients without malignant disease (BICHEL, 1952; BENYESH-MELNICK, FERN-BACH, and LEWIS, 1963; TROWELL, 1965; MOORE, ITO, ULRICH, and SANDBERG, 1966; CLARKSON, STRIFE, and DE HARVEN, 1967; PONTÉN, 1967, POPE, 1967). This was interpreted to indicate that the established lymphoblast lines took their origin from neoplastic cells. More recently, however, it has been shown that cultures of lymph nodes, spleen or peripheral blood undergo lymphoblastoid transformation regardless of whether the donor has a malignant condition or not (MOORE, GERNER, and FRANKLIN, 1967; PONTÉN, 1968; GERBER and MONROE, 1968; NILSSON, PONTÉN, and PHILIPSON, 1968). No consistent qualitative dif-ference has been documented between established lymphoid cell lines from normal or neoplastic lymphoid tissue although MOORE and McLIMANS (1968) have suggested that certain differences may exist.

A. General Description of Lymphoblastoid Transformation

1. Lymphatic Tissue and Bone Marrow

A long and intimate cell-cell contact seems to be a prerequisite for a high rate of lymphoblastoid transformation. This has best been accomplished by a modified Trowell organ type of culture (TROWELL, 1954, 1959; JENSEN, GWATKIN, and BIGGERS, 1964) where the original explant is preserved intact for several months on top of a grid placed in a Petri dish (Fig. 5). In this simple system normal human lymph nodes almost regularly form established lines (PONTÉN, 1968; NILSSON, PONTÉN, and PHILIPSON, 1968) in contrast to the low rate of success obtained with dispersed tissue (JENSEN E., KOROL, DITTMAR, and MEDREK, 1967). The process can be divided into three stages:

a) A primary lymphoid cell production phase lasting 2—5 weeks during which decreasing numbers of small lymphocytes and larger lymphoid elements are released into the medium. Some of these survive for a short time and display such typical lymphoid behavior as peripolesis and emperipolesis (see below). Eventually, however, the number of lymphoid cells decreases almost to zero and the culture gradually passes over to the second stage.

b) A fibroblast production phase which partially overlaps the preceding and succeding stages. It is characterized by growth of fibroblast-like cells individually similar to those produced by other human organs, but sometimes arranged in an irregular manner suggestive of some impairment of normal contact inhibition of cell movement (PONTÉN, 1967). These enter a degenerative phase (phase III, page 15) and show no tendency to infinite growth—typical features of human fibroblasts of any normal origin (HAYFLICK and MOORHEAD, 1961). Fibroblast production lasts until a maximum of about 40 weeks after explantation if Eagle's medium and calf serum are used as nutrients.

c) A secondary lymphoid cell production phase commencing 9—18 weeks after explantation. This is the "lymphoblastoid transformation" and it is charac-terized by production of dividing cells morphologically resembling lymphoblasts, plasmoblasts or "immunocytes" (ANDRÉ, SCHWARTZ, MITUS, and DAMESHEK, 1962). These readily form established cell lines which produce various immuno-globulins. This lymphoblastic transformation occurs in a large proportion if not all normal lymph nodes (PHILIPSON and PONTÉN, 1967; NILSSON, PONTÉN, and PHILIPSON, 1968 and unpublished) from adults.

A few lines established from non-malignant human spleens have shown the same development as lymph nodes (SINKOVICS, SYKES, SHULLENBERGER, and HOWE, 1967; LEVY, VIROLAINEN, and DEFENDI, 1968; NILSSON and PONTÉN, unpublished). In cultures of leukemic bone marrow, where the number of lymphoid cells is small, the primary lymphoid cell production phase was difficult to observe. The fibroblastic and secondary lymphoid cell production phases seemed, however, to be identical to those described above (BENYESH-MELNICK, FERNBACH, and LEWIS, 1963).

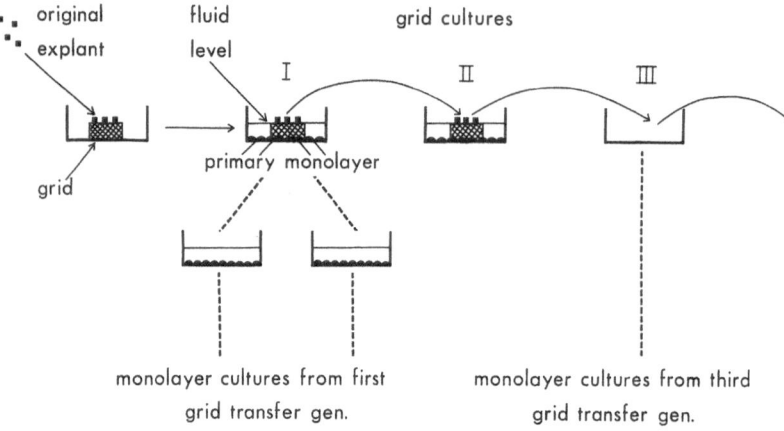

Fig. 5. The modified Trowell type of culture method used to produce established lymphoid cell lines from human organs. Small pieces of tissue are kept intact on top of stainless steel grids covered by tissue paper or gelatin foam. They will shed cells according to a fixed scheme with three phases described in the text. The third — lymphoblastoid transformation — requires the original tissue to be kept intact for many weeks. From PONTÉN, 1967

Lines developed from African lymphoma (BURKITT, 1958) tissue have evolved in a more variable fashion, possibly because the biopsies were often transported over long distances. Already after 2—3 weeks there was evidence of rapid growth with a fall in the pH and rapid proliferation of lymphoblasts which to a large extent grew in suspension (EPSTEIN and BARR, 1964; PULVERTAFT, 1964; EPSTEIN and ACHONG, 1965; EPSTEIN, BARR, and ACHONG, 1965; EPSTEIN and BARR, 1965; STEWART, LOVELACE, WHANG, and NGU, 1965; O'CONOR and RABSON, 1965; RABSON, O'CONOR, BARON, WHANG, and LEGALLAIS, 1966; MINOWADA, G. KLEIN, CLIFFORD, E. KLEIN, and MOORE, 1967; POPE, ACHONG, EPSTEIN, and BIDDULPH, 1967). A similar rapid appearance of lymphoblasts from a lymph node of a patient with lymphocytic lymphoma was described by SABIN (1965) and by TRUJILLO, LIST-YOUNG, BUTLER, SHULLENBERGER, and GOTT (1966) in a patient with reticulum cell sarcoma.

2. Peripheral Blood

MOORE and collaborators developed a method by which a large number of established lymphoid cell lines have been generated and carried as suspension cultures (MOORE, ITO, ULRICH, and SANDBERG, 1966). In this technique it is

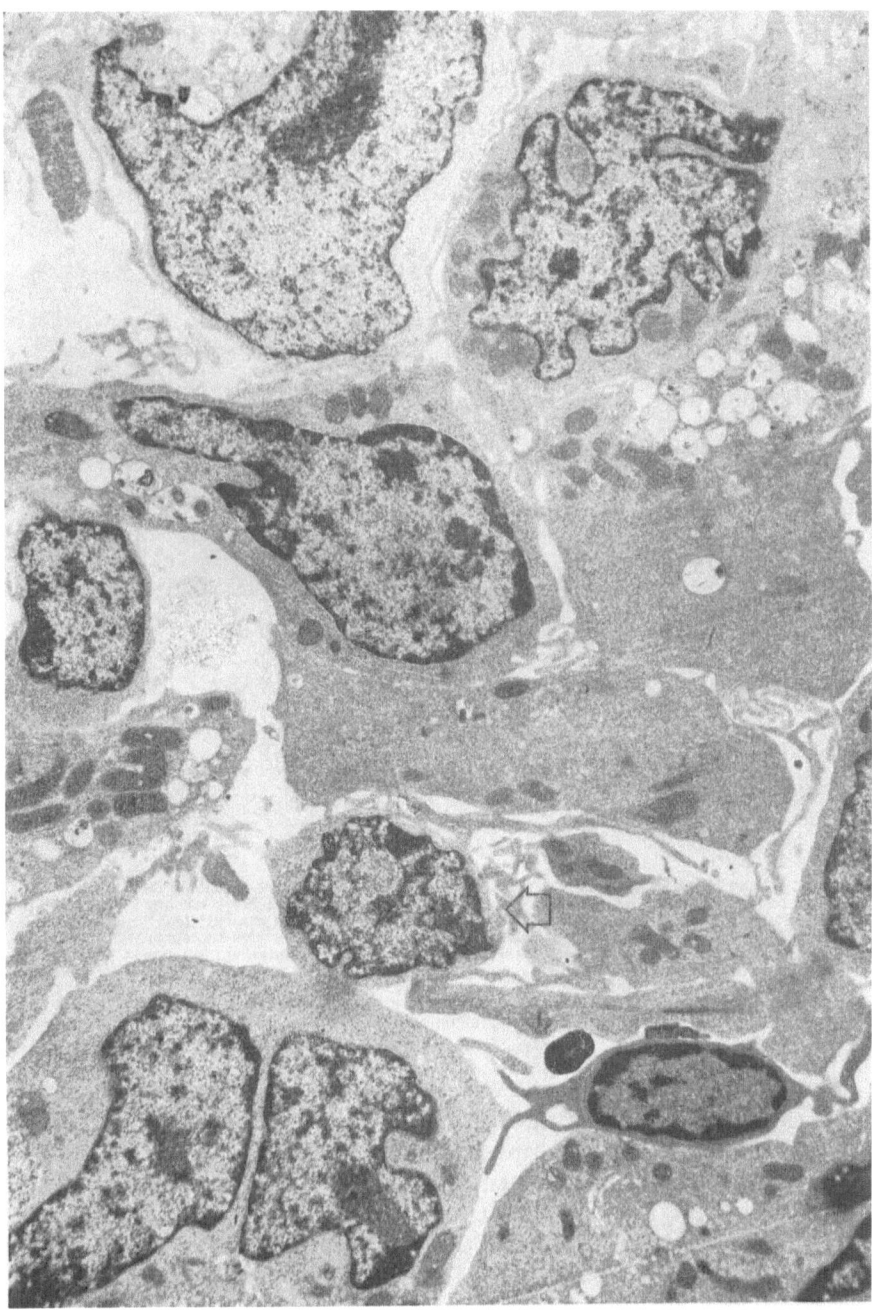

Fig. 6. Low magnification survey of an established human lymphoid line. Note pleomorphism with a small lymphocyte (arrow), large lymphoblasts and a binucleated cell (bottom left). Magnification 9,000 ×. (DE THÉ, NILSSON, and PONTÉN, in preparation)

necessary to maintain a high concentration of lymphoid cells. The lines usually start to emerge after an interval of about 3—14 weeks during which the number of cells have remained stationary or decreased. The lines from peripheral blood were originally from patients with leukemia and lymphoma and it was believed that their malignant condition was essential until Moore, Gerner, and Franklin (1967) obtained lines from the peripheral blood of 2 out of 16 and Gerber and Monroe (1968) from 6 out of 8 normal persons. Recently lines have been

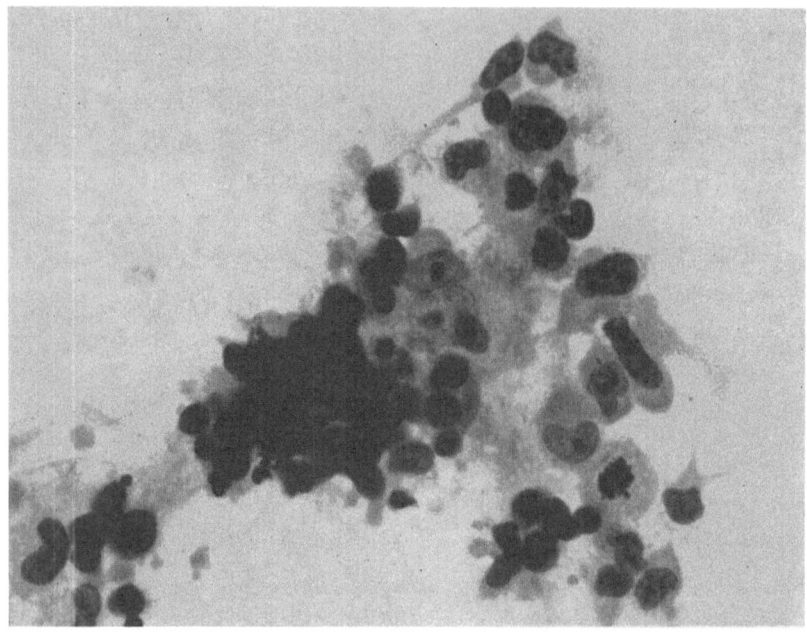

Fig. 7. Appearance of an established lymphoid line carried serially on homologous fibroblasts. Only two fibroblasts are clearly recognized near center of picture. Both have assumed the appearance typical of long and intimate contact with the lymphoid cells. Instead of many small nucleoli one large nucleolus is present. The nucleus contains finely dispersed chromatin and the cytoplasm is abundant. These cells resemble "reticulum cells" as seen in smears from human lymph nodes. The lymphoid cells are of two types. One has a small compact nucleus and scant cytoplasm and tends to aggregate as dense clumps. The other has pleomorphic, less dense nuclei and a moderate amount of cytoplasm. Time-lapse cinematography would show these actively engaged in peripolesis. A mitotic figure in the latter type of cell is seen near the lower right corner. May-Grünwald-Giemsa. Appr. magn. 1,200 ×

established from a large number of patients with mononucleosis (Pope, 1967; Glade, Kasel, Moses, Whang-Peng, Hoffman, Kammermeyer, and Chessin, 1968; Moses, Glade, Kasel, Rosenthal, Hirshaut, and Chessin, 1968).

Regardless of the circumstances under which the established lymphoid cell lines have evolved, their general cytology seems similar (Moore, Kitamura, and Toshima, 1968). The predominant cell shows moderate pleomorphism. It has an average diameter of about 7 μ, a moderate amount of dense strongly basophilic cytoplasm with numerous free ribonucleoprotein bodies, strong pyroninophilia and fluoresces orange with acridin orange. The nucleus contains a few large nucleoli and may be excentrically placed. It has generally been concluded that the cell is a lymphoblast (Fig. 6) (Benyesh-Melnick, Fernbach, and Lewis, 1963; Epstein and Achong, 1965; Epstein, Barr, and Achong, 1966; Rabin,

BENYESH-MELNICK, and BRUNSCHWIG, 1967; SINKOVICS, 1968). Its morphology resembles the phytohemagglutinin stimulated lymphocyte (for a review see LENNERT, 1966). A few lines also contained multinucleated elements or promyelocytes (MOORE, GERNER, and FRANKLIN, 1967).

3. Relationship between Lymphoblasts and Fibroblasts

Intimate contact with fibroblasts or reticulum cells often enhances the growth and survival of fresh lymphoid cells *in vitro* (review by TROWELL, 1965). The lymphoid-fibroblast interaction takes one of two related forms. Peripolesis, the adherence of mobile lymphoid cells to the outer surface of the fibroblast (SHARP and BURWELL, 1960); or emperipolesis, the entry of lymphoid cells into the fibroblast (PULVERTAFT and HUMBLE, 1962). It has been suggested that reticulum cells nurse lymphoid cells by energy transfer (TROWELL, 1961) but no experimental evidence has been supplied.

Some established lymphoid lines also require the presence of fibroblast-like or reticulum-like cells. BENYESH-MELNICK, FERNBACH, and LEWIS (1963), TRUJILLO, LIST-YOUNG, BUTLER, SHULLENBERGER, and GOTT (1966), PONTÉN (1967) and NILSSON, PONTÉN, and PHILIPSON (1968) were unable to grow their lymphoblasts in the absence of fibroblasts. SINKOVICS, SYKES, SHULLENBERGER, and HOWE (1967) and SINKOVICS, SHULLENBERGER, and HOWE (1967) obtained dependent as well as autonomous lines. POPE, ACHONG, and EPSTEIN (1968) reported fibroblast dependence of 2 sublines from a Burkitt lymphoma during an initial phase of 8—10 weeks. Only thereafter could fibroblast independent growth be recorded. In contrast, lines from peripheral blood have always grown as suspension cultures in the absence of fibroblasts even initially.

The reasons for the dependence are unknown. Time-lapse cinematography (WESTERMARK and PONTÉN, unpublished) has shown a very active peripolesis and that the fibroblasts seem to succumb slowly in the process (Fig. 7).

B. Production of Immunoglobulins after Lymphoblastoid Transformation

The morphology of the transformed lymphoblastoid cell suggests that they should be capable of specific lymphoid functions (Fig. 8). This was shown by FAHEY, FINEGOLD, RABSON, and MANAKER (1966), TANIGAKI, YAGI, MOORE, and PRESSMAN (1966), WAKEFIELD, THORBECKE, OLD, and BOYSE (1967) and MATSUOKA, TAKAHSHI, YAGI, MOORE, and PRESSMAN (1968); immunoglobulins are produced in suspension lines derived from leukemic peripheral blood or Burkitt lymphomas. Most produced IgG and a few IgA or IgM. This has also been found in fibroblast-dependent lymphoid lines of normal and neoplastic origin. Eleven lines have been shown to synthesize either IgG (9 cases), IgA (1 case) or IgM (1 case). Each seemed to produce only a single molecular species (PHILIPSON and PONTÉN, 1967; NILSSON, PONTÉN, and PHILIPSON, 1968 and unpublished). Similarly, one or only a few species of immunoglobulins were produced in each of 19 lines from patients with various malignancies (mainly leukemia-lymphoma) (FINEGOLD, FAHEY, and GRANGER, 1967).

Immunoglobulin production by established leukemic as well as normal lymphoid lines violates the "rule" that differentiated functions cannot be maintained

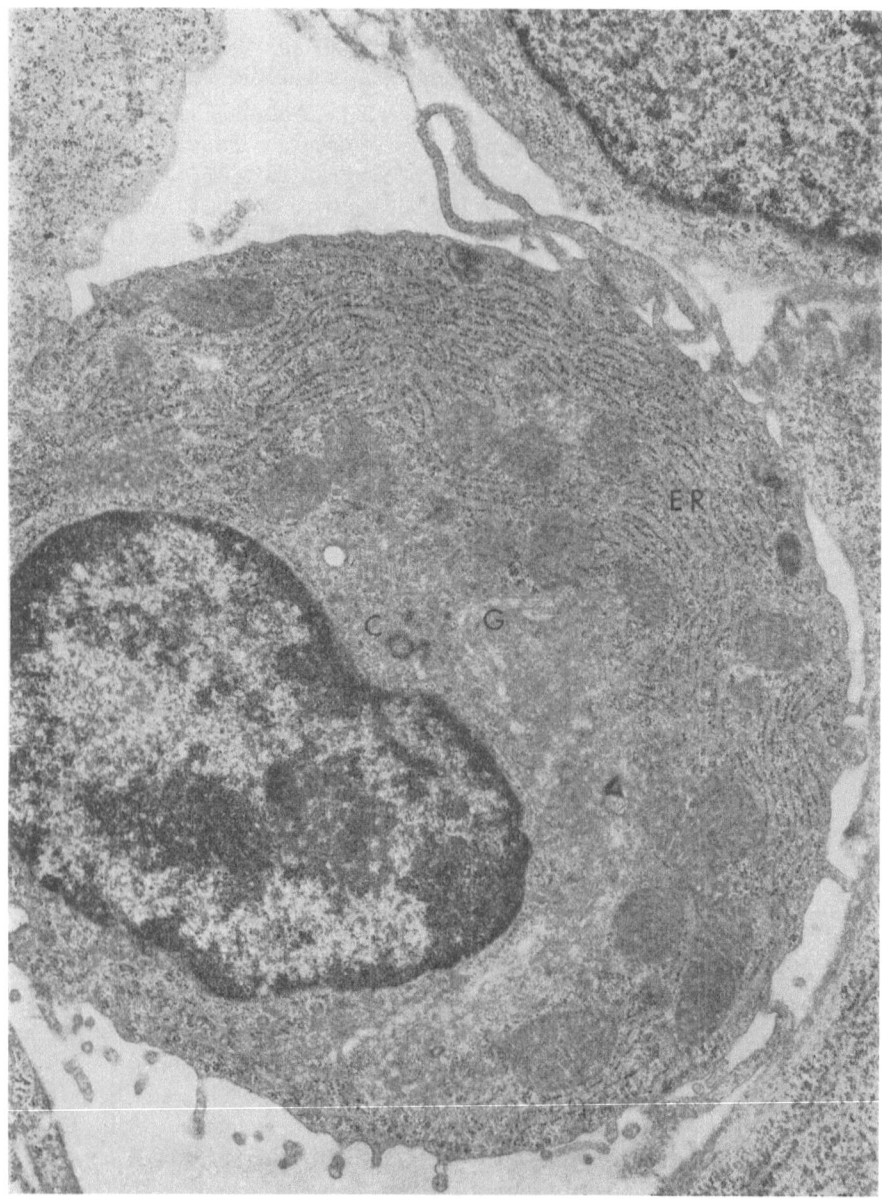

Fig. 8. A plasmocyte from an established human lymphoid cell line. Note well organized endoplasmic reticulum (ER). The centriole (c) is located close to a shallow indentation of the eccentrically placed nucleus in the vicinity of the Golgi zone (G). Morphologically identifiable plasmoblasts or plasmo-cytes form only a minority of the various cell types of a lymphoid line. Magnification 22,500 × (DE THÉ, NILSSON, and PONTÉN, in preparation)

in tissue culture. It should now be possible to study the control of immunoglobulin synthesis under well defined conditions and possibly even the induction of the primary antibody response *in vitro* (KAMEI and MOORE, 1967).

C. Presence of Virus-like Particles in Lymphoid Lines

Since the first report of BENYESH-MELNICK, FERNBACH, and LEWIS (1963), interest has centered around the significance of virus-like particles observed in transformed lymphoblasts from the leukemias or lymphomas used as starting material. The confused present situation illustrates the pitfalls and difficulties in the search for human oncogenic viruses. The following impressive list of agents has either been observed in the electron microscope or directly isolated. Myxovirus-like particles (BENYESH-MELNICK, FERNBACH, and LEWIS, 1963; BENYESH-MELNICK, SMITH, and FERNBACH, 1964), herpes-like particles, (synonyms; HLV, leukovirus, EBV) (EPSTEIN, ACHONG, and BARR, 1964; EPSTEIN, BARR, and ACHONG, 1965; EPSTEIN, G. HENLE, ACHONG, and BARR, 1965; STEWART, LOVE-LACE, WHANG, and NGU, 1965; O'CONOR and RABSON, 1965; HUMMELER, G. HENLE and W. HENLE, 1966; RABSON, O'CONOR, BARON, WHANG, and LEGALLAIS, 1966; TOPLIN and SCHIDLOVSKI, 1966; ZEVE, LUCAS, and MANAKER, 1966; TOSHIMA, TAGAKI, MINOWADA, MOORE, and SANDBERG, 1967; MINOWADA, G. KLEIN, CLIF-FORD, E. KLEIN, and MOORE, 1967; JENSEN, KOROL, DITTMAR, and MEDREK, 1967; POPE, ACHONG, EPSTEIN, and BIDDULPH, 1967; EPSTEIN, ACHONG, and POPE, 1967; OBOSHI, 1969), vaccinia virus (DALLDORF, LINSELL, BARNHART, and MAR-TYN, 1964), herpes simplex (SIMONS and ROSS, 1965; WOODALL, WILLIAMS, SIMPSON, and HADDOW, 1965; GRIFFIN, WRIGHT, BELL, and ROSS, 1966), reovirus type 3 (BELL, MASSIE, ROSS, SIMPSON, and GRIFFIN, 1966), ECHO virus type 11 (MUNUBE and BELL, 1967), mycoplasma (DALLDORF, BERGAMINI, and FROST, 1966) and an unidentified agent causing encephalitis in hamsters (STEWART, LANDON, LOVELACE, and PARKER, 1965; STEWART, GLAZER, BEN, and LLOYD, 1968). None of these agents has consistently been demonstrated in established lymphoid lines, but EBV has been most commonly observed and merits special consideration (review by HARRIS, 1967).

"Leukovirus" (EBV) or the apparently identical "herpes-like particle" (HLV) has a fairly characteristic structure. In thin sections, mature 110—115 mµ particles may be seen containing two membranes which enclose a dense 45—50 mµ nucleoid. Presumed precursors measure about 75 mµ and are limited by a single membrane. The particles typically exist in low numbers in the nucleus and the cytoplasm of a small minority of cells (Fig. 9).

EBV particles were reported in 9/26 lymphoid cell lines derived from peripheral blood by MOORE and collaborators. Six of the positive lines were from lymphomas (Burkitt and other types) or leukemia (MOORE, GRACE, CITRON, GERNER, and BURNS, 1966) whereas 2 were isolated from normal subjects (MOORE, GERNER, and FRANKLIN, 1967) and one from a patient with malignant melanoma (MOORE and PICKREN, 1967). Similar particles have been observed in 5/6 thoroughly investigated lines of Burkitt tumor origin but in one (Raji) no such particles could be demonstrated (EPSTEIN, ACHONG, BARR, ZAJAC, G. HENLE, and W. HENLE, 1966).

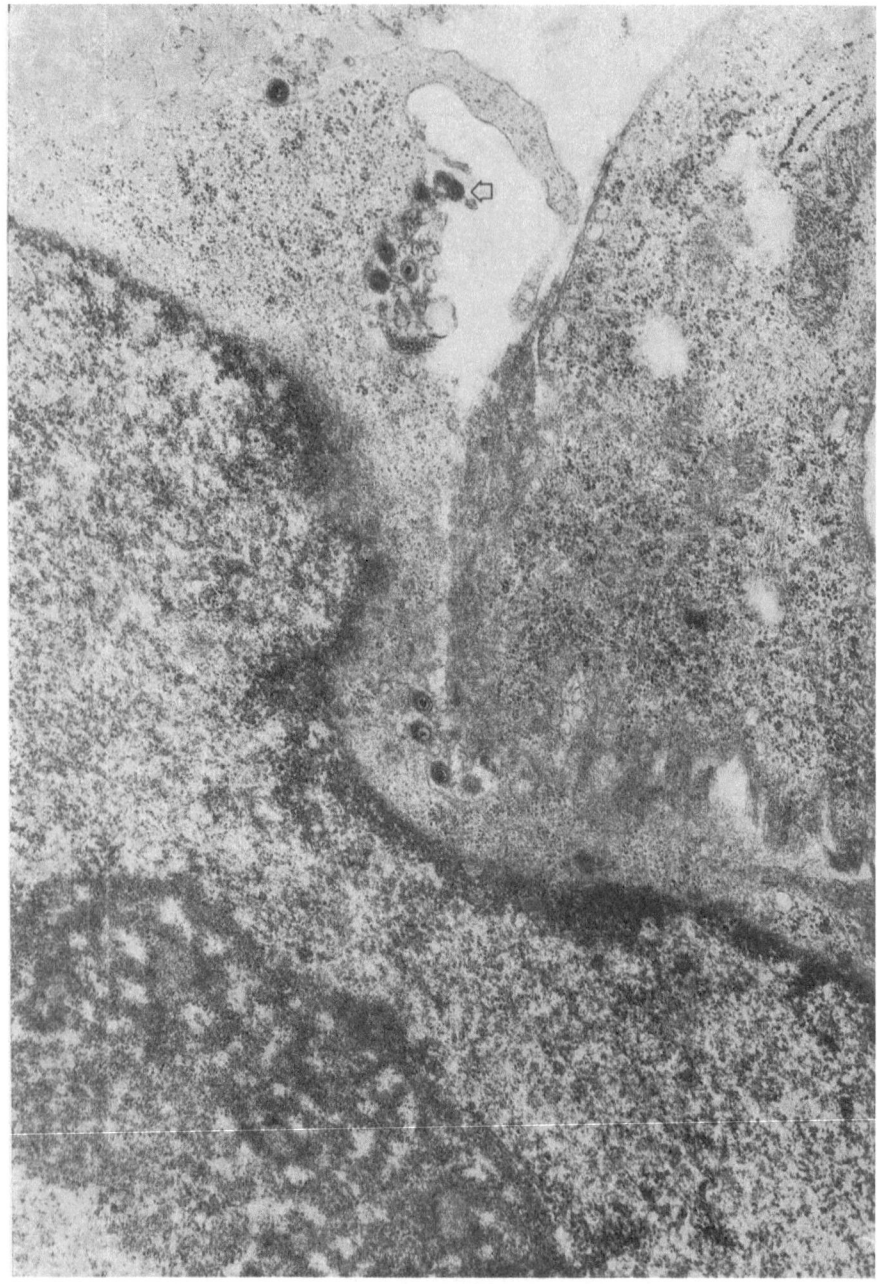

Fig. 9. Herpes-like virions (EBV) in the extracellular spaces of the lymphoid line illustrated in Fig. 8. In all except one particle (arrow) a dense nucleoid is present. The cell line was originated from a patient without neoplastic disease or mononucleosis. Magnification 30,000×, (DE THÉ, NILSSON, and PONTÉN, in preparation)

All attempts at serial passage of EBV under experimental conditions have failed and no cytopathic effects have been noted in tissue culture (e.g. EPSTEIN, HENLE, ACHONG, and BARR, 1965; MINOWADA, G. KLEIN, CLIFFORD, E. KLEIN, and MOORE, 1967). The HENLE's and their collaborators have attempted to demonstrate indirectly the presence of a virus in Burkitt cell lines by their resistance to challenge by vesicular stomatitis virus (VSV). Several lines containing an interferon-like substance have been partially resistant to VSV and capable of transmitting this resistance to normal human and simian cells. Apparently, however, this was unrelated to the presence of EBV as seen in the electron microscope, since two EM-negative lines (Raji and Ogun) were also resistant to VSV. At present it is concluded that the method has failed to provide a reliable means for the detection of the herpes-like viral carrier state in lymphoblastoid lines (G. HENLE and W. HENLE, 1966a). Recent attempts to select lymphoblasts yielding a rather large number of EBV particles by cloning may open a more fruitful avenue towards isolating EBV as an infective entity (HINUMA and GRACE, 1967, 1969), as may the enhanced virus yield obtained after deprivation of cultures of arginine (HENLE W. and HENLE G., 1968). Virus particles have been concentrated by centrifugation (MINOWADA, CHAI, and MOORE, 1969), the only usable indication of infection may be a rather minor chromosome alteration (W. HENLE, DIEHL, KOHN, ZUR HAUSEN, and G. HENLE, 1967).

An important clue to the nature of EBV was provided by the findings of G. HENLE, W. HENLE, and DIEHL (1968) that antibodies directed against EBV appeared in patients with mononucleosis, strongly suggesting that EBV is the etiological agent of this disease. Serological data (G. HENLE and W. HENLE, 1966 a, b; LEVY and G. HENLE, 1966; OLD, BOYSE, OETTGEN, DE HARVEN, GEERING, WILLIAMSON, and CLIFFORD, 1966; G. KLEIN, CLIFFORD, E. KLEIN, and STJERNSWÄRD, 1966; E. KLEIN, CLIFFORD, G. KLEIN, and HAMBERGER, 1967; GERBER, HAMRE, MOY, and ROSENBLUM, 1968) indicate that infection with EBV (or a serologically related agent) is common in many parts of the world. The proportion of positive sera among healthy persons increases with age and at least 80% of persons older than 35 have antibodies indicating the existence of numerous subclinical infections. Until better evidence is available it is quite possible that EBV represents a ubiquitous "passenger" occasionally causing clinically manifest mononucleosis but of unknown etiological significance for leukemia or lymphomas. The observation that lymphoblastoid transformation is more easily obtained in peripheral blood cultures from patients with mononucleosis than from healthy donors (POPE, 1967; MOSES, GLADE, KASEL, ROSENTHAL, HIRSHAUT, and CHESSIN, 1968; DIEHL, G. HENLE, W. HENLE, and KOHN, 1968) may not depend on the presence of EBV *per se* but rather on the large number of circulating immature cells (GLADE, HIRSHAUT, STITES, and CHESSIN, 1969). MOSES, GLADE, KASEL, ROSENTHAL, HIRSHAUT, and CHESSIN (1968) could identify EBV electronmicroscopically only in four of fourteen lines from patients with infectious mononucleosis.

D. Search for a Specific Antigen in Burkitt Lymphoma Lines

All experimental virus induced tumors contain "new" antigens. The circumstantial evidence that the Burkitt African lymphoma may be caused by an insect-born virus (BURKITT, 1958, 1962a, 1963a and b; EPSTEIN and BARR, 1964) has motivated an intensive search for specific antigens. HENLE and collaborators (G. HENLE and W. HENLE, 1966a, b; LEVY and G. HENLE, 1966; W. HENLE, HUMMELER, and G. HENLE, 1966; have used indirect and direct immunofluorescence tests. The results have been some-what difficult to interpret and are of at least three different types (G. HENLE and W. HENLE, 1965). 1. A low proportion of lymphoma cells show a brilliant coarsely granular fluorescence when treated with certain human sera mainly from patients with Burkitt's disease, and then stained indirectly with labeled rabbit anti-human γ-globulin. This could also be elicited directly with fluorescein-isothio-cyanate tagged positive sera. 2. Cells derived from various leukemias and of normal leukocyte cultures showed a moderately strong cytoplasmic fluorescence directly with labeled rabbit anti-human globulin. This reaction is reportedly blocked by unlabeled human globulin and considered non-specific. 3. Cells show a fine surface stippling in direct tests with labeled human γ-globulin, and this also is thought to be non-specific.

With the first type of staining, not only were Burkitt sera positive but also 67 and 83% of sera from normal African and American adults, respectively. The test was considered likely to be specific for EBV antibody because the only positive target cell lines were those which in the electron microscope showed the characteristic virus-like particles.

The KLEIN's and collaborators have also employed immunofluorescence in their search for a Burkitt specific antigen. Their original membrane immuno-fluorescence method (MÖLLER, 1961) employed fresh cells from Burkitt lympho-mas, bone marrow or lymph node biopsies, which were incubated with undiluted test serum for 30 minutes at 37°C. After washing, the cells were exposed to fluorescein-labeled rabbit anti-human globulin. Cells showing fluorescence around the membrane were scored as positive and a fluorescence index (FI) is computed, based on the difference between the proportion of negative cells obtained after exposure to control serum and test serum. This test was later replaced by a blocking technique, where a positive standard serum was conjugated with fluores-cein-isothiocyanate. Serum under test was assessed by its ability to block sub-sequent staining with the labeled standard serum (G. KLEIN, PEARSON, G. HENLE, W. HENLE, GOLDSTEIN, and CLIFFORD, 1969).

In general, sera from Burkitt patients gave high FI values only with Burkitt lymphoma cells but not with normal bone-marrow cells from the Burkitt patients or allogeneic lymph node cells from Swedish patients with or without malignant lymphoma. The results have cautiously been interpreted to mean either a) a Burkitt tumor-specific reaction or b) non-specific binding to Burkitt lymphoma cell surfaces of some serum globulin fraction (G. KLEIN, CLIFFORD, E. KLEIN, and STJERNSWÄRD, 1966; E. KLEIN, CLIFFORD, G. KLEIN, and HAMBERGER, 1967). Most Burkitt lines reacted in parallel in the Klein and Henle test, but the existence of a few exceptions (G. KLEIN, CLIFFORD, E. KLEIN, SMITH, MINOWADA, KOU-RILSKY, and BURCHENAL, 1967) suggests that different antigens may be involved.

The membrane antigen demonstrated by the Klein procedure is probably different from antigens of EBV virions because certain sera react differently in tests designed to distinguish between these two antigens (G. KLEIN, PEARSON,

J. S. NADKARNI, J. J. NADKARNI, E. KLEIN, G. HENLE, W. HENLE, and CLIFFORD, 1968; PEARSON, G. KLEIN, G. HENLE, W. HENLE, and CLIFFORD, 1969; G. KLEIN, GEERING, OLD, G. HENLE, W. HENLE, and CLIFFORD, 1970).

OLD, BOYSE, OETTGEN, DE HARVEN, GEERING, WILLIAMSON, and CLIFFORD (1966) tested a 5-fold concentrated saline extract of the Jiyoye line of Burkitt lymphoma, which is unusually rich in herpes-like particles in the electron microscope, by immunoprecipitation in agar against 352 different sera. They found that 56% of Burkitt lymphoma patient's sera were positive and, surprisingly, not less than 85% of sera from patients with nasopharyngeal carcinoma were also positive. A third group of patients with various malignancies and also with non-malignant disease had an average of 11% positive sera. Two other Burkitt lines which contained a smaller number of virus particles visualized in the electron microscope, as well as fresh cells from acute or chronic leukemia were negative, even when run against strongly positive sera.

A complement-fixation procedure for EBV has been described (ARMSTRONG, G. HENLE, and W. HENLE, 1966; GERBER and BIRCH, 1967; McCORMICK, STENBACK, TRENTIN, G. KLEIN, J. S. NADKARNI, J. J. NADKARNI, and CLIFFORD, 1969).

No definite interpretation of the immunological data in terms of a specific "tumor" antigen seems possible at present. The field is in a state of rapid flux and such artifacts as interference by the immunoglobulins produced by the cell lines or reactions with normal isoantigens have been difficult to exclude. TRUJILLO, BUTLER, AHEARN, SHULLENBERGER, LIST-YOUNG, GOTT, ANSTALL, and SHIVELY (1967) have interpreted the positive fluorescence obtained after staining of an established line from lymphocytic lymphoma by rabbit anti-human gamma-globulin to indicate synthesis of gammaglobulin rather than presence of a tumor or virus specific antigen.

E. Chromosomes and Lymphoblastoid Transformation

Established lines of non-lymphoid origin are characterized by mixoploidy and unstable karyotypes while the majority of established lymphoid lines do not show any extensive chromosome alterations (McCARTHY, JUNIUS, FARBER, LAZARUS, and FOLEY, 1965; STEWART, LOVELACE, WHANG, and NGU, 1965; MOORE, ITO, ULRICH, and SANDBERG, 1966; KOHN, MELLMAN, MOORHEAD, LOFTUS, and G. HENLE, 1967; MILES and O'NEILL, 1967; BISHUN and SUTTON, 1967; MILES, O'NEILL, ARMSTRONG, CLARKSON, and KEANE, 1968). Only minor alterations have been observed with trisomies involving various chromosomes, recurrent marker chromosomes and pseudodiploid karyotypes. The changes have remained stable for many generations and in several lines they can be attributed to the proliferation of one or only few cell clones. With extended cultivation slight indications of random chromosome damage may be detected (SAKSELA and PONTÉN, 1968). Only one contrary report claims the development of considerable mixoploidy (COOPER, HUGHES, and TOPPING, 1966) in lines derived from Burkitt lymphomas.

Of particular significance is the fact that the chromosomal characteristics of lymphoblastoid lines from normal lymphoid tissue or peripheral blood were essentially the same as those in lines originated from leukemias or lymphomas

(MOORE, GERNER, and FRANKLIN, 1967; SAKSELA and PONTÉN, 1968; MILES, O'NEILL, ARMSTRONG, CLARKSON, and KEANE, 1968).

Attempts to correlate alterations of certain chromosome regions with the production of a specific immunoglobulin have failed (SAKSELA and PONTÉN, 1968). More precise correlative studies with cloned cell populations are, however, required before it can be excluded that certain chromosome lesions specifically involve genetic determinants for the synthesis of immunoglobulin subunits.

F. Mechanism for Lymphoblastoid Transformation—Selection or Induction

It has not been determined whether the lymphoblastoid transformation is due to a cell change induced *in vitro* or a selection of preexisting variants in the original explant. The argument is similar to that concerning the spontaneous infinite growth transformation of rodent cells (page 28). The weight of evidence seems to favour selection. SINKOVICS, SYKES, SHULLENBERGER, and HOWE (1967) studied the fibroblastic stage and observed a few emperipoletic lymphoid cells alive after the majority of the primary lymphoid cells had died. The transformation originated from such cells which suddenly proliferated into dominance. The authors speculated that the lymphoblast may not originally be equipped for autonomous growth but that during its intrafibroblastic stay it might acquire genetic information essential for independent proliferation. It seemed quite clear that the large majority of the lymphoid cells contained in the original explant did not survive and that the recurrent growth seen after many weeks originated from only one or a few cells. It is an intriguing problem whether these latter cells are selected at random or represent preexisting cells fit for infinite survival *in vitro*.

"Monoclonal" immunoglobulin production (FINEGOLD, FAHEY, and GRANGER, 1967; PHILIPSON and PONTÉN, 1967; NILSSON, PONTÉN, and PHILIPSON, 1968) and the defined neardiploid karyotypes (SAKSELA and PONTÉN, 1968) of the transformed lines strongly supports that only a few survivors are responsible for the lymphoblastoid transformation.

Induction is, however, by no means excluded. The interesting finding by W. HENLE, DIEHL, KOHN, ZUR HAUSEN, and G. HENLE (1967) that a lethally irradiated male Burkitt lymphoma cell line (Jiyoye) harboring EBV induced lymphoblastoid transformation in female peripheral blood leukocytes obtained from healthy infants supports such a contention. Transfer of EBV was suggested by positive immunofluorescence in some of the target cells. A virus-free Burkitt line (Raji) used as control failed to cause a lymphoblastoid transformation. Essentially similar results were obtained by POPE, HORNE, and SCOTT (1968), who induced lymphoblastoid transformation in fetal lymphocytes by a filtrate from an EBV positive lymphoid line.

G. Relation between Lymphoblastoid Transformation and Neoplasia

The early studies on lymphoblastoid transformation used mainly lymphoid tissue or peripheral blood from patients with malignant lymphoma or leukemia. It was generally assumed that the lines obtained were direct descendents of the patient's neoplastic cells and that a malignant condition was a necessary prerequisite. This must, however, be questioned since it has recently been found that the incidence of lymphoblastoid transformation is the same when malignant lymphomas, lymph nodes draining carcinomas or lymph nodes from patients without any malignancy are cultivated under identical conditions (PONTÉN, 1968; NILSSON, PONTÉN, and PHILIPSON, 1968 and unpublished). The incidence of positive cultures from peripheral blood under different conditions is not accurately known, but it is significant that MOORE, GERNER, and FRANKLIN (1967) and GERBER and MONROE (1968) obtained lymphoblast lines from normal subjects. Furthermore, no systematic differences have been found in established lines of lymphomatous and normal origin with respect to general pattern of development, morphology, immunoglobulin production, chromosomes or the presence of herpes-like particles.

Two main hypotheses for the biological significance of lymphoblastoid transformation may be offered.

1. Lymphoid tissue is different from other somatic tissues because it contains a proportion of cells (lymphoblasts) capable of infinite proliferation. The immunologic function of lymphoid cells requires rapid and prolonged proliferation for instance in a secondary response, and this may be connected with the peculiar ease by which established lines are formed. The capacity for unlimited proliferation may thus be entirely unrelated to neoplasia (MOORE and McLIMANS, 1968). According to this hypothesis the lines obtained from Burkitt lymphomas and leukemia would be derived from non-neoplastic lymphoid cells present among the tumor cells and thus cannot be expected to shed light on the malignant process. Indirect support for this is suggested by work of IWAKATA and GRACE (1964), TANIGAKI, YAGI, MOORE, and PRESSMAN (1966), ZEVE, LUCAS, and MANAKER (1966) and FINEGOLD, HIRSHAUT, and FAHEY (1968), in which established immunoglobulin producing lymphoid lines were obtained from patients with *myeloid* leukemia, a condition where the lymphatic system generally is believed not to be part of the neoplastic process. It is also difficult to understand why malignant lymphomas such as Burkitt's disease should produce cell lines synthesizing immunoglobulin — a feature not displayed by the tumor cells *in vivo*.

2. Lymphoblastoid transformation reflects the common occurrence in human lymphoid tissue of a small proportion of latent neoplastic cells capable of infinite proliferation. In the normal patient all these cells are kept under control by unknown mechanisms. *In vitro* this repression is absent and the cells are free to form established lines having a tumorigenic potential. In the lymphomatous patient the control mechanism is deficient and a proportion of the lymphoid cells are permitted to proliferate already *in vivo*. The cell lines obtained from lymphomas may either be derived from latently neoplastic cells which are released from control by being submitted to *in vitro* conditions, or from the neoplastic cells which have already escaped the control mechanism.

Chromosome studies lend some support to the second hypothesis. Principally the same alterations have been found in certain leukemias and lymphomas *in vivo* as in lymphoblastoid cell lines, even when the latter are derived from normal subjects (MOORE, GERNER, and FRANKLIN, 1967; SAKSELA and PONTÉN, 1968 and unpublished). This would be compatible with the existence of clones of

Table 2. *Transformation in vitro by Oncogenic Viruses*

Virus	Natural host	Size (mμ)	Nucleic acid — Type	Nucleic acid — Appr. M. W.	Species range for transformation *in vitro* (irregular and/or unrestrained growth)
Papova group polyoma SV 40	mouse monkey	45	DNA	3×10^6	mouse, rat, hamster, calf mouse, rat, mastomys, hamster, guinea pig, rabbit, pig, calf, horse, monkey, man
Papilloma canine rabbit bovine human	dog rabbit cattle man	55	DNA	6×10^6	mouse, calf
Adeno-group avian canine bovine simian human	chicken dog calf monkey man	70—80	DNA	23×10^6	no transformation reported hamster hamster mouse, rat, hamster rat, hamster
Pox-group fibroma-myxoma Yaba	rabbit monkey	230 × 300	DNA	160×10^6	no transformation reported no transformation reported
Avian leukemia complex	chicken	80—120	RNA	10×10^6	mouse, rat, hamster, calf, monkey, duck, chicken, turkey, quail, man[1]
Murine leukemia complex	mouse	80—200	RNA	10×10^6	mouse, hamster[1]

[1] Transformation only by sarcomagenic members of the complex.

abnormal cells, which are kept in a latent neoplastic state in normal persons *in vivo* but which are free to proliferate in patients with malignant lymphoma or in tissue explanted *in vitro*.

The hypothesis fails to explain why the latently malignant cells become producers of immunoglobulin *in vitro*. Malignant lymphomas do, with rare exceptions, not produce any myeloma-like proteins *in vivo*, in sharp contrast to the lines obtained *in vitro*, which almost always have synthesized mono- or oligoclonal globulins.

IV. Virus Induced Transformation in Cell Culture

The interaction between a tumor virus and its target cell can be studied with a precision and refinement impossible *in vivo* and experimental viral carcinogenesis derives much of its basic knowledge from studies *in vitro*. Table 2 lists oncogenic viruses whose capacity to transform cells *in vitro* according to the criteria of Table 1 has been investigated. They encompass a wide spectrum with respect to their size and amount and type of nucleic acid. Papovaviruses, adenoviruses and poxviruses contain DNA. The papova viruses are grouped together on the basis of their physico-chemical similarities (MELNICK, 1962) and they may be divided into two subgroups one of which gives rise in several species to contagious, naturally occurring and usually benign papillomas of skin and mucuous membrane and the second of which comprises polyoma virus (PV) and Simian virus 40 (SV 40), which induce non-contagious sarcomas and other tumors experimentally in rodents but are of no known natural oncogenic importance. Polyoma and SV 40 are the two DNA viruses most widely employed for studies on transformation *in vitro*.

Certain types of adenoviruses induce tumors under artificial conditions for example when a non-natural rodent host is infected. Their transforming capacity in cell culture has only recently been extensively investigated. The pox group causes benign mesenchymal lesions ("cutaneous nodules", histiocytomas) of doubtful neoplastic character *in vivo* but no transformations *in vitro* have been reported.

As a group, DNA tumor viruses have been shown to give rise to a variety of benign and malignant solid mesenchymal or epithelial neoplasms but not to leukemias and lymphomas. Transformation in cell culture has with some exceptions (MONTAGNIER, MACPHERSON, and JARRETT, 1966; WELLS, WURTMAN, and RABSON, 1966; SHEIN, 1967a, b; PAULSON, RABSON, and FRALEY, 1968; DEFTOS, RABSON, BUCKLE, AURBACH, and POTTS, 1968; ALBERT, RABSON, GRIMES, and VON SALLMAN, 1969) only involved fibroblast-like cells.

The RNA tumor viruses form a more homogenous group and are similar to each other in size, shape, composition and general biological character, regardless of the species of origin. Their morphology resembles, but is not identical to the myxoviruses. Typically they induce solid mesenchymal tumors, leukemias and lymphomas, but not pure epithelial tumors. The RNA tumor viruses form two distinct "complexes", avian and murine, with principally the same characteristics. Within each, certain members give rise to sarcomas or other solid mesenchymal neoplasms whereas others induce leukemias. Only sarcomagenic members of the avian complex have definitely been shown to effect irregular and unrestrained growth transformation in cell culture.

The mammary tumor agent (MTV), also believed to contain RNA, represents a special case. It is not well characterized chemically and is only effective as a "co-carcinogen" against a certain genetic and hormonal background. Its capacity to induce transformation *in vitro* is questionable (LASFARGUES and MOORE, 1966) and MTV is not considered further in this review.

A. Transformation by DNA Viruses
The Papova- and Adeno Virus Groups
1. Polyoma Virus and Simian Virus 40

Soon after GROSS (1951a, b, and 1952) in his pioneering studies had discovered the transmissibility of mouse leukemia by cell-free extracts of leukemic tissue, it became apparent (GROSS, 1951a, 1953a, b; DULANEY, 1956; DULANEY, MAXEY, SCHILLING, and GOSS, 1957) that some of the injected mice developed salivary gland tumors and/or fibrosarcomas in the absence of leukemia. GROSS realized (1953b) the possibility that he was dealing with two agents — a leukemia virus and a "parotid tumor agent", but could not achieve a clear separation between the two.

The isolation of the "parotid tumor agent" was accomplished by STEWART and EDDY (STEWART, EDDY, GOCHNOUR, BORGESE, and GRUBBS, 1957; STEWART, EDDY, HAAS, and BORGESE, 1957; STEWART, EDDY, and BORGESE, 1958; EDDY, STEWART, YOUNG, and MIDER, 1958), who, after passage through monkey kidney or mouse embryo cultures obtained a potent material free from leukemic activity. The virus was named SE (Stewart-Eddy) polyoma virus because of the large variety of histologically different tumors it induced in mice and hamsters (STEWART and EDDY, 1959). Later the prefix "SE" has been dropped and the "parotid tumor agent" of Gross is now generally called polyoma virus (PV).

The first transformations in cell culture by PV were described by M. VOGT and DULBECCO (1960) and by SACHS and MEDINA (1961). Irregular growth and neoplastic properties were induced in hamster and mouse cultures.

Simian virus 40 or "vacuolating virus" (SWEET and HILLEMAN, 1960) was regarded a harmless "virus in search of disease" until EDDY, BORMAN, BERKELEY, and YOUNG (1961) found that Rhesus monkey kidney cultures spontaneously contaminated with the agent induced sarcomas if inoculated into newborn hamsters (review by EDDY, 1964).

Unrestrained growth transformation *in vitro* by SV40 was first described by KOPROWSKI, PONTÉN, JENSEN, RAVDIN, MOORHEAD, and SAKSELA (1962) and SHEIN and ENDERS (1962b) in human cells.

a) Physico-chemical Properties

Comparisons between PV and SV40 particles have revealed a great similarity in general structure and composition (L. CRAWFORD and E. CRAWFORD, 1963).

Highly purified preparations have been obtained by density gradient centrifugation of lysed cells and tissue culture fluid. Infective, complete particles from murine cells infected by PV have a density in CsCl or RbCl of about 1.32—

1.33 g/ml (L. CRAWFORD, E. CRAWFORD, and WATSON, 1962; ABEL and CRAWFORD, 1963; WINOCOUR, 1963). The same density has been reported for SV40 (MAYOR, JAMISON, and JORDAN, 1963; BLACK, E. CRAWFORD, and L. CRAWFORD, 1964) isolated from infected African green monkey kidney cultures. Both types of particles have a diameter of about 42—45 mμ in the electron microscope by negative staining (KAHLER, ROWE, LLOYD, and HARTLEY, 1959; WILDY, STOKER, MACPHERSON, and HORNE, 1960; SHEININ, 1962; L. CRAWFORD and E. CRAWFORD, 1963; MAYOR, JAMISON, and JORDAN, 1963; WINOCOUR, 1963; BLACK, E. CRAW-FORD, and L. CRAWFORD, 1964; KOCH, EGGERS, ANDERER, SCHLUMBERGER, and FRANK, 1967). They are indistinguishable icosahedrons containing, according to several early authors, fortytwo morphological units (capsomeres) (WILDY, STOKER, MACPHERSON, and HORNE, 1960; HOWATSON and ALMEIDA, 1960; MAYOR, JAMI-SON, and JORDAN, 1963; HOWATSON and CRAWFORD, 1963).

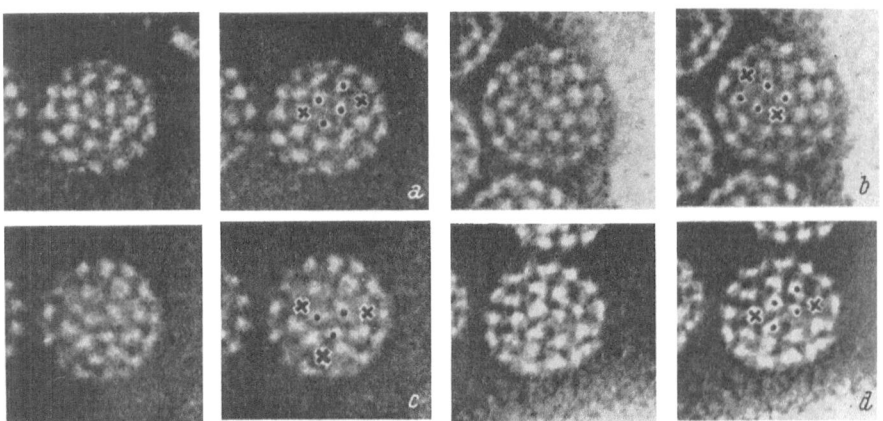

Fig. 10. Selected "one side" particle images of negatively stained SV40 particles together with duplicates where the 5-coordinated morphological units are marked X and the 6-coordinated units which determine the "72nd path" are marked with a black dot. Magnification × 380,000. From ANDERER, SCHLUM-BERGER, KOCH, FRANK, and EGGERS (1967)

MATTERN, TAKEMOTO, and DE LEVA (1967) have recently suggested that the polyoma virus has a different and somewhat more complicated structure. By centrifugation in CsCl, small-plaque PV showed four bands at 1.28, 1.29, 1.33 and 1.34 g/ml. Electron microscopy revealed different structures interpreted alto-gether as indicating that PV has a 22 mμ core of 12 capsomeres, a thin shell of DNA, a 38 mμ shell of 32 capsomeres and outermost a 48 mμ protein shell of 72 capsomeres. The possibility remains that the smaller morphological units may represent artifacts and this interesting model needs confirmation. The 72 capso-mere interpretation was first proposed by MATTERN (1962) while a 92 capsomere model has also been proposed by CASPAR and KLUG (1962).

The large-plaque variant of SV40 (KOCH and EGGERS, 1967) studied by KOCH, EGGERS, and collaborators represents the most highly purified SV40 prepara-tion. These authors report 72 capsomeres (ANDERER, SCHLUMBERGER, KOCH, FRANK, and EGGERS, 1967) rather than 42 (see Fig. 10).

The virion is 13% nucleic acid on a weight basis (WINOCOUR, 1963; ANDERER, SCHLUMBERGER, KOCH, FRANK, and EGGERS, 1967). Chemical analyses of PV and SV40 report only DNA and protein constituents, however, small amounts of carbohydrates have not been excluded.

The nucleic acid of PV was identified as DNA by Dı MAYORCA, EDDY, STE-WART, HUNTER, FRIEND, and BENDICH (1959), SMITH, FREEMAN, M. VOGT, and DULBECCO (1960) and WEIL (1961). The group around EDDY and STEWART also established that phenol extracted DNA is infectious. The double stranded mole-

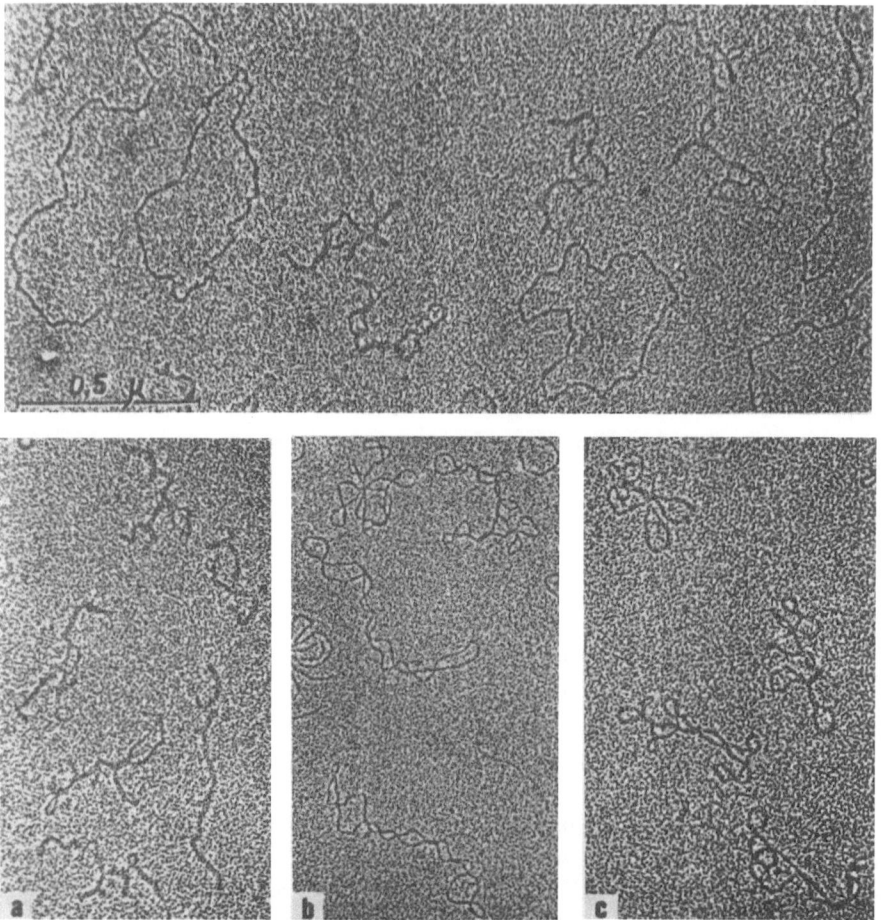

Fig. 11. Electron micrographs of SV40 DNA showing non-circular, open circular and twisted circular molecules. From ANDERER, SCHLUMBERGER, KOCH, FRANK. and EGGERS (1967)

cule has a molecular weight of about 3×10^6. From its unusual behavior in heat and alkali denaturation experiments, it was deduced that a large proportion of the isolated DNA molecules has a ring structure (DULBECCO and M. VOGT, 1963; WEIL and VINOGRAD, 1963). Thorough analysis in the analytical ultracentrifuge has revealed three forms of the DNA with sedimentation velocities of about 20, 16 and 14 Svedberg units (WEIL and VINOGRAD, 1963; CRAWFORD, 1964). The double-stranded (CRAWFORD, 1963) twisted cyclic molecules correspond to the 20S fraction (form I) and make up 80—90% of the total DNA. The artificial creation of a break in one of the strands of the DNA ring increases its infectivity and changes in to the open, nickend ring molecules of the 16S (form II) species (review by WEIL, 1964). The 20S—16S DNA molecules are believed to represent interchangeable forms of the same molecule (VINOGRAD, LEBOVITZ, RADLOFF, WATSON, and LAIPIS, 1965). Both have the same molecular weight and low content of guanine + cytosine (about 49%) (WEIL, 1964). The size corresponds to some 5,000 nucleotide pairs. The 14S linear DNA (form III) molecule is made up of a cellular rather than viral nucleic acid (DULBECCO, 1964; WEIL, MICHEL, and RUSCHMANN, 1965; WINOCOUR, 1967 a, b, 1968; MICHEL, HIRT, and WEIL, 1967). Infection by the circular 20S or 16S molecules is capable of inducing either lysis or transformation (DULBECCO and M. VOGT, 1963; CRAWFORD, DUL-BECCO, FRIED, MONTAGNIER, and STOKER, 1964).

The DNA from SV40 shows great physico-chemical similarities to that of PV. Its molecular weight is estimated to about 3×10^6 (CRAWFORD and BLACK, 1964) or $2.25 \times 10^6 \pm 15\%$ (ANDERER, SCHLUMBERGER, KOCH, FRANK, and EGGERS, 1967). A guanosine + cytosine content of only 41% has been reported and native SV40 DNA is also presumed to exist in a circular form (CRAWFORD and BLACK, 1964; ANDERER, SCHLUMBERGER, KOCH, FRANK, and EGGERS, 1967) (Fig. 11).

The empty capsids of both viruses are formed naturally in the course of virus synthesis (WILDY, STOKER, MACPHERSON, and HORNE, 1960; L. CRAWFORD, E. CRAWFORD, and WATSON, 1962; BLACK, E. CRAWFORD, and L. CRAWFORD, 1964). These are lighter than complete virions and band at a density of about 1.29 g/ml. The polyoma capsid protein agglutinates red blood cells from many species in the cold (EDDY, ROWE, HARTLEY, STEWART, and HUEBNER, 1958; SACHS, FOGEL, and WINOCOUR, 1959; FOGEL and SACHS, 1959; ROIZMAN and ROANE, 1960). Guinea-pig erythrocytes have been used in routine procedures for viral hemagglutinin (HA) or hemagglutinin inhibition (HI) antibodies (SACHS, FOGEL, WIN-OCOUR, HELLER, MEDINA, and KRIM, 1959; FOGEL and SACHS, 1959). PV capsid antigens may be detected in neutralization tests (ROWE, HARTLEY, ESTES, and HUEBNER, 1959; SACHS, FOGEL, WINOCOUR, HELLER, MEDINA, and KRIM, 1959), by immunofluorescence (G. HENLE, DEINHARDT, and RODRIGUEZ, 1959; MALM-GREN, RABOTTI, and RABSON, 1960; WILLIAMS and SHEININ, 1961; LEVINTHAL, JAKOBOVITS, and EATON, 1962; FRASER and GHARPURE, 1962) or in the electron microscope by ferritin-labeled antibody (BIBERFELD and RINGERTZ, 1966).

SV40 does not hemagglutinate. It induces neutralizing antibody in various species, including man (MORRIS, JOHNSON, AULISIO, CHANOCK, and KNIGHT, 1961).

Highly purified SV40 preparations have yielded three rather small different polypeptide chains (A, B and C), in an estimated proportion of 45.5:45.5:9.0 with average molecular weights of $16,350 \pm 10\%$ (ANDERER, SCHLUMBERGER, KOCH, FRANK, and EGGERS, 1967). One strand of viral DNA with an estimated molecular weight of about 1.5×10^6 can be calculated to carry information for some 1,200 amino acids. The A, B and C polypeptides, which have been separated and analysed with respect to their amino acid composition represent about 450 amino acids. This implies that over one third of the genetic code of each strand of viral DNA is used for the synthesis of "structural" proteins found in the mature virion (SCHLUMBERGER, ANDERER, and KOCH, 1968).

Each of 420 structural subunits of the particle shell is made up of one A and one B polypeptide. The C protein, on the other hand, is located in the interior of the particle where it seems to act as a link between the DNA and the capsid and play a role in orienting the DNA and stabilizing the structure of the virion. The basic internal C protein does not react with antisera against intact SV 40 particles (ANDERER, KOCH, and SCHLUMBERGER, 1968).

FINE, MASS, and MURAKAMI (1968) detected two different proteins in polyoma virions; one internal basic protein and one capsid protein.

b) Variants

Polyoma virus. GOTLIEB-STEMATSKY and LEVENTON (1960) infected primary mouse embryo cultures with the wild strain PV isolated by STEWART and EDDY and obtained plaques of two different sizes. Stable variants were propagated and designated large plaque (Lp) and small plaque (Sp). Variants with the same characteristics were observed by MEDINA and SACHS (1960) and DULBECCO and M. VOGT (1960). The Sp variant adsorbs to hamster cells at least 36 times more efficiently than the Lp variant (DIAMOND and CRAWFORD, 1964; DIAMOND, 1964). It has been suggested that the large negative charge of Lp polyoma counteracts adsorption to the negatively charged surface of the cell (THORNE, HOUSE, and KISCH, 1965). Sp virus contains more lysine and serine and less methionine than Lp virus (MURAKAMI, FINE, HARRINGTON, and BEN SASSAN, 1968).

In the terms of input multiplicity the Sp variant appears to be the more efficient transforming agent (MEDINA and SACHS, 1963); however, in terms of number of adsorbed plaque-forming units, the large plaque variant is superior (DIAMOND, 1964). Lp usually multiplies to higher final virus yields in mouse cells than Sp (GOTLIEB-STEMATSKY and LEVENTON, 1960; KOCH and EGGERS, 1967), possibly because of a slower cell lysis.

Another variant of PV, M-polyoma, was described by LAW, RABSON, and DAWE (1961). It was developed by passing PV on mouse lymphoma cells in a medium containing 20% defatted human serum. It showed a considerably reduced oncogenicity in mice (RABSON and LAW, 1963), possibly because of its increased capacity to induce synthesis of interferon (FRIEDMAN, RABSON, and KIRKHAM, 1963; FRIEDMAN and RABSON, 1964a and b). The same correlation between strong stimulation of interferon production and low tumorigenicity was observed by GOTLIEB-STEMATSKY, ROTEM, and KARBY (1966) in two PV variants.

HARE and collaborators (HARE and MORGAN, 1963; HARE, 1964a, 1967a, b; HARE and GODAL, 1965) have described variants of PV defective in transplantation antigen induction. One large plaque (Lp-D) and small plaque (Sp-D) variant failed to induce transplant immunity against cells transformed by several other PV strains.

A temperature sensitive mutant (Ts-a) able to transform hamster cells only at decreased temperature (31.5°C) was first described by FRIED (1965). Recently many new Ts mutants have been isolated (ECKHART, 1969b; DI MAYORCA, CALLENDER, MARIN, and GIORDANO, 1969; OZER and TAKEMOTO, 1969). Preliminary complementation results suggest that a genetic mapping should be possible since these mutations fall into defined groups.

Simian virus 40. RIGGS and LENNETTE (1965) isolated a large plaque variant which produced "red" plaques in contrast to the wild strain's "white" plaques because of a difference in the uptake of neutral red. TAKEMOTO, KIRSCHSTEIN, and HABEL (1966) isolated three variants designated Lp (large), Sp (small) and Mp (minute) according to the plaque size in primary African green monkey kidney cultures. Lp was the only variant which failed to yield infectious virus at temperatures above 39°C. Lp and Sp variants can be immunologically separated (OZER, TAKEMOTO, KIRSCHSTEIN, and AXELROD, 1969). All three could transform human fibroblasts but the Lp variant was slightly more oncogenic in hamsters than the other two. Minute plaque virus was considerably more efficient in transforming mouse and human cells than the large plaque variant with the small plaque variant in an intermediate position (TODARO and TAKEMOTO, 1969). Minute plaque virus has the unusually low ratio of 1,000:1 between lytic plaque forming units and transforming hamster cells (KOCH and EGGERS, 1967).

Successive undiluted passage of SV40 through GMK cultures results in the production of heterogenous defective virions incapable of starting a lytic sequence (UCHIDA, WATANABE, and KATO, 1966; SAUER, KOPROWSKI, and DEFENDI, 1967; UCHIDA, YOSHIIKE, WATANABE, and FURUNO, 1968). Such defective virions are as efficient tumor formers in hamsters as complete virions (UCHIDA and WATANABE, 1968).

c) Virus-cell Interactions of PV and SV40

Evidently, an oncogenic DNA virus effects cells in two different ways. One leads to the well known cytopathic effect with release of large numbers of newly synthesized virions which most "ordinary" viruses show (ROBBINS, ENDERS, and WELLER, 1950). The other leads to the opposite of cell lysis — an increased growth rate, prolongation of the life span of the cultures and a capacity for growth under conditions where normal cells are inhibited (M. VOGT and DULBECCO, 1960). No or very little infectious virus is produced in this transformative type of interaction. Much effort has been made to understand precisely what determines whether an interaction between an oncogenic DNA virus and a target cell will lead to transformation or to lysis.

Apart from lysis or transformation, PV and SV40 may also give rise to a so-called abortive infection. This ill-defined state is characterized by the association of virus with cytoplasmic structures, where it remains for several days before it is slowly expelled into the medium without any consequences for the exposed cell. It has never been convincingly demonstrated that this is not merely a passive adsorption to the outer cell membrane or intracytoplasmatic vacuoles without decoating and exposure of viral genome to the synthetic cell machinery. In the SV40-human fibroblast interaction abortive infection may, however, involve not only adsorption but also temporary disappearance of virus infectivity (CARP and GILDEN, 1966).

d) Lytic Effects of Polyoma and SV40

Polyoma virus. The cytopathic effect of PV is most conveniently studied in cultures of embryonic mouse cells where almost all infected cells undergo lysis (Table 3). This cytopathic effect (EDDY, STEWART, and BERKELY, 1958), which becomes visible after 4—7 days, is used to titrate the virus either in tubes or by a plaque assay (DULBECCO and FREEMAN, 1959; SACHS, VOGEL, and WINOCOUR, 1959; SHEININ, 1961).

4*

A one-step growth curve of PV in mouse cells shows no principle difference to that of ordinary cytopathic viruses. One infectious unit is about 100 physical particles and the eclipse phase is long, about 24 hours (WINOCOUR and SACHS, 1960). Regardless of the cell type employed, about 800—1,000 PFU are released from one lysing cell. The infectious cycle is strongly asynchronous (WEISBERG, 1963; WEIL, MICHEL, and RUSCHMANN, 1965; WEIL, PÉTURSSON, KÁRA, and DIGGELMANN, 1967; PÉTURSSON and WEIL, 1968), and cell destruction is slow and newly synthesized PV is bound to the cells for some time before it is released into the medium (WINOCOUR and SACHS, 1960).

Table 3. *Effects of Polyoma Virus in Different in vitro Systems*

Target cells	Percentage of exposed cells showing			Virus synthesis
	Transformation	"Abortive infection"[1]	Lysis	
Mouse				
embr. mix. tissue	<1	n.t.	>90	+++
embr. kidney	<1	n.t.	>90	+++
3T3 line	low	n.t.	moderate	++
Rat				
embr. mix. tiss.	0.3-0.6	>90	<0.1[2]	(+)
Hamster				
embr. mix. tissue	0.003-0.01	>90	0	0[3]
embr. kidney	1	n.t.	0	0
BHK 21 line	1-5 (-10)	95	0	0[4]
Bovine				
embr. mix. tissue	very low	n.t.	0	0
Human	0	100	0	0

[1] No morphologic change, unimpaired viability. Uptake of virus established by fluorescent antibody, electron microscopy, radioactive virus etc.
[2] Based on fraction of cells synthesizing virus.
[3] A few exceptions with continuously PV releasing lines have been reported (DULBECCO and M. VOGT, 1960; STANNERS, TILL and SIMINOVITCH, 1963; DEFENDI and LEHMAN, 1965).
[4] An early phase with limited PV synthesis has been suggested (BOURGAUX, 1964; FRASER and CRAWFORD, 1965).
n.t. = not tested.

Adsorption of PV proceeds rapidly, but is to a large extent reversible suggesting an equilibrium between free and adsorbed virus. The receptor destroying enzyme from *Vibrio cholerae* releases the virus from cell debris (CRAWFORD, 1962).

BOURGAUX (1964) followed the early fate of P^{32}-labeled small plaque PV in mouse embryo cells. Crude nuclear and cytoplasmic fractions were prepared, the acid soluble and insoluble radioactivity was estimated and samples were subjected to RbCl density gradient centrifugation. The adsorbed radioactivity followed three paths. 1. Rejection as whole virus particles back into the medium. This was still observed 24 hours post infection. 2. Breakdown of virions with loss of low molecular weight material to the medium. 3. Transfer to the nuclear fraction.

Autoradiography of mouse embryo cells infected with H³-thymidine labeled PV showed grains exclusively in the cytoplasm at 3 hour post infection, at 6 hour they were found in the perinuclear region beginning to penetrate into the nucleus. At 18 hours virtually all the radioactivity was intranuclear (KHARE and CONSIGLI, 1965).

These fragmentary studies indicate that PV is adsorbed to mouse embryo cells with great efficiency. The details of the process are unknown and a large proportion of PV may be unproductively associated with membrane structures. From this temporary association it may be released again, apparently intact, or it may be uncoated in which case the DNA enters the nucleus, the site for virus synthesis.

Some insight has been gained in the details of PV replication during recent years. A schematic outline is found in Fig. 12; there is still uncertainty about some of the steps. A certain confusion has been caused by the different responses

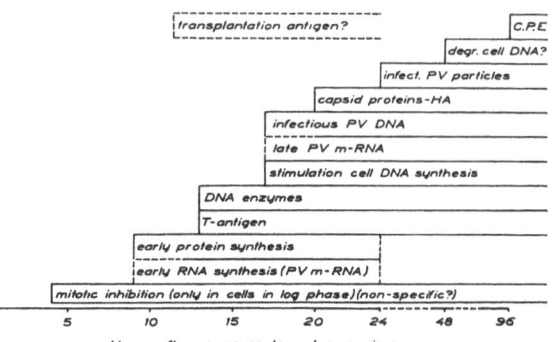

Fig. 12. Temporal relationship between various events in a lytic polyoma virus — mouse cell interaction. The indicated times can only be regarded as approximate

of cells in exponential growth phase and those whose cell cycle is subjected to cell density dependent depression (MINOWADA, 1964; SHEININ and QUINN, 1965; CRAMER and FEINENDEGEN, 1966).

The earliest recorded event is the inhibition of mitosis. It is observed already 4 hours after infection of exponentially multiplying mouse embryo cultures where the mitotic index decreases from 5 to about 1.5%. By 10—12 hours it is virtually zero (SHEININ and QUINN, 1965). It is not known whether this action is indeed under the direct influence of the viral genome. It may have represented a non-specific effect from constituents in the unpurified virus pool used for infection, or due to the extremely high multiplicity of infection (1,000—2,000 PFU *i.e.* $1-2 \times 10^5$ virus particles/cell) used.

BENJAMIN (1966a, b) was able to follow the time course of the production of virus specific RNA in mouse kidney cells by using the sensitive method of hybridization of H³-thymidine labeled virus DNA with P³² pulse labeled cell RNA. A gradual increase was noted from 4—28 hours after PV exposure (Fig. 13); the exact point of time at which the production of virus specific RNA starts cannot be determined. The gradual increase suggests an early initiation and in Fig. 12 the start has been placed at 9 hours. If the reasonable assumption is made that the RNA represents messenger molecules transcribed from the polyoma DNA (BENJAMIN, 1966b) it becomes natural to accept it as one of the earliest

events during the replicative polyoma cycle. This fits with the observation by
GERSHON and SACHS (1966) that an RNA required for PV development was
present about 9 hours after exposure to PV. The "early" RNA of BENJAMIN
(1966a) only hybridized with part of the virus DNA. At a latter stage, tentatively
placed at 15—21 hours, a "late" RNA was detected which hybridized with the
entire virus DNA molecule.

The T-antigen (neo-antigen) is defined by its reactivity in complement-
fixation tests with immunsera from tumor-bearing hamsters (page 71). Detect-

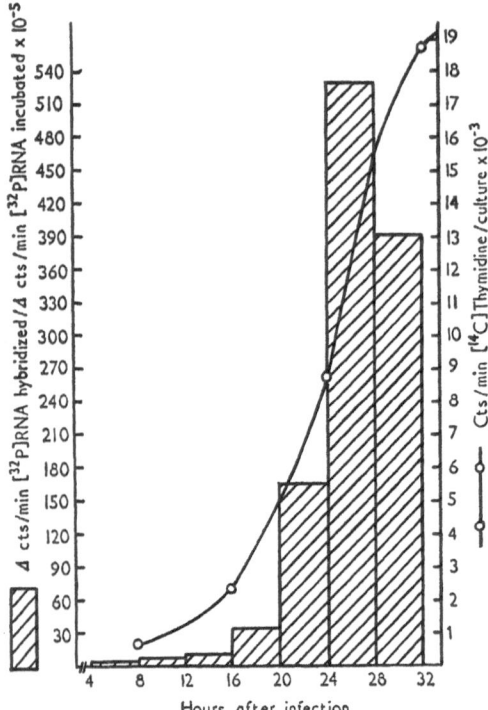

Fig. 13. Relative amounts of virus-specific
RNA at different times after infection of
mouse kidney cell by PV. From BENJAMIN
(1966a)

able amounts have been reported about 9 hours post infection after which there
is a rapid increase in the proportion of positive nuclei (TAKEMOTO, MALMGREN,
and HABEL, 1966a). Synthesis of early proteins necessary for production of infectious PV starts 8—9 hours after infection (GERSHON and SACHS, 1964) as
determined by the response of infected cells to puromycin.

Simultaneously with early protein synthesis, the activity of thymidine kinase,
DNA polymerase and at least 7 more enzymes involved in DNA synthesis begins
to rise steeply (HARTWELL, M. VOGT, and DULBECCO, 1965; DULBECCO, HARTWELL,
and M. VOGT, 1965; FREARSON, KIT, and DUBBS, 1965; KÁRA and WEIL, 1967).
The relatively small size of PV DNA precludes that all 9 enzymes are coded
by the viral genome. With the suggested exception of thymidine kinase (SHEININ,
1966a) the increase probably represents the stimulation of synthesis of normal
cellular enzymes (KIT, DUBBS, and FREARSON, 1966; KÁRA and WEIL, 1967).
Polyoma virus failed to increase the thymidine kinase activity in sublines of

3T3 selected for absence of enzyme activity (LITTLEFIELD and BASILICO, 1966). Vaccinia virus known to contain genes for thymidine kinase did under the same conditions induce enzyme activity (BASILICO, MATSUYA, and GREEN, 1969). This seems to establish that cellular genes code for thymidine kinase.

Stimulation of cell DNA synthesis (Fig. 12) was independently described by three groups (DULBECCO, HARTWELL, and M. VOGT, 1965; WEIL, MICHEL, and RUSCHMAN, 1965; WINOCOUR, KAYE, and STOLLER, 1965). DULBECCO and his collaborators employed "terminal density" kidney cells obtained from 10—14 days old mice. A few days after plating such cultures showed pronounced cell cycle inhibition. When purified Lp polyoma was added, a 10-fold increase in DNA synthesis, as measured by incorporation of 3H-thymidin, was observed after 16—24 hours. The newly formed DNA was fractionated into cell and virus DNA on a methylated albumin kieselguhr column and about two thirds of it eluted in the same manner as mouse cell DNA. Its probable cellular nature was further substantiated by its ability to hybridize with normal mouse DNA. The appearance of infectious DNA coincided with the time course of the virus DNA isolated by chromatography. Under slightly different conditions the other two groups obtained essentially the same results.

Later studies by WEIL, PÉTURSSON, KÁRA, and DIGGELMANN (1967) reveal complexities which make the above interpretations less certain. So-called stationary control cultures are in fact composed of epithelial-like cells with only little DNA synthesis but also contain fibroblast-like cells with a high rate of DNA synthesis even under crowded conditions. It was furthermore observed that up to 30% of the cells already are lost 35 hours after PV infection, and it may be objected that this triggers the cell cycle of these remaining in a non-specific manner by decreasing the local cell density.

By using Eagle's medium with only 0.5% calf serum FRIED and PITTS (1968) could reduce the proportion of mouse embryo cells participating in DNA synthesis to about 8% of the population in a dense culture. PV infection will then lead to a marked increase in total DNA synthesis. This system may have some advantage by its lowered background of DNA synthesis. It is on the other hand difficult to exclude that the low serum concentration prevents the full expression of the PV genome.

The stimulation of DNA synthesis by PV has usually been interpreted as a release from cell cycle inhibition. The discovery that differentiated multinucleated rat muscle fibers, which do not show any DNA synthesis in culture, are activated by exposure to PV (YAFFE and GERSHON, 1967) as are mouse cells whose DNA synthesis has been depressed by X-irradiation (GERSHON, SACHS, and WINOCOUR, 1966), suggest that the stimulation of DNA synthesis may have another explanation. The stimulation of cellular DNA synthesis by PV (and SV40) has been considered a result of the operation of a specific viral gene [cf. DULBECCO (1966) and ECKHART (1969a) for reviews]. A possibility not usually taken into account is, however, a direct viral action during adsorption and penetration on the cell surface, which counteracts the density dependent cell cycle inhibition of confluent target cells.

It should also be noted that PV causes inhibition rather than stimulation of host cell DNA synthesis in exponentially replicating mouse embryo cells

(SHEININ and QUINN, 1965; BIRNIE and FOX, 1965; SHEININ, 1966b; 1967) at least after a high multiplicity of infection (BRANTON and SHEININ, 1968).

The significance of a PV induced cellular DNA synthesis in murine cells is not clear. It has been regarded as an expression of the mechanism which causes transformation, an idea compatible with induction of cellular DNA synthesis in rat embryo cells (SHEININ, 1966c), where less than 0.1% of the exposed cells lyse. However, the rate of transformation in the mouse or the rat system is too low to permit definite conclusions. The large majority of infected mouse cells will undergo lysis, which clearly excludes that induction of cell DNA synthesis is sufficient to cause transformation. Combined use of autoradiography and immunofluorescence (M. VOGT, DULBECCO, and SMITH, 1966) has also established that the cells whose DNA synthesis is enhanced also produce PV capsid antigen and therefore probably lyse in a productive infection.

The newly synthesized DNA is broken down during a later stage of the infectious cycle (BEN-PORAT, CATO, and KAPLAN, 1966; BEN-PORAT and KAPLAN, 1967) and no histone synthesis was recorded in connection with the DNA synthesis (GERSHON, HAUSEN, SACHS, and WINOCOUR, 1965). This may only concern X-irradiated target cells since non-irradiated mouse embryo cells were recently found to increase their synthesis of histone as well as DNA after PV infection (SHIMONO and KAPLAN, 1969). Phase-contrast microscopy and Feulgen microspectrophotometry of PV infected mouse kidney cells have shown that DNA stimulation probably entails synthesis without mitoses (WEIL, PÉTURSSON, KÁRÁ, and DIGGELMANN, 1967). These facts suggest that, although the DNA shows physico-chemical similarities to normal cell DNA, its assembly, stability and function may be quite different. Before PV stimulation of DNA synthesis is accepted as a step leading to transformation it has to be shown that the very cells whose DNA synthesis is stimulated are those which are transformed. Stimulation of cell DNA has been considered unique for oncogenic viruses because of a failure to detect it after infection by pseudorabies virus of stationary rabbit kidney (BEN-PORAT and KAPLAN, 1963). More work on stationary cells infected by non-oncogenic DNA viruses multiplying in the nucleus seems, however, necessary before any generalizations can be accepted.

During the last part of the infectious cycle, capsid protein and viral DNA are assembled to complete particles and the cultures gradually lyse. The DNA of the infecting particles does not seem to be directly incorporated into the progeny virions (MAURER, WERTMAN, and YALL, 1969). In addition to infective and empty virus particles, there are also found viral capsids containing cell DNA. It is not known whether the incorporation of cell DNA is random or prefers certain genes and no function has been attributed to the encapsidated cell DNA (MICHEL, HIRT, and WEIL, 1967; WINOCOUR, 1968). Transformation seems incompatible with rapid synthesis of PV which invariably leads to cell destruction (M. VOGT and DULBECCO, 1962; HARE, BALDUZZI, and MORGAN, 1963; G. HENLE, HINZE, and W. HENLE, 1963).

Simian virus 40. Exposure of either primary African green monkey cells (GMK) or such established lines as BS-C-1 or CV-1 to SV40 leads to a productive infection and lysis in the vast majority of the cells (SWEET and HILLEMAN, 1960; HSIUNG and GAYLORD, 1961; GAYLORD and HSIUNG, 1961; MEYER, HOPPS,

Rogers, Brooks, Bernheim, Jones, Nisalak, and Douglas, 1962; Mayor, Stinebaugh, Jamison, Jordan, and Melnick, 1962; Levinthal and Shein, 1963; Diderholm, 1963). The general features are similar to those of the PV-mouse system with a long eclipse phase of about 20 hours, commencement of a cytopathic effect after 3—4 days and considerable asynchrony in virus maturation (Mayor, Stinebaugh, Jamison, Jordan, and Melnick, 1962). The events have, as far as they have been worked out, followed the same time course as already outlines for polyoma virus (Fig. 12). It is particularly noteworthy that a stimulation of cell DNA synthesis also can be observed in the lytic GMK-SV40 system (Werchau, Westphal, Maass, and Haas, 1966; Hatanaka and Dulbecco, 1966; Frearson, Kit, and Dubbs, 1966). Suppression of the stimulation of cellular DNA synthesis did not reduce synthesis of viral DNA (Werchau, Kaukel, Maass, Brandner, and Haas, 1968). The increased DNA synthesis is preceded by a pronounced increase in thymidine kinase, DNA polymerase, dihydrofolate reductase and thymidylate synthetase activity (Kit, Dubbs, de Torres, and Melnick, 1965; Kit, Dubbs, Frearson, and Melnick, 1966; Hatanaka and Dulbecco, 1966; Kit, Dubbs, Pierkarski, de Torres, and Melnick, 1966; Kit, 1967). Differences in the Michaelis constant between the partially purified SV40 induced thymidine kinase and the natural GMK enzyme have suggested that the former may be a "new" enzyme coded for by the SV40 genome (Kit, 1967; Hatanaka and Dulbecco, 1967). Immunologic differences between enzyme from infected and normal cells also point in the same direction (Carp, 1967). Again the stimulatory effect was only observed in cell-cycle inhibited cultures. It should be stressed that the normal cell cycle inhibition in the GMK system is only relative. Kit (1967) had a background of 10—20% nuclei incorporating thymidine after a 2 hour long pulse label, which signifies considerable DNA synthesis in the "stationary" control cultures.

Under conditions of active growth, cell DNA synthesis was instead depressed (Munk, Fischer, Goedemans, and Sauer, 1965; Sauer, Fischer, and Munk, 1966). This again is not absolute; Carp and Gilden (1966) reported continued cell replication after infection of exponentially growing AGMK cells.

The SV40 induced T-antigen (tumor antigen, CF-antigen, ICFA, neoantigen) (review by Black, 1966c) was first demonstrated by Black, Rowe, Turner, and Huebner (1963) in hamster sarcoma tissue reacted in a complement fixation test with sera from tumor bearing hamsters. The antigen can be detected in individual cells using FA techniques (Pope and Rowe, 1964a; Rapp, Butel, and Melnick, 1964; Gilden, Carp, Taguchi, and Defendi, 1965). Antibodies to T-antigen are found in monkeys infected experimentally or naturally with SV40 in the absence of any tumor formation (Vonka, Závadová, Kutinová, and Řezáčová, 1967; Rapp, Tevethia, Rawls, and Melnick, 1967; Shah and Hess, 1968; Shah, Willard, Myers, Hess, and Di Giacomo, 1969). The SV40 specific T-antigen is intranuclear and probably associated with the chromatin (Leduc, Wicker, Avrameas, and Bernhard, 1969).

The T-antigen, originally believed to be specific for neoplastic cells, was first observed in a lytic interaction by Pope and Rowe (1964a) and Gilden, Carp, Taguchi, and Defendi (1965). In a study by Carp and Gilden (1966) it was noted 18 hours post-infection in about 3% of the nuclei. The proportion increased rapidly with time to reach 80% at 48 hours and about 90% at 72 hours. BLACK and ROWE (1965) have indicated a minimum latent time of 6—10 hours.

The development of the T-antigen is influenced by the growth state of the infected cells. Stationary phase AGMK cells 4 days after infection showed more than 80% positive nuclei in contrast to replicating cells which only showed 1—10% fluorescent nuclei.

Synthesis of T-antigen occurs under conditions where all DNA synthesis has been blocked by cytosine arabinoside or FUDR (BUTEL and RAPP, 1965; RAPP, MELNICK, and KITAHARA, 1965; GILDEN, CARP, TAGUCHI, and DEFENDI, 1965; ROWE, BAUM, PUGH, and HOGGAN, 1965). The messenger RNA coding for the T-antigen must then be transcribed from the original infecting SV40 DNA. T-antigen can also be induced by purified SV40 DNA (BLACK and ROWE, 1965). T-antigen induction is inactivated by a pretreatment of virus with UV light according to a first order dose-response curve while infectivity is inactivated at a faster rate, indicating that only part of the SV40 genome codes for the T-antigen (CARP and GILDEN, 1965). Non-infectious defective SV40 particles carrying information for T-antigen have been described (UCHIDA, WATANABE, and KATO, 1966; ALTSTEIN, SARYCHERA, and DODONOVA, 1967; UCHIDA, YOSHIIKE, WATANABE, and FURUNO, 1968). The finding that defective particles contain a DNA molecule 12% shorter than that of complete infectious SV40 virions strongly suggests that the defectiveness is actually caused by physical loss of a piece of DNA (YOSHIIKE, 1968a, b).

Synthesis of the T-antigen requires DNA-dependent RNA synthesis and protein synthesis (RAPP, BUTEL, FELDMAN, KITAHARA, and MELNICK, 1965; GILDEN and CARP, 1966). In the first of these two reports puromycin did not reduce synthesis of T-antigen. Later GILDEN and CARP (1966) explained this as an artifact caused by an inability of puromycin to affect protein synthesis in GMK cells; another protein synthesis inhibitor, cycloheximide, is fully effective. Arginine-deficient medium curtails synthesis of infectious virus but has no or only a slight effect on the T-antigen (GOLDBLUM, RAVID, and BECKER, 1968).

The T-antigen is destroyed by trypsin and precipitates in half-saturated ammonium sulfate at pH 5 (DEFENDI, 1966) supporting that at least part of the antigen is a protein. Purified preparations had protein-like properties. GILEAD and GINSBURG (1968a, b) considered the T-antigen a single protein, whereas TOCKSTEIN, POLASA, PIÑA, and GREEN (1968) could resolve their material into 3—4 components.

The function of the T-antigen during the infectious SV40 cycle is not understood.

All available evidence, as reviewed by DEFENDI (1966), indicates that the T-antigen is not part of the virion. Neither whole nor disrupted SV40 particles react with T-antigen reactive antiserum and antiviral sera do not appear to react with transformed cells (DEFENDI, 1966). DNA inhibitors affect the viral antigens but not the T-antigen. Their time courses in different systems is not identical and a thermal separation of the synthesis of the T- and virus-antigen can be achieved (KITAHARA and MELNICK, 1965). Defective SV40 virions which induce synthesis of either T- or viral (V)-antigen can be obtained by serial passage of undiluted virus (UCHIDA, YOSHIIKE, WATANABE, and FURUNO, 1968).

Lytic infection of monkey cells by SV40 is characterized by cytoplasmic vacuoles (SWEET and HILLEMAN, 1960), intranuclear inclusion bodies (HSIUNG and GAYLORD, 1961; GAYLORD and HSIUNG, 1961; VENDRELY, KASTEN, TOURNIER, and WICKER, 1962; MAYOR, STINEBAUGH, JAMISON, JORDAN, and MELNICK, 1962) and an early involvement of the nucleous (MAYOR, STINEBAUGH, JAMISON, JORDAN, and MELNICK, 1962; BERNHARD and TOURNIER, 1962; LEVINTHAL and SHEIN, 1963; CRANBOULAN, TOURNIER, WICKER, and BERNHARD, 1963; PRUNIERAS, CHARDONNET, and SCHIER, 1964; GRANBOULAN and TOURNIER, 1965; KASTEN, VENDRELY, TOURNIER, and WICKER, 1965; LOVE and FERNANDES, 1965). These features do not seem to be an equally striking part of the cytopathic response to polyoma virus. A nucleolar involvement reminiscent of SV40 infection has, however, been described in polyoma infected mouse and hamster cells (BERNHARD, FEBVRE, and CRAMER, 1959; LOVE and RABSON, 1960; BERECZKY, DMOCHOWSKI, and GREY, 1961; BERECZKY and DMOCHOWSKI, 1962; FRASER and CRAWFORD, 1965).

Cytopathic effects of SV40 in systems other than GMK have only been incompletely analysed but seem to follow a similar pattern in those individual cells infected. Since such cells constitute only a small minority of the entire population, the events of the level of the mass population will be entirely different.

2. Transformation by Polyoma Virus

PV has a rather narrow spectrum of transformation *in vitro* and only rodent or bovine cultures are susceptible (Table 3). The virus replicates to high titers only in cells from its natural host — the mouse. The virus yield is directly related to the degree of cell lysis. Hamster cultures usually show no virus synthesis and no cytopathic effects; in contrast, murine cells which produce large quantities of virus react with widespread lysis. In all systems the number of transformed cells is far less than the number of cells exposed to virus.

The types of transformations obtained by polyoma virus are summarized in Table 4. In inherently stable bovine cells PV produces the irregular and unrestrained growth transformation but only rarely infinitely growing cell lines. This is based on long term observation of 6 individual calf embryo lines infected with PV around passage 10. All underwent a characteristic morphologic transformation and reacted positively in CF and FA tests with sera from polyoma tumor bearing hamsters (DIDERHOLM and WESSLÉN, 1965). At 50—60 passages after irregular growth transformation, the cultures showed degenerative changes with formation of vacuolated multinucleated cells, and gradual loss of proliferative capacity terminating in total lysis in all but one line, which progressed to infinite growth transformation (PONTÉN, unpublished). These results have been reproduced on several occasions with various frozen or fresh sublines and are interpreted as a "growth crisis" similar to that encountered in the SV40-human system (page 81) (DIDERHOLM and PONTÉN, unpublished). That the infinitive growth transformation is only regularly obtained against an unstable rodent background suggests that PV infection in itself is insufficient for the creation of established cell lines.

PV infected embryonic mouse cells have been difficult to analyse. Typically a majority will undergo lysis leaving a few survivors behind, which eventually give rise to established lines with an irregular and unrestrained growth pattern. Because of the tendency for murine cells to undergo spontaneous transformation, it has not been possible to determine if PV plays a direct role in the transformation process or merely acts as a selective agent eliminating the normal cells. As pointed out by DEFENDI (1966), the fact that PV induced antigens are commonly found in the surviving population does not prove a direct role of the

Table 4. *Transformation in vitro by Polyoma Virus*

Cell line	Target background			Lysis	Target response		
	Transformation				Transformation		
	Irreg. growth	Unrest. growth	Infinite growth		Irreg. growth	Un-restr. growth	In-finite growth
Mouse embryonic	tendency to spontaneous trans-formation			+++	yes	yes	yes
3T3 estab. line	absent	absent	present	++	yes	yes	1
Rat embryonic	tendency to spontaneous trans-formation			—	yes	yes	n.t.
Hamster embryonic	tendency to spontaneous trans-formation			—[2]	yes	yes	yes
BHK 21 estab. line	absent	present	present	—[3]	yes	1	1
Bovine embryonic	no tendency to spontaneous trans-formation			—	yes	yes	yes[4]

[1] Already present before exposure to PV.
[2] A few lines with small production of PV have been reported (DULBECCO and M. VOGT, 1960; STANNERS, TILL, and SIMINOVITCH, 1963; DEFENDI and LEHMAN, 1965).
[3] Slight PV synthesis early after infection has been suggested (FRASER and GHARPURE, 1962; BOURGAUX, 1964).
[4] Not regularly, most cultures succumb in a growth crisis (PONTÉN, unpublished).
n.t. = not tested.

virus in the transformation. It is equally possible that spontaneously transformed cells have become secondarily infected and that PV only acts as a passenger. The 3T3 established mouse line also shows a high grade of lysis (TODARO, GREEN, and GOLDBERG, 1964) and has not been used for any extensive work on PV transformation.

Embryonic hamster and rat cells are also inherently unstable (page 22). Their superiority to mouse cells stems from the rarity of infectious virus synthesis and cell lysis so that the transformation process can be studied with only a minimal risk of secondary infection and increased chances that the observed effects are due to the primary encounter with the virus. These advantages are particularly prominent in the embryonic Syrian hamster kidney line BHK-21.

Work with hamster cells will therefore be discussed in detail, whereas other systems will be only briefly touched upon.

The chain of events in hamster cell populations have been studied by M. VOGT and DULBECCO (1963), SANFORD and HOEMANN (1967) and others, and reviewed by DEFENDI (1966).

a) Individuality of Transformed Lines

DEFENDI and LEHMAN (1965) exposed 10 individual Syrian hamster embryonic cell lines to a high multiplicity of PV from the same pool. Each line subsequently showed a characteristic morphology and karyology and a characteristic rate of progression towards increasing transplantability. Also, the morphology of the sarcomas produced by implantation was characteristic for the individual transformed lines. One of the 10 lines synthesized infectious virus continuously, the others rapidly lost this ability, when tested by immunofluorescence, capacity to induce HA inhibiting antibody or directly for infective virus. Transplantation antigens were shown to be present in some, but not all lines, by an increase of the number of cells required for a 50% take (MTD) in animals pretreated with PV. In two originally positive lines the transplantation antigen disappeared after prolonged passage. Isolated colonies of transformed cells had previously been found heterogenous with respect to morphology and plating efficiency (STANNERS, TILL, and SIMINOVITCH, 1963). Clones of mouse embryo cultures which transformed after exposure to PV also displayed a broad spectrum of individual characters (WEISBERG, 1964).

A recent example of variability between PV transformed cells has been published by RABINOWITZ and SACHS (1968, 1969a, b). Clones of two types — epithelioid or fibroblastoid — could be isolated from hamster cells transformed by LpPV. The growth of the former was selectively promoted by the presence of glutaraldehyde fixed normal cells. The epithelioid clones showed a decreased saturation density and a diminished colony forming capacity in soft agar but augmented tumorigenicity. The frequency of epithelioid variants was increased by growing cells at low density.

The high tumorigenicity of clones with decreased saturation density suggests that the correlation between unrestrained growth transformation and tumorigenicity claimed by AARONSON and TODARO (1968b, Fig. 34) may not be generally applicable. This question cannot be settled until the density dependent proliferation control of the PV cells has been investigated with a more reliable technique than the dye marking method used by RABINOWITZ and SACHS (1968).

The variability and individuality of the transformed cultures is highly reminiscent of *in vivo* tumors which often have characteristic minor features, which differ even within the same tumor category.

b) Induction of Irregular and Unrestrained Growth Transformation

The variation between lines should not obscure their common features outlined in Fig. 14. All PV exposed mass cultures have shown irregular growth transformation (cf. Fig. 2) after a variable time. In the study of DEFENDI and LEHMAN (1965) this varied between 5 and 15 passages and the shortest time corresponded to about 10 cell generations. The irregular growth transformation is a diffuse change, if the cells are subcultivated, or a focal alteration if they are allowed to remain in the original culture vessel. The cell orientation becomes completely random, but on the basis of whether the cells tend to spread out or pile up on top of each other, the transformed colonies may either be classified as "thin" or "thick" (M. VOGT and DULBECCO, 1962). The proportion of the latter increased with serial passage and it was suggested that they represented a later step in a progression (M. VOGT and DULBECCO, 1962). This suggestion is indicated in Fig. 14 by the insertion of unrestrained growth transformation after irregular growth transformation although it is not definitely known if a thick pattern indeed corresponds to unrestrained growth. This sequence has not been confirmed by STANNERS, TILL, and SIMINOVITCH (1963) or DEFENDI and LEHMAN (1965), but has apparently been observed in mouse cells by WEISBERG (1964). The lack of contact inhibition of cell movement is readily detected in time-lapse cinemato-

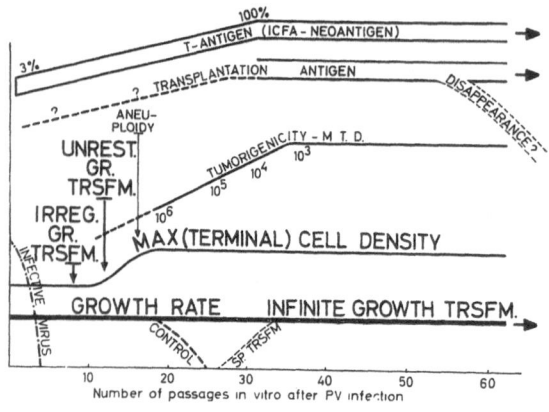

Fig. 14. Typical sequence of changes in a mass population of hamster cells exposed to polyoma virus. An early alteration involves the appearance of T-antigen in an increasing proportion of the nuclei. Infective virus disappears but may persist in an occasional line. Cytopathic changes are absent. The latent time until morphological alterations with irregular growth is long. Aneuploidy and capacity to produce tumors after animal inoculation are late steps. The tendency to spontaneous transformation unrelated to PV infection is indicated by the interrupted lines in the low r part of the chart. This unstable background is probably responsible for the variation between individual lines and the depicted events do not necessarily apply to inherently stable cells

graphy, where the cells move over each other without any restriction or formation of adhesions (WESTERMARK and PONTÉN, unpublished). It has been suggested that this transformation results from a surface change caused by an increased membrane content of sialomucins (DEFENDI and GASIC, 1963).

Cells which have undergone irregular growth transformation remain elongated ("fibroblast-like") but are cytologically "atypical" with increased cytoplasmic basophilia, nuclear pleomorphism and a high nucleo-cytoplasmatic volume ratio (DEFENDI, 1966).

These findings in hamster cells with respect to the morphology, and latent time of the PV induced transformation, and the absence of overt cytopathogenicity have also been registered with bovine (DIDERHOLM, 1967; THOMAS and LE BOUVIER, 1967) and rat cells (SACHS and MEDINA, 1961; MEDINA and SACHS, 1961; WILLIAMS and TILL, 1964) (Fig. 15).

The long latent time in a mass culture has a complex not completely clarified background. One alternative explanation — that all cells exposed to PV undergo a long series of alterations before they simultaneously express the irregular growth transformation — has been excluded by cloning experiments. In these a cell population is exposed to PV and then plated out after dilution to permit analysis of the colonies formed by the individual cells. In non-established lines the method is seriously hampered by the low plating efficiency. More than 99% of the colonies formed were normal but transformed colonies appeared so as soon as they could be evaluated (STOKER and MACPHERSON, 1961). Clones of the established BHK-21 line also showed a great excess of untransformed colonies after exposure to PV (MACPHERSON and STOKER, 1962). Similar results were published by SACHS, MEDINA, and BERWALD (1962). From these facts it is clear that at least a partial explanation for the long latent time in mass cultures is provided by the presence of a large fraction of untransformed cells which have the same or only a slightly lower growth rate than those which are transformed (DEFENDI and LEHMAN, 1965).

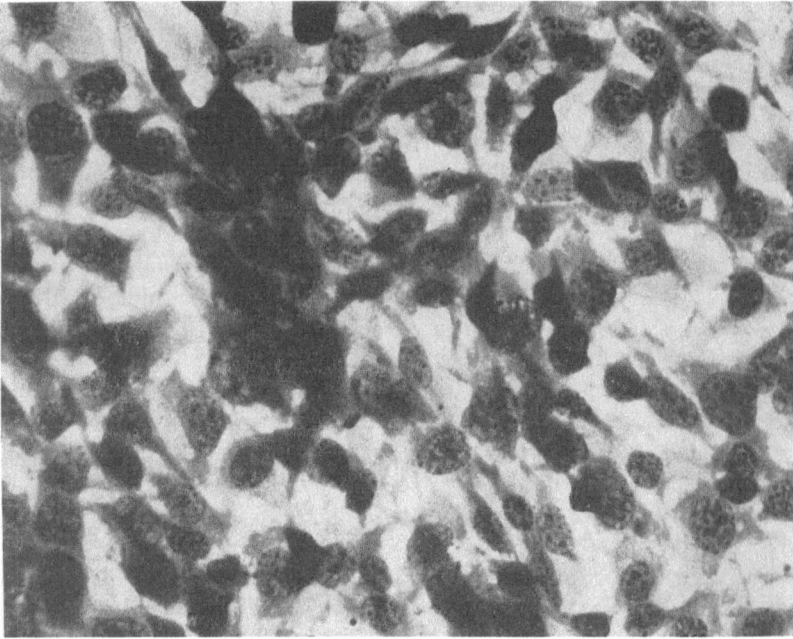

Fig. 15. Embryonic mouse cells after transformation by polyoma virus. The picture is from an established line several years after exposure to virus. Note the irregular arrangement of overlapping polygonal or spindle-shaped cells with pleomorphic nuclei. The culture shows irregular, unrestrained and infinite growth transformation. Magnification ×1,200

An approximate estimate of the true latent time can be achieved by observing the clone morphology in populations plated out at different times after infection. M. VOGT and DULBECCO (1963) found a low proportion of mainly "thin" colonies in hamster embryo cell populations seeded for colony formation two days after exposure to PV. The colonies consisted exclusively of cells displaying typical irregular growth transformation strongly indicating that induction of lack of contact inhibition is a rapid effect of polyoma virus possibly present already in the primarily infected cell. Further support for this view has been delivered by STOKER, MACPHERSON, and collaborators. In their extensive, quantitative studies they used a clone (C13) of the embryonic hamster kidney line BHK-21, which spontaneously has undergone infinite and unrestrained growth transformation (Table 4). The original stock of BHK-21/C13, which has a high plating efficiency, responds to PV infection by the formation of colonies with irregular growth transformation. These foci can be clearly identified already after 7 days of incubation, when they are composed of about 100 cells which all seem to show lack of contact inhibition of locomotion (MACPHERSON and STOKER, 1962; STOKER and ABEL, 1962; STOKER, 1963).

These results provide very strong evidence for an early and direct induction of irregular growth transformation by PV. That all transformations need not occur immediately after the primary interaction is, however, indicated by the occurrence of "delayed transformation". This manifests itself as sectors of transformed cells occupying parts of an otherwise normal colony. It has been interpreted as a lag of an estimated number of 7 or 8 cell generations before the transformed character is expressed (STOKER, 1963a; STOKER and SMITH, 1964).

The presence of only a few transformed cells among a multitude of untransformed cells is compatible with the long latent time before transformation is detected in a mass culture. A final confirmation would be desirable and could perhaps be obtained in the bovine system where the normal background is highly stable and uniform excluding the possibility that PV infection acts by favoring growth of preexisting variant cells which have acquired a tendency to lose their contact inhibition of locomotion abnormally soon after infection by polyoma virus.

Several factors probably contribute to determine the point of time when transformation is detected in a mass culture. One, which only operates in nonestablished lines, is the finite life span of normal cells. When the proliferation rate of the latter begins to decline, the transformed cells maintain their original growth rate and will thus have a selective advantage (DEFENDI and LEHMAN, 1965). Another is a possible progression among the transformed cells from irregular to unrestrained growth transformation. Only after acquisition of the latter property will the transformed cells have a strong selective advantage in a crowded culture.

A third mechanism may be a restraint by normal cells on neighboring PV transformed cells. STOKER (1964) and STOKER, SHEARER, and O'NEILL (1966) found that replication and movement of PV transformed BHK-21 cells were strongly inhibited in the presence of non-dividing normal mouse or hamster embryo fibroblasts. A similar cell cycle inhibition of PV transformed mouse and hamster cells has been found by BOREK and SACHS (1967). PONTÉN and MACINTYRE (1968) repeated these experiments in a syngeneic or allogeneic stable bovine system. The PV transformed cells were prelabeled with radioactive thymidine so that their fate could be autoradiographically determined. When seeded on stationary layers of normal cells, the transformed cells attached normally and remained viable for at least 96 hours but showed no division.

This could explain a long latent time in mass cultures, where isolated transformed cells are surrounded by normal cells. Their eventual breakthrough might occur in the course of subcultivation by the creation of small nests of transformed cells, where contact by normal cells is eliminated.

c) Rate of Transformation

Even in the highly susceptible BHK-21/C13 line only a maximum of about 5% of the originally infected cells will become transformed (irregular growth) colonies. The transformation frequency is linearly related to the virus dose up to a multiplicity of about 10^3 PFU/cell and further increase does not lead to increased transformation (STOKER and ABEL, 1962).

An elegant methodological improvement for the determination of the transformation rate was introduced by MACPHERSON and MONTAGNIER (1964) in an English and French (MONTAGNIER and MACPHERSON, 1964) version. They found that PV transformed but not normal colonies of BHK-21 were capable of growth in soft agar, permitting an easy monitoring of large number of cells. With this sensitive method the existence of a ceiling at about 5% transformed cells could be confirmed and a plateau was also seen when transformation was induced by PV DNA (BOURGAUX, BOURGAUX-RAMOISY, and STOKER, 1965). The same type of dose-response curve has been observed in rat embryo cells (WILLIAMS and TILL, 1964).

STOKER (1968) studied the behavior of BHK-21 cells exposed to PV in detail. After infection with purified virus the cells were suspended in a viscous medium containing carboxymethyl cellulose, where they could form isolated colonies by cellular multiplication. From colony size the number of divisions could be estimated. Individual colonies were picked and their irregular growth transformation evaluated from the microscopic emigration pattern. Unexposed cells were, with occasional exceptions, unable to multiply more than once. 30—50% of the PV exposed cells, on the other hand, formed colonies, *i.e.* were not dependent on anchorage to a solid surface for replication. The large majority of these colonies stopped growing after a maximum of 6—8 divisions if kept in suspension. The early emigration pattern of cells from such colonies suggested a loss of contact inhibition of locomotion but after continued incubation the population resembled unexposed cells whose movement was controlled by cell to cell contacts. It was concluded that the PV genome may induce a temporary loss of contact inhibition of locomotion with great efficiency. This is, however, obscured in all but a few cells either because the cells fail to go on dividing or because the "transformed" property is lost. The maximal 5% transformation frequency only reflects these cells which have permanently lost their contact inhibition of locomotion.

d) Possible Causes for Cellular Resistance to Transformation

About 10^6 physical PV particles are necessary to produce one transformed cell (MACPHERSON and MONTAGNIER, 1964). This could suggest that a contaminant virus was responsible. Such facts as transformation by purified circular or linear PV DNA (CRAWFORD, DULBECCO, FRIED, MONTAGNIER, and STOKER, 1964), the unaltered properties of plaque purified virus and the similarity of the biological activity of PV isolated from lytic systems or virus yielding transformed lines provide very strong evidence against this speculation.

A low transformation rate could be explained on a statistical basis, if the virus-cell interaction leading to transformation is due to a random event of extremely low probability. This hypothesis is contradicted by the absence of a direct proportion between dose and response at high multiplicities of infection.

ALLISON (1961) has suggested that interferon plays a role in transformation by restricting the lytic effect of PV. This is contradicted by the observation that PV variants with a low capacity for induction of interferon are more oncogenic *in vivo* than those with a high capacity (FRIEDMAN and RABSON, 1964a, b).

The most likely interpretations assume that only a proportion of the cells are competent to react with permanent irregular growth transformation after exposure to PV. The competence could either be genetically determined or due to the physiological state of the cell at the time of infection. The different competence does not involve the initial cell-virus interaction because all cells adsorb PV, which also becomes associated with the cytoplasm (FRASER and GHARPURE, 1962).

The genetic competence was investigated by MACPHERSON and STOKER (1962) by infection of four fibroblast clones of BHK-21. Since the same low transformation rate was found as in the parental population, it was concluded that the transformation was not limited to a stable genetically competent subpopulation. Similarly, 19 clones of BHK-21 origin reacted to approximately the same extent (BLACK, 1964). Because of the inherent genetic instability in established lines such as BHK-21 (page 26), these conclusions should not be generalized. A desirable complement would be a similar experiment with stable bovine fibroblasts, if a method could be found to increase their plating efficiency.

The physiologic competence of cells for transformation was tested by BASILICO and MARIN (1966). A non-synchronized population of BHK-21/C13 cells was infected with a high multiplicity of an Sp variant of PV. Cells in mitosis were selectively harvested at different times after infection and scored for growth in soft agar according to MACPHERSON and MONTAGNIER (1964). Early harvest cells, which can be deduced to have been in the G2 phase (page 10) when exposed to PV, had a 2.4-fold increase in transformation rate over cells in G1. Cells synthesizing DNA during adsorption of PV were intermediate.

These results suggest that the competence to undergo transformation is partly determined by the physiologic state of the cell and that the G2 period is particularly sensitive. The maximal transformation frequency was, however, only 20% and the experiments could thus not entirely account for the resistance seen in mass populations.

The interpretation of the experiments by BASILICO and MARIN is rendered difficult by the observations of FRASER and GHARPURE (1962) and FRASER and E. CRAWFORD (1965) who showed by immunofluorescence, that all BHK-21 cells exposed to PV ingest roughly equal amounts of virus in the cytoplasm, where it persists for at least six successive cell cycles. A controlled pulse of PV infection is therefore impossible to administer and the cells of BASILICO and MARIN must have gone through several complete cycles with the virus associated with cytoplasmic structures.

Another type of physiologic competence is suggested by the results of PV infection of differentiating embryonic tissue. VAINIO, SAXÉN, and TOIVONEN (1963) observed a decreased production of PV during embryonic differentiation in mouse kidney mesenchyme. In submandibular salivary glands cultivated *in vitro*, DAWE, MORGAN, and SLATICK (1966) have shown an increased responsiveness to the neoplastic PV action with increasing embryonic age. Transformation required the "interaction" of primitive mesenchyme and salivary epithelium. These findings are difficult to apply to the cell culture systems usually employed in

PV transformations, but indicate that it may be necessary to consider physiologic differences in competence other than those connected with the cell cycle.

To sum up, the reasons for the resistance which the large majority of exposed cells show to permanent irregular growth transformation by PV remain unknown. A temporary physiologic competence rather than genetic susceptibility on the part of the cell is perhaps the most likely explanation. The sensitive stage is possibly related to the cell cycle but also to other factors. As will be discussed later, a similar situation exists with other transforming DNA viruses, but not necessarily with the RNA group.

e) Chromosomes and Polyoma Transformation

The subject has recently been reviewed (DEFENDI, 1966). The low transformation rate of all systems and particularly those where stable euploid cells are employed (Table 3) has rendered it difficult to assess the importance of karyologic changes for transformation. For adequate analysis a large number of metaphases are required and before a sufficient population size is reached the critical steps in the transformation process have already occurred.

All transformed lines, regardless of species origin, eventually become aneuploid. The pattern of aneuploidization of mass cultures of embryonic hamster origin has differed between individual lines. Most lines become hyperdiploid, but subtetraploid modes are also on record (FORD, BOQUSZEWSKI, and AUERSPERG, 1961; MACPHERSON, 1963; DEFENDI and LEHMAN, 1965). Around the time of diffuse irregular growth transformation a small proportion of cells showed abnormal chromosomes but the counts remained diploid in the majority of the cells. Only later when the tumorigenicity began to increase did a significant degree of aneuploidy develop (DEFENDI and LEHMAN, 1965). The same lag between irregular growth transformation and aneuploidy was observed in the PV exposed C13 clones of BHK-21 cells (MACPHERSON, 1963).

It is not possible to interpret the significance of the occurrence of some chromatid breaks and minor rearrangements at the approximate time or irregular growth transformation. In some experiments the chromosomal abnormalities may have occurred normally during the aging of the cell in "phase III" (page 15).

It is quite possible that the aneuploidization after transformation by PV is an epiphenomenon akin to the aneuploidy seen in spontaneously established lines (page 25). No specific pattern of chromosome damage after PV infection has been established and there seems to be no karyological difference, in principle, between spontaneously established lines or established lines derived from cultures exposed to polyoma virus. Both undergo continuing karyological rearrangements and highly tumorigenic variants are rapidly selected by implantation *in vivo* (MACPHERSON, 1963).

M. VOGT and DULBECCO (1963) have reported an increased frequency of anaphase bridges early after infection of hamster cells by PV. This suggested that the virus directly induced many chromatid breaks and is at variance with the largely negative findings in early metaphases analysed by other workers (see DEFENDI, 1966). The reasons for this discrepancy are unknown. VOGT and DULBECCO's experiments contained a few control lines with a remarkably high frequency of anaphase bridges and it is not excluded that an undetected contamination by PPLO, which is known to be capable of inducing chromosome breaks (J. Fogh and H. Fogh, 1965), may have been responsible.

It seems clear that no early and direct effect of PV on chromosomes has been convincingly demonstrated. The slow but inevitable unfolding of aneuploidy can only be detected long after exposure to virus and it has not been possible to analyse the relationship between infection with PV and the evolution of a cell population with mitotic instability and aneuploidy. The effect of PV may be non-specific in the sense that it produces a general increase in mitotic errors and genetic rearrangement.

f) Tumorigenicity of PV Transformed Cells

All adequately tested established lines derived from populations exposed to PV will form sarcomas after implantation in appropriate hosts. This capacity is gradually evolved as shown by the decrease in cell numbers necessary for a 50% take (Fig. 14). It is a late step preceeded by irregular growth transformation by at least several weeks (Todaro, Nilausen, and Green, 1963; M. Vogt and Dulbecco, 1963; Defendi and Lehman, 1965).

Many attempts have been made to correlate tumorigenicity with morphological or other characteristics *in vitro* (Sanford, Dunn, Covelesky, Dupree, and Earle, 1961; Macpherson, 1963; Stanners, Till, and Siminovitch, 1963; Sanford, Barker, Woods, Parshad, and Law, 1967; Hare, 1967b).

In the BHK-21/C13 line colonies of cells with parallel orientation were found to be less tumorigenic than colonies of unoriented cells by a factor of $10^2—10^3$ (Macpherson and Stoker, 1962; Macpherson, 1963). The claim by Stoker (1962) that a "change in colonial morphology is a satisfactory indication of neoplastic transformation" has, however, had to be modified. Even low numbers of BHK-21 cells carried as normal controls in other laboratories have repeatedly been found to induce tumors in hamsters (Defendi, Lehman, and Kraemer, 1963; Gotlieb-Stematsky and Shilo, 1964; Rapp, Khera, and Melnick, 1964; Defendi and Lehman, 1965; Defendi, 1966) in spite of preservation of contact inhibition of cell movement. Jarrett and Macpherson (1968) calculated that only about 0.001% of the cells of clone 13 of BHK-21 were highly tumorigenic 50 generations from cloning. With passage *in vitro* this proportion can reach 90% or more.

Some reports have claimed (M. Vogt and Dulbecco, 1963; Weisberg, 1964) that "thick" colonies are transplantable in contrast to "thin" ones. However, in other experiments no such consistent difference was established (Stanners, Till, and Siminovitch, 1963; Defendi and Lehman, 1965; Defendi, 1966). In view of the large variations between transformed lines (Stanners, Till, and Siminovitch, 1963) it is furthermore likely that the division of transformed colonies into "thin" and "thick" represents an oversimplification. Sanford, Barker, Woods, Parshad, and Law (1967) recently attempted to select indicators *in vitro* which would permit prediction of the *in vivo* tumorigenicity of mouse and hamster cells. No such indicators were found and features like morphologic growth pattern, karyology, growth rate and glycolysis were found unsuitable. Cytological "atypia" (page 27) was possibly the best sign of tumorigenicity.

The difficulty of interpreting implantation tests has already been discussed (page 16). The selection pressure *in vivo* and *in vitro* is necessarily widely different and a perfect correlation between the appearances scored as transformations *in vitro* and tumorigenicity *in vivo* should hardly be expected. It is noteworthy that tumors induced *in vivo* by PV, need to undergo changes after explantation with selection of fit variants before progressive growth is established *in vitro* (TODARO, NILAUSEN, and GREEN, 1963; DEFENDI and LEHMAN, 1966).

g) Antigens in PV Transformed Cells

Two antigens not present on the surface of intact PV particles are known to be induced in polyoma cells.

The transplantation antigen. The PV transplantation antigen was first discovered *in vivo* by SJÖGREN, I. HELLSTRÖM, and G. KLEIN (1961 a, b) and HABEL (1961). Its detection is based on the fact that virus infection of newborn or adult rodents induces a state of resistance to subsequent grafting of transformed cells (Sjögren, 1964a).

The degree of resistance can be estimated from the increment of the cell dose necessary to obtain the same tumor response in immunized as in control animals. In a strict sense, this test measures the sensitivity of a cell population to a resistance induced by polyoma virus rather than antigenicity of the transplantation antigen *per se*. Interpretations in terms of quantity of antigen are thus rendered difficult. A direct test for transplantation antigens *in vivo* is complicated by the low level of resistance induced and the necessity of using either repeated injections of small doses of virus-free syngeneic cells or a large dose of allogeneic virus-free cells for immunization (HABEL, 1962; SJÖGREN, 1964c). These serious shortcomings may have been overcome by IRLIN (1967) who in a preliminary report described an immunofluorescent technique for surface antigens, later employed by MALMGREN, TAKEMOTO, and CARNEY (1968). Another method — colony inhibition — for detecting PV specific tumor antigen *in vitro* has also been described (I. HELLSTRÖM and SJÖGREN, 1965). The cytotoxic effect of rabbit immune serum claimed by BASES (1964) on polyoma cells has not clearly been shown to be specific for the transplantation antigen and the usefulness of the method has not been confirmed.

The specificity of the transplantation antigen has been well established. With only occasional exceptions (which may have been due to PV contamination) spontaneously transformed cells or cells transformed by other viruses either *in vivo* or *in vitro* do not contain the PV transplantation antigen (review by SJÖGREN, 1965). Conversely, under ordinary circumstances, cells of murine, hamster or rat origin exposed to PV contain the antigen (SJÖGREN, I. HELLSTRÖM, and G. KLEIN, 1961 b; HABEL, 1961; VANDEPUTTE and DE SOMER, 1963).

No correlation exists between PV neutralizing activity and HI antibodies on one hand and transplant immunity or colony inhibition on the other (HABEL, 1962; SJÖGREN, 1964b; I. HELLSTRÖM, 1965). Tumors free of infectious virus induce transplant rejection but no antiviral antibody (SJÖGREN, 1964c). Although the possibility has not been excluded, no evidence has been presented to show that the PV transplantation antigen is an internal protein of the virus particle. Regardless of whether it might be found in the interior of polyoma virions it is firmly established that synthesis of the transplantation antigen does not require synthesis of infectious PV.

SJÖGREN (1964d) transplanted two polyoma tumors in series 42 and 19 times, respectively, in immune animals. In spite of the deliberate and high selective pressure against growth of antigenic cells, the tumors retained their full sensitivity to transplant immunity. SJÖGREN (1964e) therefore suggested that the PV transplantation antigen is indispensable for the maintenance of the neoplastic cell character. In hamster cells carried *in vitro*, similar experiments have given equivocal results (M. VOGT and DULBECCO, 1960; HABEL, 1962). The transplantation antigen was for instance found only in four of six lines tested by DEFENDI and LEHMAN (1965) and in later passages of some positive lines it disappeared. These results do not necessarily contradict those of SJÖGREN. In the unstable *in vitro* hamster system some transformed cells may have been spontaneously established and therefore not representative. The finding by SJÖGREN (1964d) that a PV transplantation antigen induced in a methylcholanthrene mouse sarcoma was strongly reduced after passage through mice with a PV specific transplantation immunity also suggests that this may be an unstable system.

The importance of the transplantation antigen for tumor growth *in vivo*, the possibilities of influencing growth by manipulations of the immune status of the host and other related problems are beyond the scope of this treatise. Reviews have been published by HABEL (1963), HARRIS (1964) and SJÖGREN (1964e, 1965).

While the possibility that the PV transplantation antigen has an important and perhaps indispensable function in tumor cells still remains likely, some evidence indicates that it is not sufficient for the transformation. PV causes an inapparent infection in adult mice, which rapidly induces a specific resistance against tumor transplants. Since this occurs in the absence of detectable tumors, it can be surmised that the PV antigen is induced without any neoplastic conversion. IRLIN (1967) and MALMGREN, TAKEMOTO, and CARNEY (1968) reported the presence of surface antigen in mouse cells during early lytic infection. The recently developed techniques for *in vitro* detection of the PV transplantation antigen (page 69) should finally resolve this important problem. That the PV transplantation antigen can be induced in cells already transformed by other oncogenic viruses is well known (SJÖGREN and I. HELLSTRÖM, 1965; TAKEMOTO and HABEL, 1966; SJÖGREN and I. HELLSTRÖM, 1967).

The lack of a sensitive and simple method for determination of transplantation antigens has made it difficult to delineate its function. Its intracellular site of synthesis is unknown; the indirect assumption that the mature antigen is associated with the cell membrane seems to have been substantiated by recent FA studies (IRLIN, 1967). Irregular growth transformation, which reflects the loss of an important membrane function, is a prominent feature of PV transformation and there may be a relationship between the new surface antigen and the altered cell growth pattern.

Although the synthesis of the transplantation antigen has almost universally been considered to be under the control of the PV genome, there is actually very little experimental support for this. No or at most only a weak cross-reactivity has been found between PV induced mouse and hamster tumors (HABEL, 1962; HARE, 1964a and b; I. HELLSTRÖM and SJÖGREN, 1966; HARE, 1967a). This is hard to reconcile with a coding by the PV genome.

With their colony inhibition technique I. HELLSTRÖM and SJÖGREN (1966) demonstrated an antigen common to BHK-21 hamster cells transformed by PV and a Moloney mouse leukemia line infected by PV. A surface antigen in transformed mouse and hamster cells reacted with the same mouse serum (MALMGREN, TAKEMOTO, and CARNEY, 1968) indicating cross-reactivity. The identity of these antigens with the conventional transplantation antigen has not been determined and the BHK-21 cells may have carried a Moloney related leukemia passenger virus. It must be concluded that it is not known if the PV transplantation antigen is coded for by the viral genome or induced via depression of cellular genes or some other mechanism primarily under cellular control.

The T-antigen. The T-antigen is defined by its reaction in complement-fixation or FA tests with sera from animals bearing transplantable virus-free polyoma tumors (DEFENDI, EPHRUSSI, and KOPROWSKI, 1964; HABEL, 1965; DEFENDI and LEHMAN, 1965; DEFENDI and TAGUCHI, 1966; TAKEMOTO, MALMGREN, and HABEL, 1966a, b; FOGEL, GILDEN, and DEFENDI, 1967; PORWIT-BÓBR, PTAK, and GARLACZ, 1967). T-antigens (neoantigen, ICFA, CF-antigen), were not readily demonstrated with PV until TAKEMOTO, MALMGREN, and HABEL (1966b) showed that a heat labile factor in fresh hamster serum was required for full antigenic expression.

The T-antigen is exclusively located in the nucleus, either as dense granules or as a diffuse fluorescence (TAKEMOTO, MALMGREN, and HABEL, 1966b; FOGEL, GILDEN, and DEFENDI, 1967). It is formed rapidly after infection of mouse, hamster or rat embryo cells and precedes the PV capsid antigen by about 6 hours (Fig. 9). It has been found even after several years in all cells of non-virus releasing tumors or transformed *in vitro* cultures (TAKEMOTO, MALMGREN, and HABEL, 1966b; FOGEL, GILDEN, and DEFENDI, 1967).

Two to four days post infection hamster and rat embryonic cells show 10—11% positive nuclei. This is higher than the proportion of lytically infected cells and also higher than the estimated proportion of colony-forming transformed cells (Table 3). It is thus possible that T-antigen is not only formed in cells undergoing lysis or permanent transformation but also in some abortively transformed cells (STOKER, 1968).

The T-antigen is almost certainly distinct from the PV transplantation antigen. Their cellular location is different and no correlation exists between transplant immunity and antibody titers against the T-antigen (HABEL, 1965; DEFENDI and TAGUCHI, 1966). HARE (1967b) isolated three transformed embryonic hamster lines positive for transplantation antigen but negative for T-antigen in FA tests. A lack of correlation between antiviral hemagglutinin antibodies and CF antibodies against the T-antigen seems to exclude that the T-antigen is part of the surface of the virion. The T-antigen is thermolabile in contrast to the viral capsid antigen (DEFENDI and TAGUCHI, 1966).

The presence of the T-antigen behaves as a stable property of transformed cells. Its function and eventual indispensability are unknown. As already pointed out the presence of the T-antigen only indicates that a cell has been infected by polyoma virus, there is no evidence that the T-antigen is specifically connected with transformation or tumor formation.

The weight of evidence strongly favors that synthesis of T-antigen is under the control of the viral genome. A complete cross-reactivity exists between the three rodent species investigated, but because of the close zoological relationship between the mouse, rat and hamster, the possibility of a PV controlled depression of an identical cellular gene cannot be excluded. Evidence that the code for the T-antigen is present in the PV DNA would be strengthened by the finding of an identical T-antigen in transformed bovine cells and abortively infected human cells.

h) Attempt to Determine the Size of the PV Genome Necessary for Different PV Functions

Complete virus particles induce a number of effects in cells. It is possible to determine the dependence of these effects on the size of the viral genome. The genome is exposed to graded doses of irradiation or certain inactivating chemicals and the specific dose-response for each effect is determined, the rate of loss is then related to the size of the gene target. If one direct hit within a defined target area is sufficient to destroy its function irreversibly, an exponential rate of decay is obtained; the slope is used to calculate the target size.

BENJAMIN (1965) used UV light, X-rays, P^{32}-decay and nitrous acid to inactivate an Lp variant of PV. Infectivity in mouse cells was compared to the capacity of infected BHK-21 hamster cells to form colonies in soft agar. The dose-response curves for all four types of inactivation were of the one-hit type and indicated that the part of the PV genome required for colony formation was approximately 60% of that required for infectivity. A similar study using gamma-rays, gave the same relative target size (BASILICO and DI MAYORCA, 1965). LATARJET, CRAMER, and MONTAGNIER (1967) used the soft agar assay for colony formation of BHK-21/C13 cells and deduced from their experiments with X- and gamma-rays that only one-fifth of the PV genome was required for transformation.

Induction of cell DNA synthesis in X-irradiated rat cells (GERSHON, HAUSEN, SACHS, and WINOCOUR, 1965) and synthesis of T-antigen (DEFENDI, JENSEN, and SAUER, 1967) require a smaller part of the PV genome than does infectivity in mouse cells.

The experiments designed to determine the genome size for transformation are not easily interpreted. Colony formation in soft agar by unstable BHK-21 hamster cells means, as already pointed out, that both infinite and unrestrained growth transformation exist before exposure to PV. Strictly speaking, the data have therefore only determined that the size of the genome required to superimpose anchorage independent growth (= equivalent of irregular growth transformation?) on cells already partially transformed is smaller than the intact genome. It is still not known if the entire PV genome is required when a normal cell is used as target.

i) Persistence of Polyoma Genome in Transformed Cells

Even the first theories for transformation invoked the possibility that the polyoma virus genome persisted indefinitely in the infected cells, and that this persistence was the very cause of the altered cellular state (review by DULBECCO, 1966). The viral genome was transmitted to the daughter cells thereby ensuring that reversion to normal did not take place.

Such an integration would then be comparable to bacterial lysogeny, where infected organisms carry a prophage occupying a specific site of the bacterial DNA. The expression of the prophage genome is repressed but may be activated spontaneously in a small proportion of the infected bacteria. This frequency can be increased by certain physical or chemical agents. If the incorporated phage genome is defective, the treatments will not induce the formation of complete infectious phage particles. In certain instances the prophage may direct the synthesis of a specific protein; the so-called lysogenic conversion. The presence of a prophage ordinarily makes the bacterium resistant to superinfection by the same but not other phage, although the resistance is not complete at high multiplicities of infection.

The occasional existence of cell lines or tumors releasing small amounts of virus could be interpreted in terms of bacterial lysogeny, where vegetative phage synthesis is spontaneously induced. An analysis by M. VOGT and DULBECCO (1962) of PV transformed cells made this explanation unlikely. Clones of transformed cells from virus-yielding mass cultures did not release any infectious virus, a result unexplicable on the basis of a lysogeny model with integration of the intact phage genome. FOGEL and SACHS (1969) have, however, described an unusual Lp rat line, which after cloning maintained its PV synthesis, indicating that almost every cell contained the entire PV genome in a form which could be "spontaneously" activated to instruct synthesis of complete virions.

Cell lines and tumors which never produce virus spontaneously have also been scrutinized for persistence of the viral genome. It has, however, never been possible to induce virus production by UV-light and other agents known to be effective in lysogenic bacteria (HABEL and SILVERBERG, 1960; WINOCOUR and SACHS, 1961; M. VOGT and DULBECCO, 1962). PV-free-cultures are furthermore often sensitive to superinfection by PV (I. HELLSTRÖM, K. E. HELLSTRÖM, SJÖGREN, and G. KLEIN, 1960; HABEL and SILVERBERG, 1960; I. HELLSTRÖM, K. E. HELLSTRÖM, and SJÖGREN, 1962; WINOCOUR and SACHS, 1962; I. HELLSTRÖM, 1963). Even the sensitive technique of cell fusion failed to induce synthesis of infective virus (BURNS and BLACK, 1969).

If other criteria for persistence of the PV genome are applied than those based on a complete analogy with lysogeny, there is now proof for persistence of at least part of the viral genome in transformed cells.

The existence of a specific transplantation antigen and a T-antigen has suggested the persistence of PV DNA specifically coding for these products. However, as already outlined, this evidence is only indirect and in the case of the transplantation antigen certain data even suggest that its synthesis may be directed by the cell rather than the viral genome (page 70).

Direct attempts to extract infectious DNA from PV transformed cells have consistently failed (WINOCOUR and SACHS, 1961; M. VOGT and DULBECCO, 1962). AXELROD, HABEL, and BOLTON (1964) obtained evidence for complementarity between PV DNA and DNA from murine polyoma tumor cells and to a lesser degree to normal mouse tissue or a methyl-cholanthrene-induced murine sarcoma, but not to other cell DNA's. Since such hybridization indicates the presence of complementary sequences of nucleotides, it was suggested that PV DNA was incorporated in DNA of transformed cells. However, in the light of recent findings

(WINOCOUR, 1967a, b) that cell DNA may be encapsidated and carried along together with infectious PV (page 49), these results cannot be considered final proofs of intracellular persistence of the viral genome. WINOCOUR (1965) in a carefully planned experiment looked for specific hybridization between polyoma tumor cell DNA and RNA produced *in vitro* with *E. coli* RNA polymerase and PV DNA as primer. Based on model experiments and comparisons between normal and tumor cell DNA, WINOCOUR concluded that PV DNA was not present within the sensitivity limits of the assay. The minimum amount of DNA detectable would correspond to 20 complete viral DNA strands per tumor cell and the presence of a smaller amount of viral nucleic acid could therefore not be excluded.

BENJAMIN (1966a) reported the first convincing indirect demonstration of PV DNA in transformed cells. He used P^{32} pulse labeled RNA from virus-free polyoma transformed mouse, hamster and rat lines hybridized with H^3-thymidine labeled PV DNA, and observed a highly significant degree of binding. The virus DNA was purified in a sucrose gradient and no significant binding was observed in control experiments if RNA from normal embryonic cells, cells which had undergone spontaneous infinite growth transformation or murine cells transformed by SV40 were used. Also, no hybridization was seen if SV40 DNA was substituted for PV DNA. RNA from 3T3 cells "doubly transformed" by SV40 and PV reacted with both kinds of viral DNA. Altogether 11/11 tested PV transformed cell lines contained RNA complementary to polyoma virus DNA.

These experiments do not merely establish the presence of PV DNA, they also show that the genome is active.

WESTPHAL and DULBECCO (1968) managed to establish the presence of PV DNA in transformed lines directly by hybridization of cellular DNA with specific RNA synthesized *in vitro* after priming with purified viral DNA. It was calculated that between 5—10 viral DNA equivalents per cell were present in the 3 different PV transformed lines tested. The DNA was contained in the nuclear fraction.

MARIN and LITTLEFIELD (1968) ingeniously approached the problem of the importance of persistent PV genes for maintenance of irregular growth transformation among BHK-21 hamster cells. A hybrid line was created by cocultivation of two sublines lacking thymidine kinase or inosinic acid pyrophosphorylase. These sublines are resistant either to 5-bromodeoxyuridine or 6-thioguanine, in contrast to the hybrid line which was sensitive to both analogues. If PV transformed hybrid cells were plated in the presence of 6-thioguanine resistant clones were obtained at an approximate frequency of 10^{-4}. A proportion of these clones grew in regular array indicating "reversion" to contact inhibition of locomotion. Regardless of their morphology the resistant clones had a reduced chromosome number compared to the original near-tetraploid value. "Reversion" was never seen when populations of spontaneously formed polyploid cells were studied (MARIN and BASILICO, 1967).

Two "reverted" clones were selected for further study by MARIN and MAC-PHERSON (1969). They were less tumorigenic than transformed control cells and showed a reduced colony formation in soft agar. The transplantation antigen was apparently lost. Both clones underwent irregular growth transformation after reinfection with PV.

These results show that loss of chromosomes is associated with reappearance of normal control of cell membrane movement and also a loss of viral genes, which in turn suggests that PV genes exist which specifically control the sensitivity of the cell membrane to contact with other cells. Once these genes are lost the cell is again sensitive to contact inhibition of locomotion (page 5).

To sum up, PV genetic material is present in all transformed lines regardless of species origin. It is probably not integrated with cell DNA in a manner entirely analogous to the intact phage genome in bacterial lysogeny, mainly because irradiation with UV, cell fusion etc. fail to induce synthesis of infectious particles. Only part of the PV genome is transcribed and directs the synthesis of about 3—5 proteins, of which only the T-antigen has been unequivocally identified. It has not been settled whether the absence in transformed cells of "late" viral proteins is caused by actual loss or absolute repression of PV genes. The significant problem whether the persistent genome is essential, for the transformed or tumorigenic cell state, or merely represents a functionally unimportant indicator of a passed infection by polyoma virus is unsolved, although recent data favor the first hypothesis.

3. Transformation by Simian Virus 40

The SV40 cell interaction has many features in common with the polyoma system. The range of species susceptible to SV40 infection is, however, much broader than that of polyoma (Table 2). Several simian species and particularly the Rhesus monkey seem to be natural hosts of the virus. The lytic interaction *in vitro* with green monkey cells has already been outlined (page 56). Infection of mass cultures has been studied in the systems listed in Table 5. It is seen that the transformation frequency[1] after SV40 exposure is low, only in the established 3T3 mouse line does it exceed 1%. The SV40 minute plaque variant, after passage through 3T3 mouse cells, is the most efficient stock produced. Under optimal conditions with 3T3 like cells as targets, one transforming unit (formation of colonies with unrestrained and irregular growth) corresponds to only $1.3—5.0 \times 10^2$ plaque forming units (TODARO and TAKEMOTO, 1969). It is possible to obtain transformation even among GMK cells which predominantly respond with lysis. Virus synthesis by a mass population is roughly parallel to the extent of cell lysis. Instances of synthesis of infectious virus by transformed mass cultures are common in contrast to transformed PV systems which are usually virus-free (Tables 3 and 5). The characteristics of selected systems where SV40 transformation has been studied in some detail are given in Table 6. Transformation is in many essential respects influenced by the stability of the cells infected.

a) Transformation of Stable Cells

Human cells. SWEET and HILLEMAN (1960) seem to have been the first to infect human cells by SV40. They did not, however, observe any cytopathic effect in primary amnion or a number of established lines including HeLa cells. HSIUNG and GAYLORD (1961) had the same experience with human kidney. In contrast to these negative findings, SHEIN and ENDERS (1962a) described multi-

---

[1] Transformation frequency = proportion of originally exposed cells which respond with unrestrained and/or irregular growth transformation.

Table 5. *Effects of Simian Virus 40 in Different in vitro Systems*

Target cells	Percentage of exposed cells showing			Virus synthesis
	Transformation[1]	"Abortive infection"[2]	Lysis[3]	
Chicken				
embr. mix. tissue	0	n.t.	0	(+)
Mouse				
embr. kidney	very low[4]	20	0.2—0.9	(+)
3T3 line	2—20	50	0	0
Rat				
embr. kidney	very low[4]	n.t.	0	0
Hamster				
embr. kidney	very low[4]	n.t.	1	(+)
BHK21 line	0	n.t.	0	0
Rabbit				
embr. kidney	very low[4]	n.t.	0	(+)
Guinea-pig				
embr. kidney	very low[4]	n.t.	0	0
Bovine				
embr. lung	very low[4]	n.t.	0	(+)
Porcine				
embr. kidney	very low[4]	n.t.	0	(+)
Dog				
puppy kidney[5]	very low[4]	n.t.	0	—
Monkey				
GMK	very low[4]	0	90—99	++++
Human				
diploid Ph.II	0.02—0.04	50	1	++
aneuploid Ph.II[6]	0.8	n.t.	n.t.	++
aneuploid Ph.III	low[4]	n.t.	n.t.	++

[1] Unrestrained (and irregular) growth.
[2] No morphologic change, unimpaired cell viability. Uptake of virus established by fluorescent antibody, radioactive virus etc.
[3] Inferred from proportion of cells forming infectious centers of positive for viral antigen.
[4] Inferred from behavior of mass cultures.
[5] Transformed cells were negative for T-antigen (VONKA, KUTINOVÁ and ZÁVADOVÁ, 1969)
[6] Fanconi's disease (page 82).
   n.t. = not tested.

plication of SV40 in human fetal brain, lung, skin, and muscle, heart, liver, spleen, adrenal, intestine, testes and kidney. Cells from the last organ were studied in some detail and a cytopathic effect described with piles of cells which became necrotic with nuclear pyknosis. A high mitotic rate prevailed and it was suggested that SV40 causes cell proliferation as well as degeneration. Persistent infection of the established human cancer lines HeLa and HEp-2 with virus synthesis during at least 6—8 months was reported by EDDY, GRUBBS, and YOUNG (1962), who also noted stimulation of cell proliferation.

Unequivocal transformation of human cells by SV40 was independently reported by KOPROWSKI, PONTÉN, JENSEN, RAVDIN, MOORHEAD, and SAKSELA

Table 6. *Transformation in vitro by Simian virus 40*

Cell line	Target background			Target response			
	Transformation			Lysis	Transformation		
	Irreg. growth	Unrest. growth	Infinite growth		Irreg. growth [1]	Un- rest. growth	In- finite growth
Mouse embr. kidney	high tendency to spontaneous trans- formation			0	yes	yes	n.t.
3T3 estab. line	absent	absent	present	0	yes	yes	[2]
Bovine embr. lung	no tendency to spontaneous trans- formation			0	yes	yes	yes[3]
Monkey GMK	low tendency to spontaneous trans- formation			+++	yes	yes	yes
Human diploid Ph.II	no tendency to spontaneous trans- formation			+	yes	yes	yes[3]
aneuploid Ph.III	no tendency to spontaneous trans- formation			++	yes	yes	n.t.

[1] Indirectly inferred from the existence of cell piles.

[2] Already present before exposure to SV 40.

[3] Only after a "growth crisis" in a proportion of cultures long after unrestrained growth transformation.

n.t. = not tested.

(1962) and by SHEIN and ENDERS (1962b). The former group infected adult buccal mucosa and skin and the latter perinatal kidney.

The complicated series of morphological events (Fig. 16) has been similar regardless of tissue origin, when normal diploid phase II (page 15) cells have been used, and the following description refers to such target cells. Even after infection by large amounts of virus a comparatively long silent phase of 2—7 weeks will follow during which ordinary microscopy fails to reveal any gross alterations (SHEIN, ENDERS, and LEVINTHAL, 1962; PONTÉN, JENSEN, and KOPROWSKI, 1963; GIRARDI, WEINSTEIN, and MOORHEAD, 1966). This phase is succeeded by a mildly cytopathic stage during which a high mitotic rate also becomes apparent, even in areas of very high cell density (the unrestrained growth transformation) (KOPROWSKI, PONTÉN, JENSEN, RAVDIN, MOORHEAD, and SAK- SELA, 1962; SHEIN and ENDERS, 1962b). Gradually the cytopathic alterations become less prominent and the cultures entirely composed of morphologically changed ("epithelial-like") rapidly proliferating elements. A few months later cytopathic effects reappear with loss of proliferative power. In this "growth crisis" some cultures are lost, but cultures may survive and resume their pre-crisis appearance and rapid proliferation and in such cases indefinite growth is obtained (JENSEN, KOPROWSKI, PAGANO, PONTÉN, and RAVDIN, 1964; MOYER, WALLACE, and Cox, 1964; SHEIN, ENDERS, PALMER, and GROGAN, 1964; GIRARDI, JENSEN, and KOPROWSKI, 1965; GIRARDI, WEINSTEIN, and MOORHEAD, 1966).

The long *silent phase* has been difficult to explore and the forerunners of these cells which eventually transform have not been possible to identify. The population dynamics from the time of virus exposure to the time when all cells show unrestrained growth transformation are presumable very complex. It is for instance possible that primarily infected cells are at a selective disadvantage during the silent phase, while the reverse may be true once they have become transformed. Cell to cell interactions with transfer of the SV40 genome (GERBER and KIRSCHSTEIN, 1962) and growth inhibition of transformed cells by normal cells surrounding them (SHEIN and ENDERS, 1962b; BOREK and SACHS, 1966a) may represent further complications. These circumstances influence the composition of mass cultures in a quantitatively unpredictable manner.

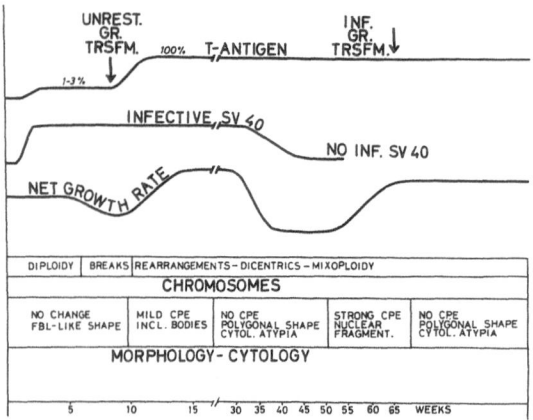

Fig. 16. Schematic outline of the events during transformation of human phase II fibroblasts by SV40. The time scale is only approximate. Note two waves with increased CPE; one partially coinciding with beginning unrestrained growth and the other constituting a late growth crisis. The interval between these waves is considerably decreased if phase III cells are exposed to SV40 (cf. KOPROWSKI, JENSEN, GIRARDI, and KOPROWSKA, 1966). Post crisis cultures have undergone infinite growth transformation, are usually virus-free and contain T-antigen in every nucleus

Early events, during the silent phase before gross CPE or transformation is observed, were studied in the human diploid cell line WI-38 by CARP and GILDEN (1966). The first cycle of growth of SV40 differed from the lytic interaction with GMK cells. Cell associated virus did not increase until 3 days post exposure and the maximum titer was not reached until day 6. In the established GMK line BS-C-1, cell associated virus rose sharply on day 2 to reach a maximum already on day 3. Virus in the supernatant of the human diploid cells only rose very slowly, in contrast to a rapid rise in the GMK system in which the virus yield on day 6 was about 2 logs higher. Another noteworthy contrast was the pronounced CPE in the GMK-cells, first noted on day 2, which lead to destruction of more than 80% of the population by day 5, and the absence of any CPE in the WI-38 cells.

The adsorption of SV40 and disappearance of infectivity followed an identical time course in the lytic simian and non-lytic human system indicating that these stages of the infectious cycle may be identical. All exposed human cells probably adsorb SV40. This finding is quite remarkable since only about 1% of the primarily infected human cells produce infectious virus (SHEIN, ENDERS, and LEVINTHAL, 1962) and only 1—3% of the cells contain SV40 specific T-antigen. It was concluded (CARP and GILDEN, 1966) that SV40, in the majority of exposed human phase II fibroblasts, goes into eclipse without altering the cell and that no further steps of the infectious cycle will take place. Whether the disappearance of infectivity reflects decoating of SV40 particles, abolishment of the infectivity by adsorption to certain cell structures or some other mechanisms has, however, not been determined. A block in the transcription process has, however, been suggested (CARP and SOKOL, 1969).

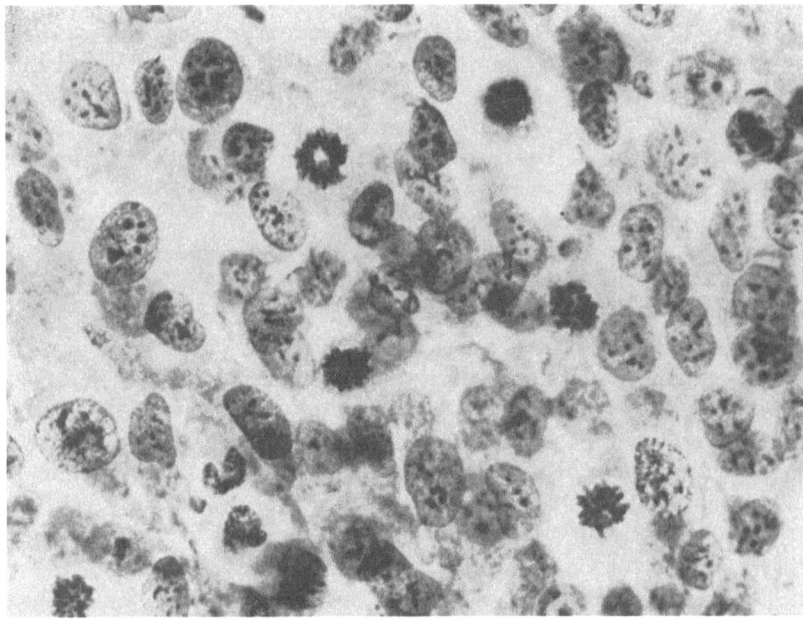

Fig. 17. Well developed transformation of human cells by SV40. The single most characteristic feature is the abundance of mitosis in spite of a high local cell density, i.e. unrestrained growth transformation. From PONTÉN, JENSEN, and KOPROWSKI, 1963. Magnification ×1,750

The silent phase preceding unrestrained growth transformation was further characterized by a persistent, moderately high yield of infectious virus from about 0.1—1% of the cells, which seemed to undergo occult slow lysis (BISSET and PAYNE, 1966). The proportion of cells having intranuclear T-antigen remains low until unrestrained transformation. Extensive chromosome analysis (MOORHEAD and SAKSELA, 1965) reveals no significant departure from normal diploidy during this phase.

A regular chain of events characterizes the *unrestrained growth transformation*, which in the case of mass cultures of phase II human embryonic fibroblasts commences 6—16 weeks after infection (PONTÉN, JENSEN, and KOPROWSKI, 1963). During the initial stage gross CPE may affect as many as 10—20% of the cells (CARP, DEFENDI, GILDEN, GIRARDI, JENSEN, and KOPROWSKI, 1966). Concomitantly the proportion of virus producing cells rises to approximately 100% but the SV40 yield per shedding cell decreases (GIRARDI, WEINSTEIN, and MOOR-

HEAD, 1966). Shortly thereafter the restraint on mitosis in dense culture is lost. At first this may seem diffuse, but soon the cells grow into thick multilayered piles (Fig. 17). The unrestrained growth transformation seems to take place in the absence of any gross cytological change in the fibroblast-like cells (GIRARDI, WEINSTEIN, and MOORHEAD, 1966). Only at a stage later do the cells assume a polygonal ("epithelial-like") form with nuclear pleomorphism, hypertrophy of nucleoli, irregular basophilia and an increased nucleo-cytoplasmatic ratio (SHEIN and ENDERS, 1962 b; JENSEN, KOPROWSKI, and PONTÉN, 1963).

WEINSTEIN and MOORHEAD (1967) analysed the relation of the karyotypic change to loss of cell cycle inhibition. In their large samples of WI-38 cells in phase II they noted an increased mitotic index even in dense cultures of fibroblastic appearance before any chromosome aberrations. They concluded that the unrestrained growth transformation may occur without apparent chromosome damage, however, extensive rearrangement is evident within approximately 2—3 cell cycles thereafter. The chromatid and chromosome breaks were randomly distributed along the chromosomes.

Unrestrained growth transformation permits at least a doubling of the maximal cell density. Even at this high density, cell division proceeds with little or no restraint. The population doubling time also seems to be shortened (MACIEIRA-COELHO, 1967 b) and a decreased rate of cell death has been recorded (NORRBY, 1970). No time-lapse observations on the movement of SV 40 transformed cells have been published, but fixed preparations containing compact cell piles suggest that the unrestrained growth transformation is accompanied by irregular growth transformation (lack of contact inhibition of cell membrane movement). The further progression of SV 40 transformation is characterized by disappearance of cell lysis, except in about 1% of cells synthesizing infectious virus. The over-all picture of a fully developed transformation is distinctive.

The proportion of cells positive for the intranuclear T-antigen rises from about 1 to 100% (Fig. 16) when unrestrained growth transformation becomes apparent (GIRARDI, WEINSTEIN, and MOORHEAD, 1966; BISSET and PAYNE, 1966). The chromosome breaks of the early unrestrained growth transformation have now led to rearrangements with dicentric and accentric fragments as well as heteroploidy (KOPROWSKI, PONTÉN, JENSEN, RAVDIN, MOORHEAD, and SAKSELA, 1962; YERGANIAN, SHEIN, and ENDERS, 1962; MOORHEAD and SAKSELA, 1963; MOORHEAD and SAKSELA, 1965).

Autochthonous or allogeneic human cells in the stage of unrestrained growth transformation have been implanted into terminal cancer patients. Seventeen homologous patients were subcutaneously inoculated with 5×10^6 cells at 53 different sites; in none did any tumors develop. Two patients who received their own skin fibroblasts back after unrestrained growth transformation developed small nodules at the sites of inoculation, which, however, rapidly regressed. Before regression the lesions were composed of pleomorphic cells in a sarcoma-like pattern. The rejection was accompanied by a violent lymphocytic reaction (JENSEN, KOPROWSKI, PAGANO, PONTÉN, and RAVDIN, 1964). RABSON, MALMGREN, O'CONOR, and KIRSCHSTEIN (1962) developed a line of human cells after exposure to SV 40 apparently equivalent to the transformed lines described above. Its tumorigenicity was tested in the hamster cheek pouch with negative results, but two cortisonized monkeys showed partly necrotic, sarcoma-like growths at the intrathalmic site of cell implantation. Negative results of hamster cheek-pouch inoculation of transformed human cells were also reported by ASHKENAZI and MELNICK (1963).

TODARO, GREEN, and SWIFT (1966) estimated the approximate transformation frequency by plating out human phase II fibroblasts on the day after exposure to concentrated, high titered SV40. If cells were seeded at an optimal density of $0.5-2.5 \times 10^3$ cells/cm^2 and grown in the presence of antiviral serum, colonies with unrestrained growth transformation were observed already within 10 days of infection. The frequency was only about 0.025%. The long silent phase in a mass culture may thus be an artifact reflecting the extended period necessary for the few transformed cells to reach a population size sufficiently large for detection.

The growth crisis. The basic features of this phenomenon — the unexpected appearance of degenerative changes and decelerated growth in previously healthy, rapidly replicating cultures with lacking cell cycle inhibition — have been noted in all human lines transformed by SV40 (JENSEN, KOPROWSKI, PAGANO, PONTÉN, and RAVDIN, 1964; MOYER, WALLACE, and COX, 1964; SHEIN, ENDERS, PALMER, and GROGAN, 1964; GIRARDI, JENSEN, and KOPROWSKI, 1965) including one hormone producing line of parathyroid gland origin (DEFTOS, RABSON, BUCKLE, AURBACH, and POTTS, 1968).

The growth crisis does not seem to be directly caused by the persistent chronic SV40 infection. No increase in virus titer or percentage of SV40 synthesizing cells has been noted (SHEIN, ENDERS, PALMER, and GROGAN, 1964). On the contrary, crisis was accompanied by a decreased virus yield (PONTÉN, JENSEN, and KOPROWSKI, 1963; JENSEN, KOPROWSKI, PAGANO, PONTÉN, and RAVDIN, 1964; GIRARDI, JENSEN, and KOPROWSKI, 1965). Cytological alterations are fairly characteristic with a marked increase in the number of cells with multilobed or fragmented nuclei sometimes taking the form of numerous micronuclei. About 10—30% of the cells showed these abnormalities (SHEIN, ENDERS, PALMER, and GROGAN, 1964). Fragmented nuclei usually signify a profoundly disturbed mitotic apparatus, but karyological data covering the crisis have not been published.

Recovery from growth crisis among SV40 transformed cultures was first observed in line WI8Va2 (PONTÉN, JENSEN, and KOPROWSKI, 1963) and an exceptional line derived from newborn human kidney (SHEIN, ENDERS, PALMER, and GROGAN, 1964). Such recovery is always accompanied by infinite growth transformation. Systematic study reveals that the probability of recovery can be augmented if the cell density is not permitted to fall below a certain level and numbers of cultures are kept for a sufficient time. Infinite growth transformation is ushered in by the focal appearance of a few rapidly multiplying cells in the midst of the stationary or degenerating cells typical of the growth crisis. The emergent transformed cells rapidly predominate (GIRARDI, JENSEN, and KOPROWSKI, 1965) and behave in all respects like classical spontaneously established lines (page 15). Infectious virus, usually but not always (GIRARDI, WEINSTEIN, and MOORHEAD, 1966), disappears and the cultures show stem line chromosome numbers in either the subdiploid or the subtetraploid range (WEINSTEIN and MOORHEAD, 1967). The T-antigen persists during years of further cultivation in apparently all nuclei.

Allogeneic implantation of cell line W18Va2 after infinite growth transformation resulted in intradermal nodules interpreted as sarcomas after 5×10^6 cells — a dose which was non-oncogenic prior to growth crisis. The transformed cells, identified by a chromosome marker, could be recovered from the sites of inoculation after reexplantation *in vitro* (JENSEN, KOPROWSKI, PAGANO, PONTÉN, and RAVDIN, 1964).

Variant cells. The induced alterations just described are typical for diploid phase II fibroblast-like cells. The sequence of events is considerably modified if other kinds of target cultures are employed.

When spontaneously degenerating, slow growing, late passage, phase III fibroblasts were exposed to SV40, unrestrained growth transformation occurred within about 1—3 weeks (JENSEN, KOPROWSKI, and PONTÉN, 1963; TODARO, WOLMAN, and GREEN, 1963) rather than the 7—10 weeks needed with phase II cells. This observation also held true for late passage amnion cells (CHANG and SINSKEY, 1968). Cellular DNA synthesis and mitosis of phase III cells were stimulated already within 2 days after infection by SV40 (SAUER and DEFENDI, 1966). The reason for the apparent speeding up of unrestrained growth transformation may have been the background of non-transformed cells practically incapable of division — a condition expected to facilitate early detection of any focal growth stimulation. This trivial explanation is rendered unlikely by the finding that 35% of the nuclei are positive for T-antigen in phase III cultures early after virus exposure in contrast to the 1—3% characteristic for phase II (CARP and GILDEN, 1966). This suggests a true difference in competence for transformation.

SV40 exposed phase III populations will only enjoy a short period of a few weeks of unrestrained growth transformation after which they enter growth crisis. It has been claimed that phase III cell lines always recover (GIRARDI, JENSEN, and KOPROWSKI, 1965), but this finding needs confirmation.

Untreated phase III human fibroblasts show chromosome abnormalities not seen in phase II populations (SAKSELA and MOORHEAD, 1963; YOSHIDA and MAKINO, 1963). Possibly certain types of aneuploidy predispose to transformation since skin fibroblasts from cases of Fanconi's anemia (TODARO, GREEN, and SWIFT, 1966), Down's syndrome or the trisomy no. 18 syndrome (TODARO and MARTIN, 1967) have a 3—30 fold increased transformation frequency over normal phase II fibroblasts. Fanconi's disease has chromosome abnormalities and an increased frequency of malignant neoplasms (GARRIGA and CROSBY, 1959). Down's syndrome (mongolism) typically shows a 21—22 trisomy and an increased frequency of leukemia (KRIVIT and GOOD, 1957). In one homozygous patient with Fanconi's disease the transformation frequency was 0.8% (Table 5) which is the highest reported for any human cell (TODARO, GREEN, and SWIFT, 1966). The increased frequency of transformation was not accompanied by acceleration of the transformation process in the individual cell, but is paralleled by an increased proportion of T-antigen forming cells (AARONSON and TODARO, 1968c). A potent small plaque variant of SV40 induced T-antigen in 7—15% of the originally exposed cells of a susceptible aneuploid human cell line (TODARO and TAKEMOTO, 1969).

Bovine cells. Cultures of stable embryonic bovine cells may be transformed by SV 40 or SV 40 DNA (DIDERHOLM, STENKVIST, PONTÉN, and WESSLÉN, 1965). The morphology resembles other systems with unrestrained and probably also irregular growth transformation. Infinite growth transformation of bovine cells occurs only in a proportion of exposed cultures and then after a growth crisis (DIDER-HOLM, unpublished). Bovine cultures exposed to SV 40 synthesize small amounts of virus (BLACK and ROWE, 1963 c; DIDERHOLM, STENKVIST, PONTÉN, and WESS-LÉN, 1965) with the exception of an unusual epithelioid line (DIDERHOLM and HERMODSSON, 1967) and become aneuploid (PONTÉN and LITHNER, 1966). Autochthonously implanted cells failed to form progressively growing tumors (DIDERHOLM and WESSLÉN, 1965).

b) Transformation of Unstable Cells or Established Cell Lines

Mouse cells. SV 40 does not seem to be oncogenic in the living mouse (BLACK and ROWE, 1963 c). In spite of this the virus caused unrestrained growth transformation and cytological "atypia" in kidney cultures from weanling mice. When tested early after their unrestrained growth transformation the cultures failed to produce tumors after syngeneic or allogeneic implantation (BLACK and ROWE, 1963 c), but late passage cells have been tumorigenic in untreated (KIT, KURIMURA, and DUBBS, 1969) or immunodepressed (TAKEMOTO, TING, OZER, and FABISCH, 1968) syngeneic mice.

Later studies on mouse cells *in vitro* have concentrated on the 3T3 line, which shows a strong cell cycle inhibition under a specified regimen, but is already aneuploid and has undergone infinite growth transformation spontaneously (page 15). Unrestrained growth transformation is detected already 10—14 days after infection by SV 40 of a mass culture with a resulting 25-fold increase of maximal cell density. The augmented cell density was not accompanied by any en-hancement of the exponential growth rate (TODARO, GREEN, and GOLDBERG, 1964).

The transformation frequency of 3T3 cells has been studied in a series of publications using morphology (irregular and unrestrained growth) of plated colonies as the criterion. 3T3 cells in exponential growth or at terminal density have a transformation frequency of 4.4 and 0.8%, respectively, under standard conditions (TODARO and GREEN, 1964a, b). Under optimal conditions, when 3T3 cells were infected in suspension by a high titer of the unusually active and concentrated SV 40 strains 776 or 777, more then 20% of the originally exposed cells gave rise to colonies with unrestrained growth transformation. Under these circumstances 10^3 PFU or 10^5 physical particles corresponded to one transforming unit (TODARO and GREEN, 1966b). With a potent minute plaque variant only $1.3—5.0 \times 10^2$ PFU were required for each transformation event, *i.e.* formation of colonies with unrestrained and irregular growth (TODARO and TAKEMOTO, 1969). It should be observed that these high figures were based not only on pure transformed colonies but also on partially transformed colonies similar to the "delayed" transformations by PV (page 64). No production of infectious virus, infectious DNA or lysis by 3T3 cells exposed to SV 40 has been reported (cf. TODARO and BARON, 1965).

The 3T3-SV 40 system resembles the BHK-21-PV system in many respects and has also been used for quantitative studies. It has not been possible, however, to apply the convenient soft agar transformation assay of MACPHERSON and MONTAGNIER (1964) (BLACK, 1966a). Transformation frequency is directly proportional to virus input (TODARO and GREEN, 1964a), up to a ceiling multiplicity of about 600 PFU/cell (BLACK, 1966a).

HENRY, BLACK, OXMAN, and WEISSMAN (1966) as well as GERSHON, SACHS, and WINOCOUR (1966) reported stimulation of cellular DNA synthesis in cell cycle inhibited 3T3 cultures exposed to 100—500 PFU/cell. The significance of the stimulation of DNA synthesis by SV40 is as obscure as that with polyoma virus (page 55). It may not be essential since it was not detected at multiplicities of infection where the frequency of transformation (and induction of T-antigen) was still high (TODARO and GREEN, 1966b; GERSHON, SACHS, and WINOCOUR, 1966).

When 3T3 cells were pretreated with the thymidine analogs IUDR or BUDR enhancement of the transformation frequency was noted (TODARO and GREEN, 1964b). The opposite effect was obtained if the 3T3 line was first transformed by PV and then exposed to SV40 (TODARO and GREEN, 1965). These results are difficult to interpret because of the selection introduced in the aneuploid cultures prior to treatment with SV40.

Hamster cells. The oncogenicity of SV40 *in vivo* was first established in Syrian hamsters (page 46). RABSON and KIRSCHSTEIN (1962) and SHEIN, ENDERS, LEVINTHAL, and BURKET (1963) reported transformation by SV40 of hamster cells in culture. Both groups infected newborn renal cell cultures and particularly the latter described the events in detail. In contrast to simian and human kidney cells, no gross CPE or inclusion bodies were noted. About one week post infection a few syncytia were observed but unrestrained growth transformation did not commence until 3—4 weeks after exposure to SV40 when disorganized cell piles similar to those described in human cells were noted. SEEMAYER (1968) could confirm this and divided the complex series of events into four stages. SHEIN (1967a, 1968) observed extensive CPE followed by complex morphological alterations in hamster brain cultures. After animal implantation a variety of histological tumor types were obtained: glioblastoma multiforme, ependymoblastoma, choroid plexus papilloma, meningioma and undifferentiated sarcoma. Evidently SV40 can transform other cell types than the fibroblast-like elements employed in most laboratories.

DIAMANDOPOULOS (1968) observed a wide variation in tumorigenicity among transformed hamster clones similar to the findings already described for the PV-hamster system (page 68). Tumorigenicity is greatly enhanced by passage *in vivo* but was not correlated with the presence of the SV40 genome (ENDERS and DIAMANDOPOULOS, 1969).

The low transformation rate of hamster cells is increased if chromosome breaks are introduced in the target cells by X-rays or IUDR prior to exposure to SV40.

In none of the hamster experiments was any growth crisis similar to that of human cells observed. Instead, infinitely multiplying cell lines were directly formed from the cultures which had undergone the unrestrained growth transformation.

Rat cells. Cultures established from excised kidneys underwent unrestrained growth transformation without any preceding cytopathic phase. Transformed cultures formed progressively growing tumors in animals, whose immune defense had been abrogated by X-irradiation. A transplant line was developed which showed a progressive increase in malignancy (DIDERHOLM, BERG, and WESSLÉN, 1966; BERG and STENRAM, 1968).

Rabbit cells. BLACK and ROWE (1963c), SCHELL and MARYAK (1966) and SCHELL (1968) reported transformation of early passage kidney cells, whereas SCHELL and MARYAK (1967) only saw a fulminant cytolytic effect with virus release in an established line.

Simian cells. The predominant effect of SV40 infection of monkey cells is either lysis (*e.g.* GMK cells) or latent infection (*e.g.* Rhesus kidney cells) (Table 6). If infection is started by small amounts of virus (FERNANDES and MOORHEAD, 1965) and particularly if the cultures are kept in a multiplying state (CARP and GILDEN, 1966), green monkey kidney cultures may undergo irregular, unrestrained and infinite growth transformation. The only intensively studied line eventually ceased to synthesize infectious virus but showed a persistent intranuclear T-antigen and aneuploidy (FERNANDES and MOORHEAD, 1965). Since the uninfected control line also became aneuploid and also underwent infinite growth transformation, it is not known if SV40 acted directly or as a selective agent.

Chronically infected Rhesus monkey cultures underwent unrestrained growth transformation after a long latent period. Transformed autochthonous cells injected back to three monkeys showed only temporary growth and were rapidly rejected (RABSON, KIRSCHSTEIN, and LEGALLAIS, 1965). Kidney cells from a pro-simian, the mongoose lemur, respond with widespread lysis, but a small proportion of exposed cells underwent unrestrained growth transformation (USHIJIMA, GARDNER, and CATE, 1966).

c) SV40 Induced Antigens

A *T-antigen* was first discovered in tumor extracts or transformed cells when these were reacted in a complement fixation test with sera from hamsters bearing SV40 sarcomas (BLACK, ROWE, TURNER, and HUEBNER, 1963). As already mentioned the T-antigen is not, as originally proposed, "tumor-specific" since it is readily demonstrated also in a lytic system (page 57). Immune hamster sera react with hamster, rabbit, mouse, pig (BLACK, ROWE, TURNER, and HUEBNER, 1963; BLACK and ROWE, 1963c), rat (DIDERHOLM, BERG, and WESSLÉN, 1966), bovine (DIDERHOLM and WESSLÉN, 1965), monkey (FERNANDES and MOORHEAD, 1965), and human (SABIN, SHEIN, KOCH, and ENDERS, 1964; HABEL, JENSEN, PAGANO, and KOPROWSKI, 1965) transformed cells indicating that the T-antigen is identical regardless of species. Therefore viral rather than cell DNA directs its synthesis. Purified SV40 DNA will also induce T-antigen production (BLACK and ROWE, 1965).

Most transformed lines have shown T-antigen in 100% of the nuclei in FA tests (POPE and ROWE, 1964a; RAPP, BUTEL, and MELNICK, 1964), a property which has remained stable, apparently indefinitely (cf. BLACK, 1966c). An exception was described by DIAMANDOPOULOS and ENDERS (1965) in hamster lung and liver lines which failed to become T-antigen positive after exposure to SV40 and transformation. It was not excluded that the apparent infinite growth transformation was spontaneous and not directly mediated by SV40. The presence of

T-antigen measured as the proportion of antigen carrying cells is positively correlated with tumorigenicity (DIAMANDOPOULOS and ENDERS, 1966).

Specific SV40 induced *transplantation antigens* were first demonstrated in hamster tumors (DEFENDI, 1963; HABEL and EDDY, 1963; KOCH and SABIN, 1964; KHERA, ASHKENAZI, RAPP, and MELNICK, 1963). Their behavior *in vivo* and occurrence in cells transformed *in vitro* have essentially been similar to the corresponding PV transplantation antigen.

Although a certain cross reactivity has been claimed with PV (KOCH and SABIN, 1964), human wart and Shope papilloma virus (MELNICK and RAPP, 1965), the general impression has been that the transplantation antigen is specific for SV40, and that the weak protection obtained after infection by other viruses than SV40 may have been non-specific.

The wide spectrum of species susceptible to transformation by SV40 has permitted investigations of the identity of the transplantation antigenicity. Non-proliferative SV40 induced hamster tumor cells administered to hamster pre-inoculated with SV40 prevented the occurrence of neoplasia (GOLDNER, GIRARDI, LARSON, and HILLEMAN, 1964) possibly by inducing immunity against a transplantation antigen. The immunizing efficiency of the tumor cells was not destroyed by exposure to IUDR, X- or γ-irradiation, but by mechanical disruption and formaldehyde (COGGIN, LARSON, and HILLEMAN, 1967), suggesting that the intact cell is required for efficient immunization. Similar results reported by GIRARDI (1965) with human cells provide further evidence that the capacity to induce resistance against SV40 virus induced neoplasia is shared by transformed cells regardless of species origin. This finding suggests that the capacity to induce tumor resistance is under control of viral rather than cellular genes. The alternative explanation that a transplantation antigen is induced in hamster cells by SV40 genome transferred from the injected xenogeneic cells has, however, not been excluded. It is still possible that SV40 acts as a specific depressor of a cellular gene controlling the synthesis of transplantation antigens (cf. page 178).

Indirectly it may be surmised that transplantation antigens are located on the cell surface (page 69). This cannot, however, be definitely established until a reliable *in vitro* test has been worked out (CARP, DEFENDI, GILDEN, GIRARDI, JENSEN, and KOPROWSKI, 1966). Direct evidence for the presence of a surface (S) antigen on cells transformed by SV40 has been published by TEVETHIA, KATZ, and RAPP (1965), TEVETHIA and RAPP (1965) and TEVETHIA, COUVILLION, and RAPP (1968). Immunofluorescent staining was seen when sera from hamsters immunized with SV40 and resistant to tumor cell challenge were employed. Sera from animals developing large tumors were non-reactive. Recently the sensitive mixed hemagglutination technique (ESPMARK and FAGRAEUS, 1965) confirmed the presence of a surface antigen in non-virus yielding transformed lines (HÄYRY and DEFENDI, 1968, 1969; METZGAR and OLEINICK, 1968). It has not been possible to prove that the surface antigens demonstrated by these two techniques are identical to those responsible for tumor transplant rejection; on the contrary, TEVETHIA, DIAMANDOPOULOS, RAPP, and ENDERS (1968) found no correlation between S-antigen and transplantation antigen. The S-antigen is contained in cells devoid of T-antigen (DIAMANDOPOULOS, TEVETHIA, RAPP, and ENDERS,

1968) and SV40 specific RNA and may therefore be coded for by cellular DNA (LEVIN, OXMAN, DIAMANDOPOULOS, LEVINE, HENRY, and ENDERS, 1969).

HÄYRY and DEFENDI (1970) used mixed hemagglutination to analyse the nature of the SV40 induced S-antigens. Surprisingly they found that low concentrations of trypsin and chymotrypsin (but not ficin, papain or neuraminidase) converted normal, spontaneously or PV transformed hamster and mouse cells to specific reactivity with antisera against SV40 induced S-antigens. The interesting implication is that SV40 alters the surface of an infected cell by exposing antigenic determinants present in the uninfected cell membrane in a masked state. Each tumor virus then has to uncover virus-characteristic normal sites to account for the clear virus-specific nature of the S-antigens. This unorthodox view suggests that the molecules responsible for virus-specific surface and by analogy also transplantation antigenicity are controlled by genes active in normal cells. The gene products are not permitted to express their antigenicity because they are too deeply buried in the cell membrane to become exposed to antibody forming cells.

Although most tumors or transformed cells have shown a positive transplant rejection response, a few reports have described the apparent loss of SV40 transplantation antigen in metastases (DEICHMAN and KLUCHAREVA, 1966). Under certain conditions sera from immunized hamsters have enhanced growth of transplantable SV40 tumors (GIRARDI, 1966).

d) The State of the SV40 Genome in Transformed Cells

The apparently ubiquitous presence of an intranuclear virus specific T-antigen strongly indicates the presence of SV40 genes in transformed cells. Further research has tried to resolve whether this reflects the presence of the whole or only part of the SV40 DNA. Attempts have also been made to define the functional state of the viral genes with particular emphasis on their possible integration with the cell's own genome.

Studies *in vivo* have been hampered by the presence of SV40 neutralizing antibody (EDDY, BORMAN, GRUBBS, and YOUNG, 1962; BLACK and ROWE, 1964) and perhaps other factors interfering with the expression of viral infectivity. It was first noted by GIRARDI, SWEET, SLOTNICK, and HILLEMAN (1962) that small amounts of infectious SV40 may be recovered from a proportion of primary or transplanted hamster sarcomas by exposing GMK cells to tumor extracts (ASHKENAZI and MELNICK, 1963; BLACK and ROWE, 1964). SABIN and KOCH (1963a) repeatedly monitored transplantable tumor lines and observed small bursts of infectious SV40 at irregular intervals. They concluded that most or perhaps all tumor cells carried the SV40 genome and that maturation of the virus occurred rarely in occasional cells — a situation resembling bacterial lysogeny (SABIN and KOCH, 1963b).

Even stronger evidence for the persistence of a complete SV40 genome was obtained by GERBER and KIRSCHSTEIN (1962) who observed that extracts of SV40 induced ependymomas, alternatively interpreted as "choroid plexus gliomas" (DUFFELL, HINZ, and NELSON, 1964), were always virus-free when tested on GMK cells. However, if intact tumor cells were seeded on GMK cells, infectivity was consistently detected. An increased probability of obtaining infectious

SV40 after cocultivation of hamster tumor cells and GMK cells was subsequently reported by SABIN and KOCH (1963 b). Further analysis (GERBER, 1963, 1964, 1966; BLACK, ROWE, and COOPER, 1963; BLACK and ROWE, 1963 a and b) has confirmed that certain transformed lines exist — which fail to yield directly demonstrable infectious SV40 in extracts, lysates or cells rendered non-viable by UV- or X-irradiation. Separation of viable transformed cells from GMK cells by a 5 μ Millipore filter prevented synthesis of infectious virus. UV-inactivated Sendai virus, which induces fusion of cells (OKADA, 1958; HARRIS and WATKINS, 1965) on the other hand, greatly enhanced SV40 synthesis in mixed GMK-transformed hamster cell cultures (GERBER, 1966). Essentially the same findings were reported for clones of SV40 transformed hamster kidney cells (BLACK, 1966 d), SV40 transformed 3T3 mouse cells (WATKINS and DULBECCO, 1967; TAKEMOTO, TODARO, and HABEL, 1968) or human SV40 transformed W18Va2 cells (KOPROWSKI, JENSEN, and STEPLEWSKI, 1967) fused or grown in intimate association with GMK cells. Five out of five tested transformed hamster clones yielded SV40 after fusion with GMK cells, although they differed with respect to the ease by which virus synthesis could be induced (BURNS and BLACK, 1968). The rare occurrence of a transformed human line which did not yield infectious SV40 even after fusion with GMK cells (KHOURY and VAN DER NOORDAA, 1969) may have been due to infection by defective virions.

Transfer of the message for T-antigen synthesis, even in the absence of infectious SV40 or viral antigen, has been shown in the spontaneous cell fusion involved in the formation of myotubes *in vitro* (FOGEL and DEFENDI, 1967). Recently these findings were extended to a heterospecific combination when transfer of T-antigen synthesizing ability from nuclei of human W18Va2 cells to rat myotubes was established by FA methods (FOGEL and DEFENDI, 1968) in the absence of any synthesis of complete virus particles.

The use of metabolic inhibitors indicates that DNA synthesis is not required for the internuclear transmission of T-antigen. Protein and RNA synthesis are on the other hand essential which excludes that the T-antigen simply diffuses to the originally negative partner of a heterokaryon. A probable mechanism is the synthesis of mRNA on the SV40 DNA template of the T-antigen positive nucleus. T-antigen is then synthesized in the common cytoplasm of the fused cells and transported to the T-antigen negative nucleus (STEPLEWSKI, KNOWLES, and KOPROWSKI, 1968). Transfer of SV40 DNA to the negative nucleus seems unlikely because heterokaryons of human T-antigen positive nuclei and T-antigen negative mouse nuclei will lose their capacity to synthesize T-antigen if all human chromosomes are lost during serial cultivation (M. C. WEISS, EPHRUSSI, and SCALETTA, 1968).

Altogether, these results strongly indicate that the transformed hamster, mouse or human cells carry the complete SV40 genome. From hybridization studies it has been calculated that about 60 SV40 viral DNA equivalents may be stably integrated in the cells of transformed lines (WESTPHAL and DULBECCO, 1968; TAI and O'BRIEN, 1969). The failure to synthesize infectious virus may be due to the integration of SV40 DNA in the cellular genome, which might alter the configuration of the molecule to render it non-infectious (KIT, KURIMURA, SALVI, and DUBBS, 1968) or even more plausibly subject it to the cells mechanism for the control of gene expression. Alternatively it may depend either on the lack of an essential cofactor or the existence of repressors (DUBBS, KIT, DE TORRES, and ANKEN, 1967). The latter alternative was rendered unlikely by JENSEN and KO-PROWSKI (1969) in experiments involving fusion of different transformed lines. Only in certain instances is there presumptive evidence that the SV40 genome is defective (TOURNIER, CASSIGENA, VICKER, COPPEY, and SUAREZ, 1967) and needs recombination or complementation to affect complete virus synthesis (KNOWLES, JENSEN, STEPLEWSKI, and KOPROWSKI, 1968). Fusion with GMK cells permits

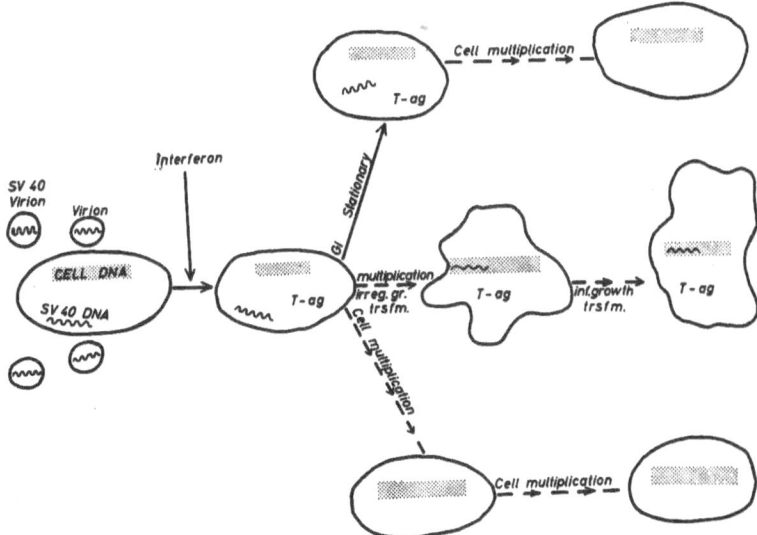

Fig. 18. Schematic outline of the effects of SV40 on 3T3 cells. Given optimal conditions the DNA of the virus probably enters all cells and gives rise to T-antigen synthesis. Interferon given in direct conjunction with virus infection can abolish the expression of the viral genes. If the cells are prevented from multiplying the SV40 genome will not become integrated and no transformation takes place. The T-antigen is lost if cell multiplication is initiated and the cells are then indistinguishable from normal 3T3 (upper row). If replication of cells starts soon after infection a majority of the cells will fail to integrate the viral DNA, remain untransformed and show no synthesis of T-antigen (lower row). A minority (middle row) will integrate the viral DNA into the cell DNA, show unrestrained growth transformation, resistance to interferon, persistence of intranuclear T-antigen but no synthesis of complete infectious virions (middle row). The integration of viral DNA into the cell genome seems to be the crucial event which makes a cell undergo unrestrained growth transformation and maintain its lack of growth control

synthesis of infectious DNA and complete virions by a process which is not yet understood (WEVER, KIT, and DUBBS, 1970). Fusion with cells which do not naturally support SV40 synthesis efficiently leads only to synthesis of T-antigen (STEPLEWSKI, KNOWLES, and KOPROWSKI, 1968; KIT, KURIMURA, SALVI, and DUBBS, 1968).

Transformed 3T3 mouse cells have been the material of choice for analysing the intracellular state of the SV40 genome. Figure 18, which is mainly based on the discussion by TODARO and GREEN (1966a), OXMAN and BLACK (1966) and OXMAN (1968) summarizes the effects of SV40 on 3T3 cells infected as stationary cultures. It should be kept in mind that certain steps are speculative and that the picture may be even more complicated if transfer of SV40 genome between cells takes place (GERBER and KIRSCHSTEIN, 1962) and/or normal cells interfere with the proliferation of transformed cells as suggested by Borek and Sachs (1966a).

Confluent 3T3 cells can be maintained in the G1 phase of the cell cycle for many days when fed with depleted medium. Infection by SV40 will lead to T-antigen synthesis which becomes maximal after 48 hours when about 50% of the nuclei are positive in the FA test. If subcultivation, which reverses the G1 block, is instituted soon after infection the transformation frequency is maximal; if the cells are kept stationary by postponing subcultivation, the frequency declines. These experiments were interpreted to show that a cell which does not proceed from the G1 phase shortly after infection with SV40 will synthesize T-antigen but will not be transformed. Evidently the SV40 genome may induce the T-antigen without causing unrestrained growth transformation or lysis. Todaro and Green (1966a) stressed the importance of DNA synthesis for expression of transformation but nothing in their experiments excludes that the G2 phase could be the crucial period, an interpretation in line with the data of Basilico and Marin (1966) in PV infected BHK-21 cells.

In cultures subcultivated 24 hours after infection the percentage of T-antigen positive nuclei decreased within the first four cell cycles from 50 to 10% indicating that about three fourths of the originally T-antigen positive cells lose this antigen. These cells are apparently unaffected by the virus infection and cannot be distinguished from uninfected 3T3 cells or the fraction of cells which did not respond to the original exposure to SV40. Nothing contradicts the assumption that the gene for the T-antigen may by diluted out during rapid cell division. The residual T-antigen positive cells will eventually undergo unrestrained growth transformation and remain T-antigen positive indefinitely.

Oxman, Baron, Black, Takemoto, Habel, and Rowe (1967) observed that T-antigen synthesis was completely unaffected in transformed cells grown for more than 100 generations under constant exposure to large amounts of interferon. This is in marked contrast to the efficiency by which T-antigen synthesis is prevented if interferon is administered prior to infection by SV40 (Oxman and Black, 1966). Interferon presumably owes its anti-viral action to a blockade of the association of viral mRNA with host cell ribosomes, an action which is specific in the sense that the interaction between cell mRNA and ribosomes is unaffected. It has been proposed (cf. Oxman, 1968) that the differential action of interferon could be explained if SV40 DNA in transformed cells is functionally integrated with host cell DNA in such a way that the T-antigens' messenger RNA could not be recognized as "viral" in origin. In cells, which do not transform, the gene for the T-antigen is not integrated. The transcribed mRNA is then perceived by interferon as "viral" and its translation is therefore averted (Fig. 18). SV40 DNA from transformed 3T3 cells also behaved as covalently bound to cellular DNA (Sambrook, Westphal, Srinivasan, and Dulbecco, 1968). Its physical state is thus fully compatible with its apparent functional integration with the cell's own DNA. The theoretical possibility that viral DNA is bound to mitochondrial DNA seems to have been excluded by Benjamin (1968).

Oda and Dulbecco (1968a) and Aloni, Winocour, and Sachs (1968) inferred from hybridization experiments that the size of the active part of the SV40 genome in transformed 3T3 cells corresponded to the size of the viral genome transcribed during the early phase of a lytic infection, *i.e.* before any infectious SV40 DNA is detected. The major part of the informational content of the viral DNA of transformed cells and that transcribed during early lytic infection was identical. The difference could be accounted for if the active part of the SV40 genome in transformed cells did not include one of the genes (induction of thymidine kinase?) transcribed during early lytic infection but instead included one of the genes transcribed during the late productive SV40-GMK interaction (ODA and DULBECCO, 1968a). SAUER and KIDWAI (1968) concluded from their hybridization experiments that about 80% of SV40 DNA was copied in transformed GMK cells in contrast to 3T3 cells where only 40% was transcribed. Actinomycin D (0.5 µg/ml) suppresses synthesis of "late" RNA (and viral coat protein) but not "early" RNA (CARP, SAUER, and SOKOL, 1969) confirming that the corresponding bits of viral DNA may have different modes of transcription. The very key to the intriguing problem of how SV40 effects cellular transformation may be the mechanism which regulates the transcription of integrated and non-integrated viral genes.

The importance of the presence of the SV40 genome for the continuous expression of the transformed phenotype is as unclear as in the PV system. It may be essential but can also be interpreted as a sign of a passed virus infection which in a non-specific manner diverted the cell to a route leading to disruption of physiologic growth controls. A specific integration site for the SV40 genome would favor the first hypothesis. But available evidence speaks against such a site. M. C. WEISS, EPHRUSSI, and SCALETTA (1968) observed that all human chromosomes were lost from their hybrids before they became T-antigen negative. No correlation was found to the exclusion of any particular chromosome. These results therefore indicate association with SV40 DNA to many different chromosomes. An important extension of these studies (M. C. WEISS, 1970) shows that T-antigen negative mouse/human hybrids with a few remaining human chromosomes can be rendered T-antigen positive by superinfection with SV40. This means that viral DNA can be associated with negative chromosomes originating from SV40 transformed cells. It strongly suggests that SV40 DNA does not favor any specific regions when it becomes integrated or associated with chromosomal DNA. The final solution of this crucial problem probably has to wait until better cytogenetic techniques have been developed.

4. Synopsis of Transformation by PV and SV40

These two small DNA viruses have similar physicochemical properties and interactions with their respective target cells also have many features in common.

Both may induce either lysis, abortive infection or transformation. Lysis is the predominant effect in cultures from the natural hosts of the respective viruses. The lytic interaction does not deviate in any significant manner from that of ordinary cytopathic viruses except possibly for the stimulation of host cell DNA synthesis and marked enhancement of cellular messenger RNA synthesis (ODA and DULBECCO, 1968b) under conditions of cell cycle inhibition. The abortive infection has only been little studied and the manner by which the SV40 or PV genome is lost remains unknown.

Transformation occurs in cells originating in the natural host of the virus but more efficiently in cells from certain non-natural hosts. It is strongly influenced by the nature of the target cells. If already partially transformed cells such as BHK-21 or 3T3 are used, irregular and unrestrained growth is rather easily superimposed. In primary lines of rodent cells, which have a spontaneous tendency to undergo infinite growth transformation, PV and SV40 will accomplish a "full" transformation including all the alterations listed in Table 1. If stable cells with no tendency to spontaneous transformation are employed, only SV40 is capable of inducing a "full" transformation with any regularity. "Infinite growth transformation" occurs then only after a growth crisis, thereby implicating severe selection factors acting on a heterogenous population.

The earliest growth control disturbance by SV40 is unrestrained growth, whereas PV first seems to induce irregular growth. These differences are sufficiently reproducible in various species to suggest different modes of action for the two viruses (TODARO, GREEN, and GOLDBERG, 1964). The predominant induction of irregular growth by PV implicates a cell membrane change as the key alteration. This would explain the lack of contact inhibition of cell movement, the susceptibility to contact with normal cells, the disturbed electrophoretic mobility and the increased acid mucopolysaccharides content of the membranes of transformed cells. As discussed in the introduction, lack of contact inhibition of membrane movement does not seem to be sufficient for the malignant cell state. A further requirement is a disturbance of the control of cell replication. Infinite growth transformation is only occasionally induced in the only stable system, where PV has been studied, but it will occur in unstable rodent cells. These facts suggest that PV needs unstable cells to accomplish complete transformation and owes part of its tumorigenic potential to a capacity to act upon an inherent tendency toward spontaneous transformation.

SV40 transformation starts with unrestrained growth in all systems and its key change may therefore involve an alteration in the control of cell division. SV40 may cause infinite growth transformation even against a stable background and probably also causes a cell surface change akin to that of PV (MARTINEZ-PALOMO and BRAILOVSKY, 1968). It is possible that the chromosome breaks and rearrangements induced by SV40 are instrumental in producing the adaptability which seems to be essential for indefinite propagability *in vitro*. For unknown reasons the wide species spectrum for transformation *in vitro* has no counterpart *in vivo* where only hamsters are susceptible to the tumorigenic action. With the exception of hamster cells (DIAMANDOPOULOS, DALTON-TUCKER, and VAN DER NOORDAA, 1969), SV40 transformed cultures usually have only a weak tumorigenic capacity after animal implantation.

The evidence that viral genetic material persists in transformed cells is almost irrefutable for both viruses. In PV transformed cells the data suggest that only part of the genome is present or at least reactivable (WATKINS and DULBECCO, 1967); in SV40 transformation the entire genome is present in many lines but in a partly non-expressed state.

The regular distribution of viral genetic information to all the progeny of a transformed cell, the insensitivity to interferon and physico-chemical binding to cellular DNA strongly suggest that viral DNA is integrated with cell DNA.

A non-integrated structure, whose rate of multiplication is sufficiently high to permit passive transmission to all daughter cells is, however, not entirely excluded, although the failure to recover infectious DNA from non-virus yielding cells speaks against this hypothesis (KIT, KURIMURA, SALVI, and DUBBS, 1968).

The significance of the persisting viral genes is unknown. It is possible that these genes and their specific products only serve as indicators of a past virus infection. This notion is favored by the failure to find a common gene product for PV and SV 40; a positive finding would have suggested the existence of information specifically responsible for the transformed cell state. No viral gene product — mRNA or antigens — regularly found in transformed cells or tumors has been shown to be consistently absent in lytically or abortively infected cells. There is thus no formal proof that such a molecule is specifically or exclusively involved in cell transformation. On the other hand recent findings that PV transformed cells regained normal control of locomotion after having lost chromosomes apparently carrying viral genes suggest that irregular growth transformation may be controlled by a specific PV gene. This important question is discussed further on page 177—181.

5. Transformation by Papilloma Virus

Bovine papilloma virus, which causes infectious fibroepitheliomas in its natural host (CREECH, 1929; BAGDONAS and OLSON, 1953; CHEVILLE and OLSON, 1964) and non-infectious fibromas or fibrosarcomas in experimental hamsters and mice (FRIEDMAN, LÉVY, LASNARET, THOMAS, BOIRON, and BERNARD, 1963; BOIRON, LÉVY, THOMAS, FRIEDMAN, and BERNARD, 1964; CHEVILLE, 1966; ROBL and OLSON, 1968), has been used to infect bovine, rabbit, murine, hamster and human cells.

Bovine cultures of various origin have shown a proliferative response and disorganized growth (BLACK, HARTLEY, ROWE, and HUEBNER, 1963; THOMAS, LÉVY, TANZER, BOIRON, and BERNARD, 1963; BOIRON, LÉVY, THOMAS, FRIEDMAN, and BERNARD, 1964). Similar alterations have also been observed in murine (THOMAS, BOIRON, TANZER, LÉVY, and BERNARD, 1964) and hamster (BLACK, HARTLEY, ROWE, and HUEBNER, 1963) but not in human or rabbit cell cultures (BLACK, HARTLEY, ROWE, and HUEBNER, 1963). The alteration presumably represents irregular and unrestrained growth transformation of fibroblast-like cells. Infectious virus is not released after transformation, which may occur in the absence of any chromosomal damage (BLACK, HARTLEY, ROWE, and HUEBNER, 1963; THOMAS, BOIRON, TANZER, LÉVY, and BERNARD, 1964). Transformed mouse cells did not cause tumors after syngenic implantation (THOMAS, BOIRON, TANZER, LÉVY, and BERNARD, 1964). Actinomycin D given early after virus exposure prevented transformation of bovine cells completely (MOULTON and GARG, 1966).

Human papilloma (wart) virus has been reported to be cytopathic for certain monkey and human cell cultures (MENDELSON and KLIGMAN, 1961; OROSZLAN and RICH, 1964; NOYES, 1965). In the last of these articles irregular growth transformation of embryonic fibroblast-like cells was also described. These interesting findings must still be regarded as preliminary and further confirmation is needed before it can be accepted that the inoculated virus is indeed directly responsible (review by ROWSON and MAKY, 1967).

6. Transformation by Adenovirus

a) General Features of Adenovirus

The well characterized adenovirus group is widespread in nature. Representatives have been isolated from humans, monkeys, cattle, dogs, mice and chickens (ROWE and HARTLEY, 1962). A recent review of their biochemistry and biology has been published by SCHLESINGER (1969). The DNA containing particles are identified by their size, icosahedral shape with 252 capsomeres and ether-resistance. Inflammatory diseases of the digestive and respiratory tract may be caused by adenovirus, but no evidence of tumor induction under natural conditions has been presented. In the laboratory, however, malignant tumors have

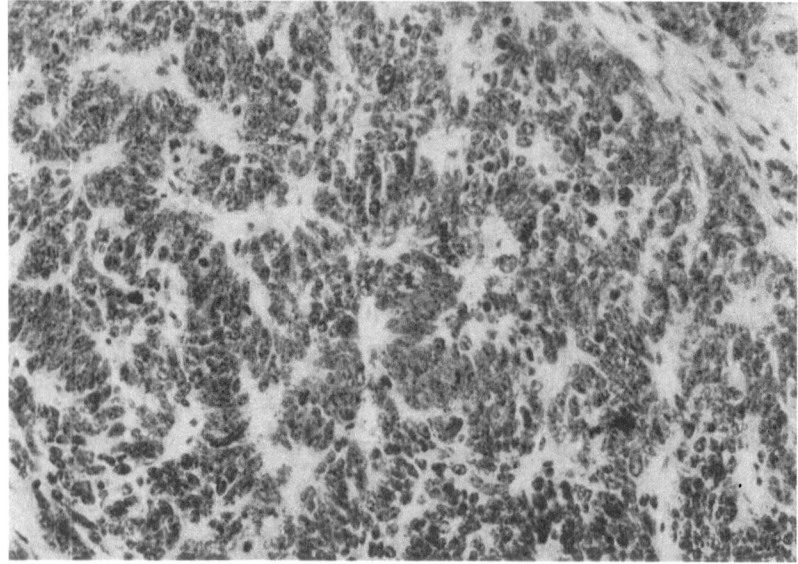

Fig. 19. Section from hamster tumor induced by adenovirus type 12. Moderately pleomorphic, columnar or cuboidal cells are arranged side by side in rather regularly arranged strands. Many mitoses may be seen. The pattern is characteristic for tumors induced by primate adenoviruses. The histogenesis has never been definitely established. Hematoxylin and Eosin. Appr. magn. ×300

been produced in rodents by human (TRENTIN, YABE, and TAYLOR, 1962), simian (HULL, JOHNSON, CULBERTSON, REIMER, and WRIGHT, 1965; ALTSTEIN, TSETLIN, DODONOVA, VASILIEVA, LEVENBOOK, and CHIGIRINSKY, 1968), bovine type 3 (DARBYSHIRE, 1966; DARBYSHIRE, BERMAN, CHESTERMAN, and PEREIRA, 1968), canine (SARMA, VASS, HUEBNER, IGEL, LANE, and TURNER, 1967) and avian (SARMA, HUEBNER, and LANE, 1965) strains. The tumors induced by primate adenoviruses have a highly characteristic uniform morphology (Fig. 19) (OGAWA, TSUTSUMI, IWATA, FUJII, OHMORI, TAGUCHI, and YABE, 1966; KIRSCHSTEIN, RABSON, PAUL, and PETERS, 1966; BERMAN, 1967; SPJUT, VAN HOOSIER, and TRENTIN, 1967) and do not produce infectious virus (HUEBNER, ROWE, and LANE, 1962). Newborn hamsters, rats, mice and Mastomys are susceptible to Ad 12, which is the most widely tested oncogenic type (TRENTIN, YABE, and TAYLOR, 1962; YABE, TRENTIN, and TAYLOR, 1962; HUEBNER, ROWE, TURNER, and LANE,

Table 7. *Subgroups of Human Adenovirus Types Arranged according to Oncogenicity and Other Characters. Only these Types where the Respective Tests have been Carried out have been Tabulated*

Sub-group	Types	Oncogenicity (hamster)	Transformation[1] in vitro (hamster, rat)	Subgroup specific T-antigen	Viral specific RNA	G+C content (types)	DNA–DNA homology
A	12, 18, 31	strong (12, 18, 31)	12	12, 18, 31	12, 18, 31	48–49% (12, 18)	70–100%
B	3, 7, 11, 14, 16, 21	weak (3, 7, 14, 16, 21)	3, 7	3, 7, 11, 14, 16, 21	3, 7, 11, 14, 16, 21[2]	50–52% (3, 7, 11, 14, 16, 21)	70–100%
C	1, 2, 5	not demonstrated	1, 2, 5	1, 2	1, 2, 5[3]	57–58% (1, 2, 5)	not done[4]
remaining types		not demonstrated	not demonstrated	not demonstrated[5]	not done	56–60% (4, 6, 9, 10, 13, 15, 17, 19, 20, 22, 28)	not done[4]

[1] Unrestrained growth with piles of cells of characteristic morphology (Fig. 20).

[2] RNA from tumor cells induced by types 3, 7, 14 and 16 hybridize specifically with viral DNA from types 3, 7, 11, 14, 16 and 21.

[3] RNA from tumor cells induced by type 2 hybridizes specifically with viral DNA from types 1, 2 and 5.

[4] Types 2 and 4 show about 35% homology.

[5] See page 96 about T-antigen induced by type 4.

1963; RABSON, KIRSCHSTEIN, and PAUL, 1964; YABE, SAMPER, BRYAN, TAYLOR, and TRENTIN, 1964). Transformation *in vitro* of hamster and rat cultures by certain human types has only been described rather recently.

Thirty-one different types of human adenoviruses have been identified by type-specific antigens. Three subgroups can be segregated on the basis of nucleic acid hybridization studies, the presence of common T- (or CF) antigens and onco-genicity in hamsters (Table 7).

Subgroup A (types 12, 18 and 31) are strongly oncogenic, *i.e.* they produce tumors within 1—2 months in a high proportion of inoculated new-born hamsters (TRENTIN, YABE, and TAYLOR, 1962; HUEBNER, ROWE, and LANE, 1962; PEREIRA, PEREIRA, and CLARKE, 1965). Tumor cells induced by these types synthesize a common viral-specific RNA not found in tumors induced by any other types (FUJINAGA and GREEN, 1966, 1967; GREEN and FUJINAGA, 1966). The viral DNA's show a high mutual homology and a low content of guanine + cytosine (LACY and GREEN, 1964, 1965; PIÑA and GREEN, 1965; GREEN and FUJINAGA, 1966). The tumors share a subgroup specific T-antigen (HUEBNER, ROWE, TURNER, and LANE, 1963; PEREIRA, PEREIRA, and CLARKE, 1965), which is also induced in a lytic interaction (HOGGAN, ROWE, BLACK, and HUEBNER, 1965).

Subgroup B (types 3, 7, 11, 14, 16 and 21) is defined by the presence of a common subgroup-specific virus induced RNA in the respective tumor cells (GREEN and FUJINAGA, 1966; FUJINAGA and GREEN, 1967, 1968). This RNA appears in the early RNA induced during lytic infections. The viral DNA:s show a high mutual homology (LACY and GREEN, 1965; GREEN and FUJINAGA, 1966). Their content of guanine + cytosine is lower than in subgroup A (PIÑA and GREEN, 1965) and they share a subgroup-specific T- (or CF) antigen which is also induced in lytically infected KB cells (HUEBNER, CASEY, CHANOCK, and SCHELL, 1965). Type 4 induces the same CF reactivity (HUEBNER, CASEY, CHAN-OCK, and SCHELL, 1965) but is not included in subgroup B because it does not show more than 33—51% DNA-DNA homology with Ad3, Ad7, Ad14 or Ad16 (GREEN and FUJINAGA, 1966). Type 4 DNA also has a higher proportion of guanine + cytosine than the members of subgroup B (PIÑA and GREEN, 1965).

Morphologically characteristic (BERMAN, 1967) hamster neoplasms have been produced by types 3 (HUEBNER, CASEY, CHANOCK, and SCHELL, 1965) and 7 (GIRARDI, HILLEMAN, and ZWICKEY, 1964). Types 14, 16 and 21 have also been reported to be oncogenic (GREEN and FUJINAGA, 1966; TRENTIN, VAN HOOSIER, and SAMPER, 1968). The tumors induced by subgroup B have a long latent time of 5—6 months and only occur in a very small proportion of animals as exemplified by type 3 which produced 1/8 and 1/17 takes in the series published (HUEBNER, CASEY, CHANOCK, and SCHELL, 1965). Type 7 seems, however, more efficient. Type 11 has to date not produced any tumors *in vivo*.

Subgroup C (types 1, 2 and 5) has been proposed by FREEMAN, BLACK, VAN-DERPOOL, HENRY, AUSTIN, and HUEBNER (1967). Previously all types not belong-ing to subgroup A or B were lumped together as a non-oncogenic subgroup. FREE-MAN *et al.* demonstrated that type 2 (and probably also type 5) induced unre-strained growth transformation in rat embryo cultures. VAN DER NOORDAA (1968a) and MCALLISTER, NICOLSON, LEWIS, JR., MACPHERSON, and HUEBNER

(1969) made the same observation with type 1. The mRNA from Ad 2 transformed cells hybridized specifically with DNA from adenovirus types 1, 2 and 5 but not with that of types 3, 4 or 12. No tumors were elicited in newborn hamsters by these types according to GIRARDI, HILLEMAN, and ZWICKEY (1964) and GIRARDI, LARSSON, and HILLEMAN (1965). TRENTIN, VAN HOOSIER, and SAMPER (1968) described 4 tumors in 60 hamsters inoculated with Ad 1. Since none of these had the characteristic "adeno-morphology" it is not clear whether they were directly caused by the virus under test. The subgroup C guanine + cytosine content is higher than that of the preceding subgroups.

The remaining types have not produced any transformation *in vitro* or tumors or typical "adeno-morphology". Whether they constitute a homogenous fourth subgroup has not been determined. Their guanine + cytosine content is in the same range as that of types 1, 2 and 5 and they may eventually also turn out to have transforming or oncogenic capacity.

In addition to the T- (or CF) antigen, the fiber (C) antigen has been demonstrated in hamster tumors induced by type 12 (HUEBNER, PEREIRA, ALLISON, HOLLINSHEAD, and TURNER, 1964).

A transplantation antigen was first demonstrated in Ad 12 mouse and hamster tumors by TRENTIN and E. BRYAN (1966). Cross reactivity between transplantation antigens induced by Ad 7 and Ad 12 has been reported (SJÖGREN, MINOWADA, and ANKERST, 1967; SJÖGREN, 1968).

b) Adenovirus Transformation in vitro

Transformation after exposure to adenovirus *in vitro* was first described by MCBRIDE and WIENER (1964). They infected mass cultures of newborn hamster kidney with type 12 and noted foci of epithelial-like cuboidal elements which tended to pile up on top of each other. Such a growth pattern was not observed in untreated controls and persisted during numerous subsequent passages. Similar findings have been recorded by POPE and ROWE (1964b), STROHL, ROUSE, and SCHLESINGER (1966), FREEMAN, BLACK, WOLFORD, and HUEBNER (1967), KUSANO and YAMANE (1967a, b), YAMANE and KUSANO (1967), BRAILOVSKY, WICKER, SUAREZ, and CASSIGENA (1967), RAFAJKO (1967), PÉTURSSON, ARMSTRONG, DE HARVEN, and FOGH (1969) and others (see below), in cultures of hamster or rat cells. The frequency of the apparent unrestrained growth transformation was extremely low and transformation only occurred unpredictably after many months. The tumorigenic potential of the altered cultures was ascertained by KUSANO and YAMANE (1967a) and BRAILOVSKY, WICKER, SUAREZ, and CASSIGENA (1967).

Studies by SCHELL and SCHMIDT (1968), SCHELL, MARYAK, YOUNG, and SCHMIDT (1968) and SCHELL, MARYAK, and SCHMIDT (1968) have attempted to define the conditions of Ad 12 transformation of hamster embryo cells. Several factors such as the quality of individual sera, the growth state of the cells at infection, the multiplicity of infection, the conditions of virus-cell adsorption and the calcium concentration in the medium influenced the ability of the virus to transform cells. A low (0.1 mM) concentration of calcium had previously been found optimal for the growth of explanted tumor cells or transformed cells (FREEMAN, HOLLINGER, PRICE, and CALISHER, 1965; FREEMAN, CALISHER, PRICE, TURNER, and HUEBNER, 1966).

Under optimal conditions colonies composed of tightly packed small poly-gonal cells could be identified 2—3 weeks post inoculation. The cells had an increased content of DNA, which permitted easy identification after Feulgen staining. Transformed cells showed an increased amount of membrane-bound mucopolysaccharides similar to that observed after transformation by papovavirus (MARTINEZ-PALOMO and BRAILOVSKY, 1968). The arrangement indicated a dis-turbed growth control of the type found in irregular and unrestrained growth transformation. When the input virus dose was just below cytopathic levels, one transforming unit corresponded to about 10^4 infectious units (HEK cultures).

Foci of cells showing unrestrained and irregular growth transformation did not always survive prolonged cultivation (RAFAJKO, 1967). Regular formation of transformed established lines was not dependent on the same environmental conditions as the primary transformation. "Normal" levels of calcium were essential for primary transformation but establishment of cell lines required a drastic reduction of the calcium concentration (SCHELL, MARYAK, and SCHMIDT, 1968; VAN DER NOORDAA, 1968b).

STROHL, RABSON, and ROUSE (1967) infected cloned-established BHK-21 cells with a high multiplicity of purified adenovirus 12 and then assayed for colony growth in soft agar (MACPHERSON and MONTAGNIER, 1964). After 21 days 0.5 — 1% of the exposed cells had formed visible colonies. After plating, these grew as a mixture of normal appearing fibroblast-like cells and cuboidal cells. The latter contained the adeno subgroup A T-antigen. 10^4 transformed cells produced tumors in 3 week old hamsters after a latent time of 7 weeks, and these had the characteristic morphology of adenovirus induced neoplasms. Uninfected control cells gave rise to tumors with a higher efficiency (only 10^2 cells were required) but these were non-descript fibrosarcomas. It was concluded that the fibroblast-like BHK-21 cell may be transformed by Ad12 and that the typical primate adenovirus tumor morphology is controlled by the viral genome and not by the nature of the infected cell.

Using the soft agar assay technique, McALLISTER and MACPHERSON (1968) reported a transformation frequency of 0.002% in another established hamster line (Nil-2), but failed to transform BHK-21.

Adenovirus types 1, 2, 3, 5 and 7 have been reported to cause unrestrained and infinite growth transformation of rat embryo cell cultures (FREEMAN, BLACK, VANDERPOOL, HENRY, AUSTIN, and HUEBNER, 1967; FREEMAN, VANDERPOOL, BLACK, TURNER, and HUEBNER, 1967; VAN DER NOORDAA, 1968a; McALLISTER, NICOLSON, LEWIS, JR., MACPHERSON, and HUEBNER, 1969). These alterations were only reproducible when the calcium content of the medium was lowered (FREEMAN, HOLLINGER, PRICE, and CALISHER, 1965; FREEMAN, CALISHER, PRICE, TURNER, and HUEBNER, 1966). Type 3 transformed cells were tumorigenic after animal implantation (FREEMAN, VANDERPOOL, BLACK, TURNER, and HUEBNER, 1967), but for unknown reasons type 1 and type 2 transformed rat cells did not give tumors after inoculation into newborn inbred rats.

A simian adenovirus (SA7) has been reported to transform hamster cells in the same way as human Ad12 (Fig. 20) (RIGGS and LENNETTE, 1967; CASTO, 1968a; WHITCUTT and GEAR, 1968; ALTSTEIN, SARYCHEVA, and DODONOVA, 1967). UV irradiation experiments suggest that the relative size of the viral genome responsible for irregular and unrestrained growth transformation is less than half of that of the total genome (CASTO, 1968b). Tumors were formed after implantation of transformed cells into animals (WHITCUTT and GEAR, 1968).

CASTO (1969) described an unusual appearance after infection by Simian virus 11 not seen with any other types of adenovirus. Within 7 days after exposure of primary hamster embryo cultures foci of T-antigen positive, very small, round,

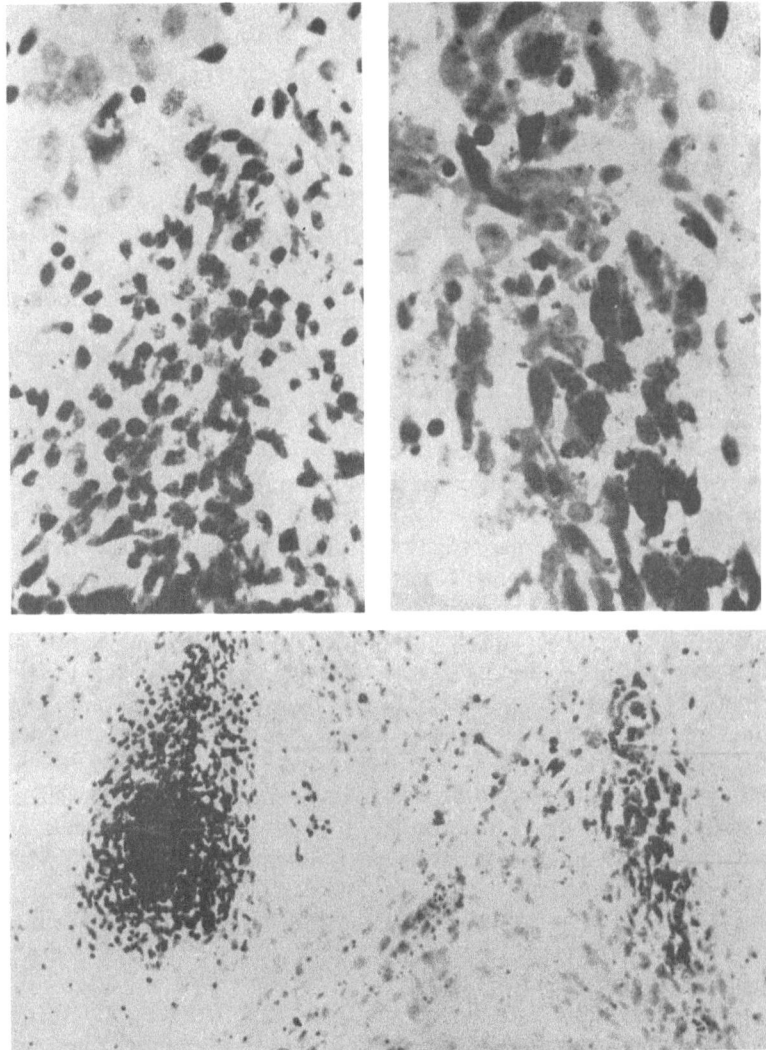

Fig. 20. Focus of hamster cells showing unrestrained growth transformation after exposure to SA7 (simian adenovirus). Note dark, lightly packed cuboidal cells contrasting against lightly stained firboblastic untransformed background. From WHITTCUTT and GEAR, 1968

loosely attached cells appeared. The cells resembled lymphoblasts (page 31) and were indefinitely propagable. Whether this means that adenovirus may be implicated in lymphomogenesis or represents an example of accidental pick-up of a simian "EBV-type" (page 37) was not discussed. Transformation of hamster and rat cultures by human and simian adenovirus must be regarded as well established. None of the experimental systems have, however, been used in detailed quantitative studies. It might be argued that the *in vitro* transformations represent selection of preexistent cells prone to undergo spontaneous transformation. However, the presence of virus specific T-antigen and mRNA constitute proof that the cells have been infected by adenovirus. Adenovirus transformed cells have a unique morphology *in vitro* and *in vivo* (KUSANO and YAMANE, 1967a) (Figs. 19 and 20), which has not been seen in spontaneously transformed rodent cells or in cells transformed by other viruses. It therefore seems most likely that the cell transformations are caused directly by the virus.

Infectious virus is not produced, and results from nucleic acid hybridization studies have suggested that a maximum of 50% of the viral genome is functioning in tumor (and by implication transformed) cells (FUJINAGA and GREEN, 1966).

DOERFLER (1968) infected BUDR-treated BHK-21 cells with H^3-labelled Ad 12 and isolated DNA in CsCl density gradients at various times after infection. Early all the label had the density of Ad 12 DNA, later increasing amounts were found at the position of heavy cellular DNA. The results were interpreted to indicate a linkage of Ad 12 DNA with cellular DNA. Since BHK-21 cells with their low transformation frequency were employed it is not known if this possible linkage is specifically correlated with transformation. It was also not ascertained whether the "linkage" is equivalent to true integration. ZUR HAUSEN and SOKOL (1969) could from a detailed study conclude that most of this "integrated" DNA was in fact reutilized degraded viral nucleic acid. Only a small fraction, mainly fragments, of viral DNA becomes integrated in a non-degraded state. ZUR HAUSEN (1968c) followed the development of subgroup specific T-antigen in the Nil-2 established line of Syrian hamster cells. Purified Ad 12, labelled with tritium, was used to infect cells at a multiplicity of 10. Within 24—48 hours 100% of the cells showed intranuclear T-antigen in the form of filaments also visible under phase contrast. If the cultures were allowed to multiply by subcultivation the percentage of T-antigen positive cells declined to 1% on day 12. If cell growth was arrested by FUDR, 100% of the cells remained positive for T-antigen. These results show that the low transformation frequency of Nil-2 cells does not depend on inefficient adsorption and penetration of virus. Expression of gene functions such as T-antigen synthesis is not sufficient for transformation. This may be considered an abortive infection analogous to that described with SV 40 in connection with Fig. 18. There is no evidence of replication of virus DNA or synthesis of infective virus in abortively infected cells, but in spite of this severe chromosome alterations occur (STICH, VAN HOOSIER, and TRENTIN, 1964) and a large majority of the cells become unable of further division (ZUR HAUSEN, 1968b, STROHL, 1969b). The data from ZUR HAUSEN (1967, 1968a, b, c) indicate that neither induction of T-antigen synthesis nor production of early chromosome lesions is responsible for transformation. It may be speculated that the rare cell which eventually transforms, does so because the viral genome becomes integrated in the cell genome as outlined for SV 40 in Fig. 18. The probability is increased if multiplying cells are used as targets (STROHL, 1969a).

Stimulation of hamster embryo cellular DNA synthesis by human types 12, 31, 3, 7, 2 or 5 has been reported by SHIMOJO and YAMASHITA (1968), TAKAHASHI,

OGINO, BABA, and ONAKA (1969), DOERFLER (1969) and STROHL (1969b), but it was not determined if this was a secondary effect caused by lysis of a proportion of the cells (page 55) or a direct result of the infection.

The continuous presence of subgroup specific T-antigen and viral specific RNA in the transformed cells shows that viral genes persist in the transformed cells. As with the other DNA tumor viruses it is, however, not possible to exclude that the presence and persistent functioning of the viral genome may be fortuitous and only signifies that the cell has once been infected by adenovirus.

It is possible to look for transformation of human cells by adenovirus. Although epithelium from kidney is rapidly destroyed, fibroblasts will withstand the infection for periods of several weeks (e.g. STROHL and SCHLESINGER, 1965a, b). One report (SULTANIAN and FREEMAN, 1966) has described enhanced growth and culture behavior suggestive of irregular and infinite growth transformation of human embryonic lung exposed to adeno 12. Cell contamination was not excluded and this isolated instance cannot be accepted without further confirmation, particularly since the cultures did not show the typical morphology of adeno-transformed cells of other species. An association between Ad12 and human chromosomes has been described (NICHOLS, PELUSE, GOODHEART, MCALLISTER, and BRADT, 1968; ZUR HAUSEN, 1968a) but no transformation occurred in these experiments.

7. Transformation by Combinations of SV40 and Adenovirus

Human adenovirus has no or at most only a rudimentary capacity of independent replication in simian cells. Preinfection by SV40 will, however, make them support synthesis of adenovirus to high titers (O'CONOR, RABSON, BEREZESKY, and PAUL, 1963; RABSON, O'CONOR, BEREZESKY, and PAUL, 1964). This finding has been followed by a number of experiments which have partially elucidated complicated interactions between these two unrelated DNA viruses. The single most important discovery has been that SV40 genes can be incorporated into adenovirus particles. The main concern in this review is the capacity of such "hybrid" particles to transform cells. The virological aspects have been well reviewed (BLACK, 1966b; RAPP, 1966a, b; RAPP and MELNICK, 1966; ROWE, 1967).

Enhancement of adenovirus replication by SV40 in GMK cells was originally shown for type 12 (RABSON, O'CONOR, BEREZESKY, and PAUL, 1964) but has later been extended to types 1—7, 14 and 21 (see RAPP, 1966a, b; LEWIS, BAUM, PRIGGE, and ROWE, 1966). Herpes simplex, measles and rabbit and human papilloma virus are ineffective (FELDMAN, BUTEL, and RAPP, 1966), but simian adenoviruses may also aid in the synthesis of human adenovirus in GMK cells (NAEGELE and RAPP, 1967; ALTSTEIN and DODONOVA, 1968).

Figure 21 charts some known or assumed genetic factors of SV40 and adenovirus particles. The nature and techniques for detection of the SV40 T and transplantation antigens (I and II) have been described on page 57. The factor (III) responsible for stimulation of cellular DNA synthesis is probably not identical to factor I and II, but no clear separation has been achieved. Factor IV has the specific capacity of greatly enhancing synthesis of human adenovirus in simian cells. It has not been immunologically or chemically identified but will be treated as an entity. It affects a late step in the adenovirus infectious cycle.

Factor V is recognized by its capacity to induce unrestrained growth transformation in the species listed in Table 2 and sarcomas in rodents with a fairly characteristic histology. It has not been excluded that factor V is identical with one of the other factors. Factor VI, which probably consists of three genes (page 49), directs the synthesis of "structural" viral proteins. The 6 enumerated factors constitute the only well established presumably genetically determined SV40 functions. The size of the SV40 DNA corresponds to 6—10 different genes. It is therefore possible that the factors shown in Fig. 21 reflect all the functions of the SV40 genome.

Human adenovirus DNA (mol. weight 23×10^6) would contain sufficient information to code for about 20—50 different proteins using the same calculations as for SV40. It is obvious that only a fraction of these are known (MAIZEL, WHITE, and SCHARFF, 1968) (Fig. 21). The T-antigen (factor a) may well be heterogenous

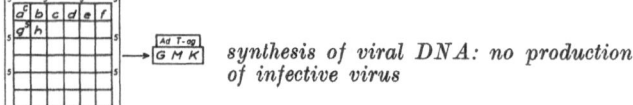

Synthesis of fully infective SV40 virions

   I induction of T antigen
  II induction of trspl. antigen
 III stimulation of cellular DNA: synthesis in stationary cultures
 IV factor potentiating synthesis of infective adenovirus in doubly infected GMK cells ('adeno helper function')
  V factor responsible for transformation of human or rodent cells to 'SV40 morphology'
 VI gene(s) for capsid protein synthesis

synthesis of viral DNA: no production of infective virus

a induction of subgroup specific T antigen
b induction of trspl. antigen
c factor(s) required for synthesis of defective hybrid particles in GMK cells ('PARA helper function')
d factor(s) responsible for transformation of rodent cells to characteristic adeno morphology
e gene(s) for hexon synthesis
f gene(s) for penton base synthesis
g gene for fiber synthesis (type specific antigen)
h gene(s) for internal protein synthesis

Fig. 21. Schematic representation of SV40 and adenovirus particles. The genomes contain 6 and 7 factors, respectively, defined by their functions. It is not established if they represent one or more genes. The adeno genome is considerably larger than the SV40 genome and contains a number of genes with undetermined specificity. Adenovirus is exemplified by a type 5 virion

(HOLLINSHEAD, BUNNAG, and ALFORD, 1967). It is defined here as the subgroup specific antigen demonstrable by sera from tumor-bearing hamsters in CF or FA tests. Its induction does not require any viral DNA synthesis. The transplantation antigen (b) is detected by the capacity to induce a specific tumor transplant rejection. Function c is operationally defined by the helper effect which allows, in mixed infections, the synthesis of hybrid particles (*i.e.* virus particles with an adeno-capsid enclosing SV40 genes, see Fig. 22) in simian cells. Factor d is analogous to the corresponding factor (V) of the SV40 genome. It induces irregular and unrestrained growth transformation *in vitro* and malignant neoplasms *in vivo* of a characteristic morphology. Factors e, f and g presumably consist of several genes including that for the adeno type specific antigen. Factor h stands for the gene(s) responsible for synthesis of internal protein (LAVER, PEREIRA, RUSSEL, and VALENTINE, 1968; PRAGE, PETTERSSON, and PHILIPSON, 1968).

The abortive cycle of adenovirus infection in GMK cells involves synthesis of T-antigen and viral DNA (FELDMAN, BUTEL, and RAPP, 1966; MALMGREN, RABSON, CARNEY, and PAUL, 1966; REICH, BAUM, ROSE, ROWE, and WEISSMAN, 1966) but no or very little capsid antigen (ROWE and BAUM, 1965; RAPP, FELDMAN, and MANDEL, 1966; SCHELL, LANE, CASEY, and HUEBNER, 1966; MALMGREN, RABSON, CARNEY, and PAUL, 1966; RAPP, JERKOFSKY, and VANDERSLICE, 1967; BLACKLOW, HOGGAN, and ROWE, 1967). The deficient replication may be due to a failure to transcribe the adeno genome or a malfunction of capsid protein synthesis. The former alternative was rendered unlikely by the demonstration of adenovirus-specific RNA in GMK cells infected by Ad7 (BAUM, WIESE, and REICH, 1968). If transcription is deficient it cannot involve the entire adenovirus genome. The deficiency is overcome if factor IV of the SV40 genome is present.

Interaction of SV40 and adenovirus in simian cells has been accomplished either experimentally (usually in GMK), or unintentionally by growing adenovirus vaccine strains in SV40 contaminated Rhesus monkey kidney cultures. Analysis of a type 7 vaccine strain (LL) grown in SV40 contaminated Rhesus monkey cells disclosed that the strain was oncogenic in hamsters and induced CF antibodies directed against the SV40 T-antigen. The latter could not be explained by contamination by SV40 particles, since the LL strain had been freed of SV40 virions by treatment with sera hyperimmune against SV40. It was soon proposed that SV40 genetic material must have been incorporated in adenovirus capsids (HUEBNER, CHANOCK, RUBIN, and CASEY, 1964; RAPP, MELNICK, BUTEL, and KITAHARA, 1964; ROWE and BAUM, 1964). This unexpected finding has later been extended to many other types and human adenovirus seems to be generally able to encapsidate part or the whole of the SV40 genome.

Figure 22 outlines the properties of the four different particles which have been identified after double infection of simian cells by SV40 and adenovirus. Two are identical to the infecting particles, whereas two represent newly created hybrids where a foreign genome is encased in an adenovirus capsid. All 4 particles are usually not detected in one system, instead a combination of 2 or 3 particles is the common outcome.

One case of SV40 genetic material encased in adeno capsids is represented by the "Ind.2" strain of Ad2 (LEWIS, BAUM, PRIGGE, and ROWE, 1966). The virus was originally isolated from a human adenoid and passed on Rhesus monkey kidney cells naturally infected by SV40. Analysis disclosed that "Ind.2" was composed of at least 3 different particles: regular adenovirus 2 particles in high titer, SV40 virus in low titer and a third particle of intermediate titer surrounded by an adenovirus type 2 capsid but containing a complete SV40 genome. The existence of the latter ('non-defective hybrid' of Fig. 22) was shown by a capacity to induce synthesis of complete SV40 virus even after pretreatment with SV40 neutralizing sera. Encasement by an adenovirus capsid was demonstrated by the antagonistic effect of adenovirus antisera. "Ind.2" may also contain a fourth particle composed of a defective SV40 genome coated by an Ad2 capsid (= defective hybrid PARA of Fig. 22). See ROWE, 1967. A fifth non-defective (one-hit kinetics) component inducing SV40-specific RNA and previously unrecognized antigen but no infectious SV40 has recently been isolated as a further proof of the heterogeneity of the SV40-adenovirus hybrid stocks (LEWIS, LEVIN, WIESE, CRUMPACKER, and HENRY, 1969).

A similar situation with apparent incorporation of the entire SV40 genome in an adenovirus capsid has been reported for strains of Ad4 (BEARDMORE, HAVLICK, SERAFINI, and McLEAN, 1965; EASTON and HIATT, 1965; EASTON, 1966), Ad3 (MORRIS, CASEY, EDDY, LANE, and HUEBNER, 1966), Ad12 (SCHELL, LANE, CASEY, and HUEBNER, 1966) and Ad5 (LEWIS, BAUM, PRIGGE, and ROWE, 1966).

"Ind.2" transforms hamster kidney cultures according to an SV40 pattern and the established cell lines contain SV40 but not any directly demonstrable

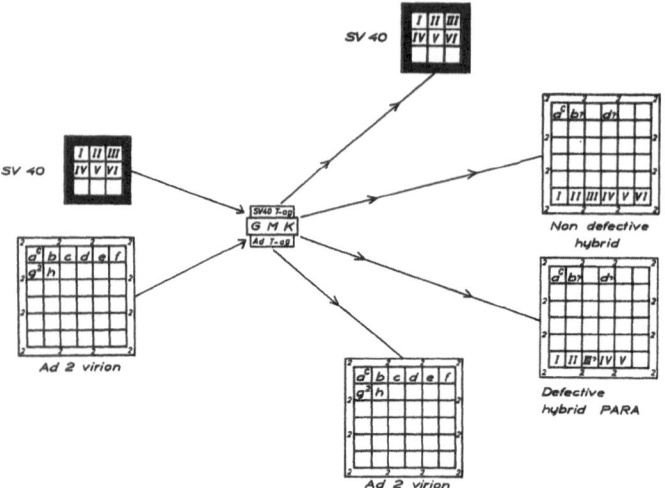

Fig. 22. The creation of hybrid particles encased in adenocapsids. Four different particles have been identified in the supernatant of simian cells accidentally or intentionally coinfected with SV40 and adenovirus. Two correspond to original infective particles. Two are coated by adenovirus capsids, but contain genetic information not found in regular adenovirus particles. The non-defective hybrid contains the entire SV40 genome and apparently an adeno-derived factor responsible for synthesis of adeno T-antigen. The defective hybrid (PARA) contains an SV40 genome lacking factor VI required for SV40 capsid protein synthesis. It retains the factor (IV) aiding adenovirus multiplication in simian cells. It contains the adeno gene(s) for the group specific T-antigen (a⁰) and probably also the factor responsible for the characteristic morphologic transformation (d). It lacks the factors required for independent replication (c) and capsid protein synthesis (e, f, g) and is therefore dependent on concomitant infection by a "helper" adenovirus for its replication and provision by an adenovirus capsid

adeno T-antigen (BLACK and WHITE, 1967). However, after animal implantation the cell lines sometimes elicited antibodies against adeno T-antigen and nests of adeno typical morphology intermingled with a predominant SV40 morphology could also be found. This suggests that factor V of the SV40 genome is of decisive importance for the transformation but that some cells have reacted under the influence of factor d of the adenovirus genome (IGEL and BLACK, 1967; BLACK, BEERMAN, and DIXON, 1969).

The composition of the non-defective hybrid particles is not known in detail. Physico-chemical data indicate that it contains adeno DNA in addition to the SV40 DNA, but this does not seem to have been directly demonstrated (LEWIS, BAUM, PRIGGE, and ROWE, 1966).

A second variation of adeno-SV 40 interaction is represented by the LL strain of adeno 7. It is composed of two kinds of particles; adeno 7 virions and particles with an Ad 7 capsid encasing a hybrid genome composed of parts of SV 40 DNA linked with adenovirus DNA. The latter particle (PARA) in Fig. 22, is defective because it is not capable of independent replication in GMK or any other cell. Only in the presence of "helper" adenovirus containing factor c will new progeny be formed.

Analysis of the PARA genome has been hampered by the failure to achieve a biological (ROWE and BAUM, 1965) or physico-chemical (ROWE, BAUM, PUGH, and HOGGAN, 1965; BOEYÉ, MELNICK, and RAPP, 1966) isolation of the particle which has the same density as adenovirions. It is nevertheless clear that it contains the following SV 40 derived pieces of information: induction of T-antigen (factor I) (HUEBNER, CHANOCK, B. RUBIN, and CASEY, 1964), induction of

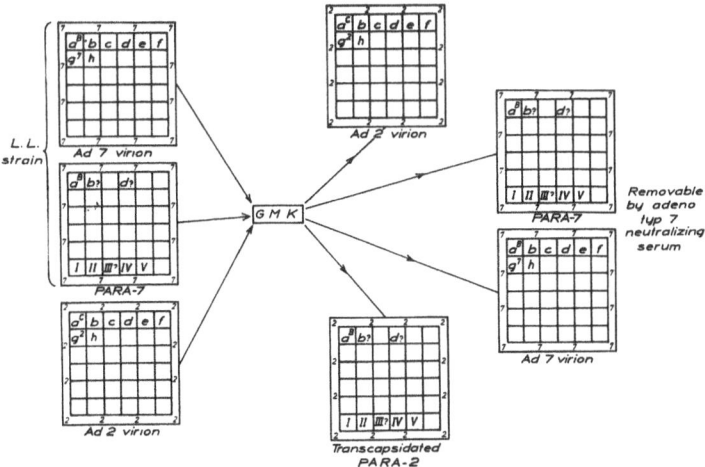

Fig. 23. Transcapsidation of a defective PARA genome. Coinfection of GMK cultures by a PARA hybrid preparation such as strain L.L. (ad 7 + PARA 7) and ad 2 gives rise to the original 3 particles *plus* a new particle PARA 2 coated by an adeno type 2 capsid. PARA 2 contains the genetic factor (a^B) for the adeno T-antigen subgroup specificity of the original PARA 7 genome indicating that the defective SV 40 genome was linked to a defective adeno 7 genome in the original particle and that this linkage persisted in the transcapsidation process

transplantation antigen (factor II) (RAPP, TEVETHIA, and MELNICK, 1966), induction of the surface (s) antigen (LAUSCH, TEVETHIA, and RAPP, 1968) (page 86), induction of the factor (IV) aiding adenovirus synthesis in GMK cells (ROWE and BAUM, 1965; BOEYÉ, MELNICK, and RAPP, 1966) and induction of "SV 40 typical" tumors *in vivo* (HUEBNER, CHANOCK, B. RUBIN, and CASEY, 1964; KIRSCHSTEIN, RABSON, and O'CONOR, 1965) or transformation *in vitro* (BLACK and TODARO, 1965; IGEL and BLACK, 1967) (factor V). It lacks the information for SV 40 capsid protein synthesis (factor VI) (HUEBNER, CHANOCK, B. RUBIN, and CASEY, 1964). Whether it is also deficient with respect to stimulation of cell DNA synthesis (factor III) is not known. A weak enhancement of thymidine kinase activity has, however, been reported (KIT, DUBBS, DE TORRES, and MELNICK, 1965).

The type specificity of the PARA capsid is determined by the adenovirus used for the original coinfection. The defective PARA genome can be made to change its capsid by coinfecting cells with an unrelated type of adenovirus (Fig. 23). It is seen that exposure of GMK to Ad 7, PARA-7 and Ad 2 will lead to production of a new particle, which differs from the original PARA-7 by the acquisition of an adeno 2 capsid. By treating the resulting virus mixture with type 7 antiserum, PARA-7 and Ad 7 can be eliminated (ROWE, 1965; RAPP, BUTEL, and MELNICK, 1965). Transcapsidation of PARA has been shown for several serotypes and seems to be a general human adenovirus property (RAPP, 1966a, b; BUTEL and RAPP, 1967; RAPP and JERKOFSKY, 1967; BLACK and WHITE, 1967).

While it has been quite clear that the PARA particle contains a defective SV 40 genome, it was only recently demonstrated that it also contains adeno genetic material. The evidence for the presence of "factor a" coding for the synthesis of the group specific T-antigen is outlined in Fig. 23. By appropriate serological methods it was possible to show that the adeno T-antigen of transcapsidated defective PARA-2 had the subgroup specificity (a^B) of Ad 7 meaning that SV 40 genes were joined by the adeno gene for T-antigen in the transcapsidation process (ROWE and PUGH, 1966). Factor d responsible for the characteristic adeno morphology of transformed cells has also been assumed to reside within the PARA particle because of apparent one-hit kinetics of the dose-response curve in transforming hamster cells. The alternative that factor d instead comes from the adenovirions mixed with the PARA particle cannot be entirely excluded because the precision of the transformation assay is insufficient (BLACK, 1968).

PARA DNA has not been isolated from mixed strain LL. Fixed-angle cesium chloride density gradient centrifugation showed a single homogenous peak, whereas an artificial mixture of SV 40 and adenovirus DNA could be resolved. The location and shape of the single peak coincided with that of adenovirus DNA indicating that the SV 40 DNA in the PARA-7 particle is contained in a molecule of approximately the same density as adenovirus DNA. Experiments with denatured DNA suggested that the linkage with SV 40 and adeno DNA is covalent (BAUM, REICH, HYBNER, ROWE, and WEISSMAN, 1966).

Transformation in vitro by defective adenovirus-SV 40 hybrids has been reviewed by BLACK and IGEL (1966) and BLACK (1968). Inadvertent hybrids such as the LL (PARA/Ad 7) strain as well as a number of intentionally produced hybrids can transform cells in culture. The main problem has been to elucidate the relative roles of the defective SV 40 and adenovirus genomes.

BLACK and TODARO (1965) first described transformation by the LL strain of baby hamster kidney cells and adult human skin fibroblasts. In both systems adenovirus CPE was noted initially. Among the survivors, unrestrained and irregular growth transformation typical of SV 40 was observed. Almost all these cells were positive for SV 40 T-antigen, but no adeno T-antigen was detected in the FA tests. The transformation was virtually identical to that obtained by pure SV 40, except that it was not prevented by anti SV 40 serum and no infectious SV 40 was produced; two findings readily explained by the defectiveness of the adeno-encased SV 40 genome (Fig. 22). These results would indicate that factor V of the SV 40 genome of PARA is of decisive importance. This interpretation was questioned by RABSON, MALMGREN, and KIRSCHSTEIN (1966), who obtained an adeno typical morphology after implantation of hamster cells transformed by the LL strain indicating that factor d derived from the adeno genome was the important one. Since the transformed cells were SV40 T-antigen positive and adenovirus T-antigen negative, the tumor morphology was not determined by the T-antigen. Other studies with the LL strain have given intermediate results with a bimorphic histology with features of SV40 and adeno transformation intermingled (WELLS, RABSON, MALMGREN, and KETCHAM, 1966; IGEL and BLACK, 1967; BLACK and WHITE, 1967).

On serial passage hamster cultures often acquire resistance to infection by SV40 presumably by selection of variant cells (DIAMOND, 1967a). The resistance to SV40 is probably caused by a failure of the virus to adsorb to the cells. If the SV40 genome is coated by an adenovirus capsid efficient adsorption takes place and infection now induces SV40 T-antigen and eventually also unrestrained growth transformation (DIAMOND, 1967b).

The artificial hybrid strains ("transcapsidants") PARA-2 and PARA-12, produced from the LL strain according to the scheme of Fig. 23, were studied by BLACK and WHITE (1967) in wealing hamster kidney cultures. The former transformed according to an "SV40 pattern" but the latter predominantly conformed with an "adeno pattern". All transformed lines contained SV40 T-antigen, but only those transformed by PARA-12 were in addition regularly positive for adeno T-antigen. In spite of the common absence of directly demonstrated adeno T-antigen, there was suggestive serological evidence of persistence of some adeno genetic material in the lines transformed by PARA-2. The adeno T-antigen in the lines transformed by PARA-12 was probably induced by the Ad12 rather than the PARA component of the virus preparation used, as indicated by the subgroup specificity.

A monkey adapting component (MAC) has been claimed (BUTEL, RAPP, MELNICK, and RUBIN, 1966). It was detected in a strain of Ad7 adapted to replication in GMK cells. MAC is encased in an adenovirion and exerts its action by enhancing adenovirus synthesis in simian cells. In contrast to PARA and SV40, which also have this property, MAC does not induce any SV40 T-antigen or SV40 transplantation resistance (RAPP, BUTEL, TEVETHIA, and MELNICK, 1967). It is therefore not established that it contains any part of the SV40 genome. MAC can be transcapsidated in a fashion analogous to PARA (BUTEL, RAPP, MELNICK, and RUBIN, 1966). Its eventual transforming or oncogenic potential remains to be determined.

In conclusion, the remarkable discovery of hybridization of two unrelated DNA genomes has opened a fruitful field for further exploitation of the mechanism of cell transformation. The aid for adenovirus multiplication provided by part of the SV40 genome constitutes an efficient mechanism by which human adenovirus can transcend its natural limits and replicate in cell species susceptible to SV40. From the point of view of the SV40 DNA a human adenovirus capsid enhances its penetrability into human and other cells. As a consequence transformation also becomes possible in otherwise resistant cells. This same principle makes it possible that oncogenic information contained in innocent adeno capsids may be widespread in nature. It should be emphasized that the true nature of such particles cannot be detected by standard serological techniques (neutralization tests etc.), which only identify the capsid.

Table 8. *Biological Properties of Viruses Belonging to the Avian Leukosis Complex*

	Virus[1]	Abbr.	Prototype strains	Lesions induced in chickens											Alterations in vitro[2]		
				LS	ML	AMC	EL	AEB	FS	END	NBL	HEP	OPT	HEM	Cytol. change	Irreg. and unrestr. growth transf.	Infinite growth transf.
Leukemia complex	Lymphomatosis	LV	RPL	+++	−		+		(+)	(+)	(+)	(+)	+	+	neg.(?)	neg.	neg.
	Myeloblastosis	AMV	BAI-A	+	+++				(+)	(+)	+	(+)	+		myelo-blasts	neg.	neg.(?)
	Myelocytomatosis	MCV	MC 29			+++	+			+	+	+			myelo-blasts	n.d.	neg.
	Erythroblastosis	CEV	R	++			+++	+							none	neg.	neg.
Sarcoma complex	Rous sarcoma	RSV	Table 11	+			+		+++					+	pos.[3]	pos.	neg.
	Fujinami	FSV							+++						pos.[3]	pos.	neg.
	Murray-Bragg	HEV	MH₂					+		++							

LS   = lymphosarcoma (visceral lymphomatosis).
ML   = myeloblastic leukemia (myeloblastosis).
AMC = aleukemic myelocytomatosis.
EL   = erythroblastic leukemia (erythroblastosis, erythro-leukemia).
AEB = aleukemic (anemic) erythroblastosis.

FS   = fibrosarcoma.
END = endothelioma.
NBL = nephroblastoma.
HEP = hepatoma and/or ovarian carcinomas.
OPT = osteopetrosis.
HEM = hemorrages and/or cysts.

[1] The listed virus stocks are mixtures of related agents. The biological effects are *inter alia* dependent on the proportion of these agents.
[2] Refers to avian cells.
[3] Basophilic round and spindle cells. Multinucleated, vacuolated cells.

Table 9. *Susceptibility to RSV in Different Species*

Species	Tumors in vivo	Transformation in vitro	References[2]
Human	n.d.[1]	++	Jensen, Girardi, Gilden, and Koprowski (1964); Stenkvist and Pontén (1964)
Monkey	+	+	Zilber (1965); Munroe, Shipkey, Erlandson, and Windle (1964)
Lamb	+		Sarma (1964)
Calf	n.d.	+	Stenkvist and Pontén (1964)
Dog	+++		Rabotti, Grove, Sellers, and Anderson (1966)
Rat	+++	+++	Svet-Moldavsky (1958); Ahlström and Jonsson (1962); Bergman and Jonsson (1962)
Hamster	+++	+++	Klement and Svoboda (1963); Ahlström and Forsby (1962)
Mouse	++++		Schmidt-Ruppin (1959); Bergman and Jonsson (1962)
Rabbit	+++	+	Zilber and Kryukova (1958); Ahlström, Jonsson, and Forsby (1962)
Guinea-pig	+++		Ahlström, Bergman and Ehrenberg (1963); Bergman and Jonsson (1962)
Steppe-lemming			Pogosianz, Bruyako and Radzikhovskaya (1966)
Chicken	+++	++	Rous (1911); Manaker and Groupé (1956); Temin and Rubin (1958)
Duck	++++		Duran-Reynals (1942); Temin (1960)
Guinea-fowl	+++		Duran-Reynals (1943)
Jap. quail	++++	++	Reyniers and Sackstedter (1960); Freeman (1963); Smida and Smidová (1968)
Turkey	+++		Duran-Reynals (1943); Prince (1962)
Pheasant	+++		Andrewes (1932)
Pigeon			Borges and Duran-Reynals (1952)
Reptiles			Svet-Moldavsky, Trubcheninova, and Ravkina (1967)
Drosophila	+		Burdette and Yoon (1967)

[1] CARR (1960) inoculated himself with RSV in 1939 without any untoward effects during an observation period of more than 30 years.

[2] For a more complete list see P. VOGT (1965).

B. Transformation by RNA Viruses

Two large families of oncogenic, physico-chemically similar RNA viruses have been identified in chickens and mice. They cause neoplastic and other diseases, collectively known as the avian and murine leukemia-sarcoma complex, respectively. Each family is defined by a common group specific (gs) antigen. An analogous feline complex may have been discovered recently (RICKARD, POST, NORONHA, and BARR, 1969).

The numerous members of these families possess biological and serological properties, which sometimes are imperfectly defined. Many virus stocks are mixtures of two or more related and interdependent strains. The viruses acquire phenotypic properties from the cells in which they mature. Phenotypic mixing of virus strains may occur, purified virus has rarely been obtained and classification and identification are difficult. Such shortcomings may in part explain the confusion and contradictory results which characterize some of the experimental data.

The Avian Leukemia-Sarcoma Complex

1. Virological Aspects — Classification of Viruses

The avian leukemia-sarcoma complex includes the following pathologic lesions: lymphosarcoma (lymphoid leukosis), myeloblastic leukemia, aleukemic myelocytomatosis, erythroblastic leukemia, aleukemic (anemic) erythroblastosis, fibrosarcoma, endothelioma, nephroblastoma, hepatoma, ovarian carcinomas, osteopetrosis, hemorrhages and hemorrhagic cysts (see ENGELBRETH-HOLM, 1938; DURAN-REYNALS, 1953; BURMESTER and WITTER, 1966; MLADENOV, HEINE, D. BEARD, and J. BEARD, 1967). None of the causative viruses produce only one of these diseases. On the basis of their predominant effect and disease spectrum they may, however, be grouped in a few rather well defined categories (Table 8). LV, AMV, MCV and CEV may be considered a separate subfamily on the basis of a predominant effect on hematopoietic cells. These leukemia inducing agents are fully capable of directing their own synthesis and have not been shown to be pathogenic for other than bird species. RSV, FSV and HEV mainly induce solid neoplasms. They infect insects, reptiles, birds and mammals (Table 9).

Marek's disease (neural lymphomatosis), originally thought to belong to the avian leukemia-sarcoma complex, is now considered a separate entity (CAMPBELL, 1961; BIGGS, 1961; BIGGS and PAYNE, 1967a, b; BURMESTER, 1967) probably caused by a herpesvirus which is cytopathic for chicken kidney cells and duck embryo fibroblasts (CHURCHILL and BIGGS, 1967; SOLOMON, WITTER, NAZERIAN, and BURMESTER, 1968; NAZERIAN, SOLOMON, WITTER, and BURMESTER, 1968). No transformation in vitro has been reported and Marek's disease will therefore not be further considered.

A comprehensive review of the avian tumor viruses has been written by P. VOGT (1965).

a) Morphology of the Virions

Thin section electron microscopy (GAYLORD, 1955) has revealed rounded often somewhat irregular particles of about 70—100 mμ. No consistent differences were found between different strains of the leukemia-sarcoma complex (BERNHARD, 1958, 1960; J. BEARD, 1963). The particles are composed of an envelope (coat) and a central nucleoid measuring about 35—45 mμ with an interposed membrane. The core is usually electron-dense but may be electron lucent in presumably immature particles (BERNHARD, OBERLING, and VIGIER, 1956; BERNHARD, BONAR, D. BEARD, and J. BEARD, 1958; HAGUENAU, DALTON, and MOLONEY, 1958; BONAR, PARSONS, BEAUDREAU, BECKER, and J. BEARD, 1959; HAGUENAU and J. BEARD, 1962; MLADENOV, HEINE, D. BEARD, and J. BEARD, 1967).

Examination of purified virus using shadow-casting or negative contrast has shown roughly spherical particles with a diameter of about 100—120 mμ (SHARP and BEARD, 1952; SHARP, ECKERT, D. BEARD, and J. BEARD, 1952; SHARP, D. BEARD, and J. BEARD, 1955). Sometimes the easily deformed particles have

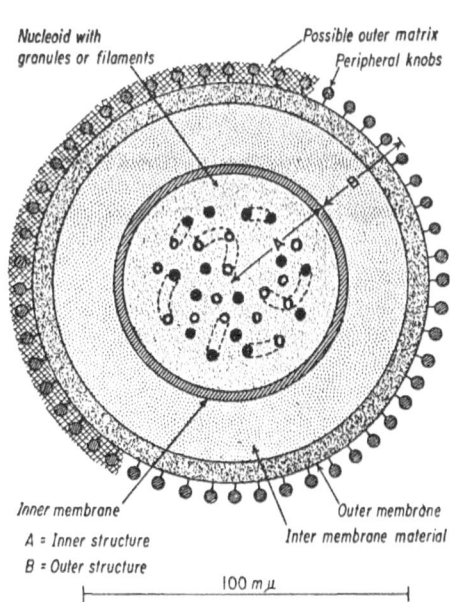

Fig. 24. Diagrammatic representation of the morphology and distribution of structural components of AMV virus.
Inner portion is shown as a spheroid consisting of a matrix with embedded entities of granular or filamentous arrangement enclosed in a membrane probably derived in particle synthesis as a part of or in association with the inner matrix and the contained entities. Inspection of thin sections of the virus at high magnification suggests that the enclosed entities, granules or filaments may be hollow.
Outer membrane is a relatively dense structure enclosing a less dense material possibly of high water content and without internal supporting structure. Some peripheral knobs have been drawn as free projections, but it has been indicated that the knobs may be embedded in an associated outer matrix. Representation of the outer membrane and other elements of the virus as homogenous indicates only that the present evidence is insufficient to establish a more detailed structure. From BONAR, HEINE, D. BEARD, and J. W. BEARD, 1963

shown tails generally considered artifacts (SHARP, ECKERT, D. BEARD, and J. BEARD, 1952; BARTH, RIMAN, and SORM, 1963; BONAR, HEINE, D. BEARD, and J. BEARD, 1963). The coat seems to be studded with small knob-like projections on its outside (BONAR, HEINE, D. BEARD, and J. BEARD, 1963; ECKERT, ROTT, and SCHÄFER, 1963). Figure 24 gives a schematic diagram of an AMV particle.

b) Chemical Composition

All leukosis-sarcoma strains contain lipid, protein, RNA and possibly also carbohydrates (BONAR and J. BEARD, 1959).

The presence of lipid was first suggested by the sensitivity to inactivation by ethyl ether (ANDREWES and HORSTMAN, 1949; FRIESEN and RUBIN, 1961; DRAYTON, 1961; WALLBANK, SPERLING, STUBB, and HUBBEN, 1962). Highly purified AMV preparations contain 35% lipid on a dry weight basis (BONAR and J. BEARD, 1959). Fractionation of the lipid showed: cholesterol 34%, neutral lipids 5% and phospholipids 58% (RAO and J. BEARD, 1964).

The localization of the lipid has been deduced from the ultramicroscopic effects of various solvents. The outer membrane is first destroyed but since the interior is also affected it has been concluded that this structure may also contain lipid (BONAR, HEINE, and J. BEARD, 1964).

Protein constitutes about 60% of dried AMV particles (BONAR and BEARD, 1959). Mild trypsinization removes the outer envelope (BONAR, HEINE, and BEARD, 1964). This fact together with the localization of enzyme activity at the viral surface *(vide infra)* prove that protein is part of the outer viral envelope. The presence of filaments and a sensitivity to proteolytic enzymes suggest the presence of protein also in the nucleoid (BONAR, HEINE, D. BEARD, and J. BEARD, 1952).

The nucleic acid of partially purified RSV was found to contain ribose and was thus identified as RNA (BATHER, 1957). Chemical analysis has shown an RNA content of about 2% dry weight corresponding to about 10^7 molecular weight units per virion in RSV (BATHER, 1957; L. CRAWFORD, and E. CRAWFORD, 1961) and AMV (BONAR and J. BEARD, 1959).

Table 10. *Base Composition of P^{32}-labeled Fast Sedimenting Viral RNA Components (From Robinson and Duesberg, 1968)*

	RSV + RAV	AMV	MLV	RSV (0)
C	24.2	23.0	26.7	24.2
A	25.1	25.3	25.5	24.8
G	28.3	28.7	25.1	29.2
U	22.4	23.0	22.7	21.7

Original attempts to isolate RNA resulted in a heterogenous mixture of low molecular weight degradation products. More recent work has produced a high molecular (about 70 S) RNA in RSV (W. ROBINSON, PITKÄNEN, and RUBIN, 1965; L. HAREL, A. GOLDÉ, J. HAREL, MONTAGNIER, and VIGIER, 1965) and AMV (W. ROBINSON and BALUDA, 1965; HUPPERT, LACOUR, J. HAREL, and L. HAREL, 1966; BAUER, 1966; ERIKSON, 1969) believed to be the intact viral nucleic acid. The isolated single stranded molecule appears as a non-branched thread with a modal length of 8.7 μ (GRANBOULAN, HUPPERT, and LACOUR, 1966). In addition the preparations have contained smaller RNA molecules of unknown significance but partly of host cell origin (BAUER, 1966; BONAR, SVERAK, BOLOGNESI, LANGLOIS, D. BEARD, and J. BEARD, 1967). One of these which sediments at 4 S is methylated and appears to be similar to host cell transfer RNA, which therefore seems to be incorporated into AMV virions (ERIKSON, 1969). One article (DUESBERG, 1968) has claimed that the fast sedimenting 70 S RNA is in fact an aggregate of smaller RNA:s; one major 36 S and some minor heterogenous components. The reported infectivity of RSV RNA by RABOTTI and COOK (1964) has never been confirmed. Not until biologically active nucleic acid has been prepared does it seem possible to resolve the physico-chemical nature of the native viral RNA.

Base ratio determinations of the RNA of purified members of the leukemia-sarcoma group have given the following values in mole per cent: cytosine 22.6 — 27.2; adenine 20.0 — 25.2; guanine 28.2 — 36.4; uracil 20.0 — 22.5 (BONAR, PURCELL,

D. BEARD, and J. BEARD, 1963; TRÁVENIČEK, BUŘIČ, ŘÍMAN, and ŠORM, 1964; ROBINSON, PITKÄNEN, and RUBIN, 1965; GRANBOULAN, HUPPERT, and LACOUR, 1966; HUPPERT, LACOUR, J. HAREL, and L. HAREL, 1966). The most recent estimates compiled by ROBINSON and DUESBERG (1968) are given in Table 10. The aberrant figures reported by THORELL (1962) for AMV are probably erroneous because of imperfections in the microtechnique employed.

The RNA is localized in the nucleoid from which it can be specifically removed by RNase (after fixation by permanganate) (EPSTEIN, 1958; YAMAGUCHI, 1962).

c) Viral Assembly

The oncogenic RNA viruses have a more complex structure and composition than the oncogenic DNA viruses. The cytoplasm is the major or perhaps exclusive site of synthesis and assembly. Final maturation occurs at the plasma membrane and is not directly accompanied by lysis of the infected cell.

All avian leukemia-sarcoma viruses show the same principal mode of development (DMOCHOWSKI, GREY, PADGETT, LANGFORD, and BURMESTER, 1964), although differences related to the cell rather than the virus exist (BONAR, PARSONS, BEAUDREAU, BECKER, and J. BEARD, 1959; HEINE, DE THÉ, ISHIGURO, SOMMER, D. BEARD, and J. BEARD, 1962; HEINE, BONAR, BECKER, and J. BEARD, 1964). The maturation at the cell membrane (GAYLORD, 1955; HAGUENAU, DALTON, and MOLONEY, 1958; VIGIER and GOLDÉ, 1959) takes place as a characteristic budding process (BENEDETTI and BERNHARD, 1958; BERNHARD, 1960). Budding is mainly observed along the outer cell surface but may also occur into cytoplasmic vacuoles. Dense "prenucleoid" material accumulates under the membrane, which reacts by forming a protrusion, which breaks off after having been connected with the cell by a thin stalk (MANNWEILER and BERNHARD, 1958; HEINE, BEAUDREAU, BECKER, D. BEARD, and J. BEARD, 1961; ZEIGEL, 1961). The sequestered cell membrane becomes the viral envelope (Fig. 25) but it has been modified to acquire regularly spaced small knob-like projections (DOURMASHKIN and SIMON, 1961; BONAR, HEINE, D. BEARD, and J. BEARD, 1963) (Fig. 24). Only after the virus particle has become free from the cell does the prenucleoid condense to become the strongly osmiophilic core of a fully mature virion (DE THÉ, 1964). DE THÉ (1964) has estimated that the budding process takes 10—20 minutes in AMV producing myeloblasts.

The stages preceding the maturation at the cell membrane have been more difficult to observe but intracytoplasmic clusters of small 44 mμ particles, possible precursors of mature virions, have been described (HAGUENAU, FEBVRE, and ARNOULT, 1962; DE THÉ, HEINE, SOMMER, ARVY, D. BEARD, and J. BEARD, 1963; DI STEFANO and DOUGHERTY, 1965). Part of the intracytoplasmic particles may, however, represent disintegrating phagocytosed virions (HEINE, BONAR, BECKER, and J. BEARD, 1964).

d) Viral Envelope Constituents

The morphological evidence that the host cell membrane is an essential part of the viral envelope has been amply confirmed by biochemical and serological investigations.

BEARD's group could show that virus, which is present in large quantity in the plasma of birds infected with AMV, showed a high enzymatic activity to dephosphorylate adenosine (and inosine) triphosphate (MOMMAERTS, ECKERT, D. BEARD, SHARP, and J. BEARD, 1952; MOMMAERTS, SHARP, ECKERT, D. BEARD, and J. BEARD, 1954). It was originally believed that this ATP:ase activity was

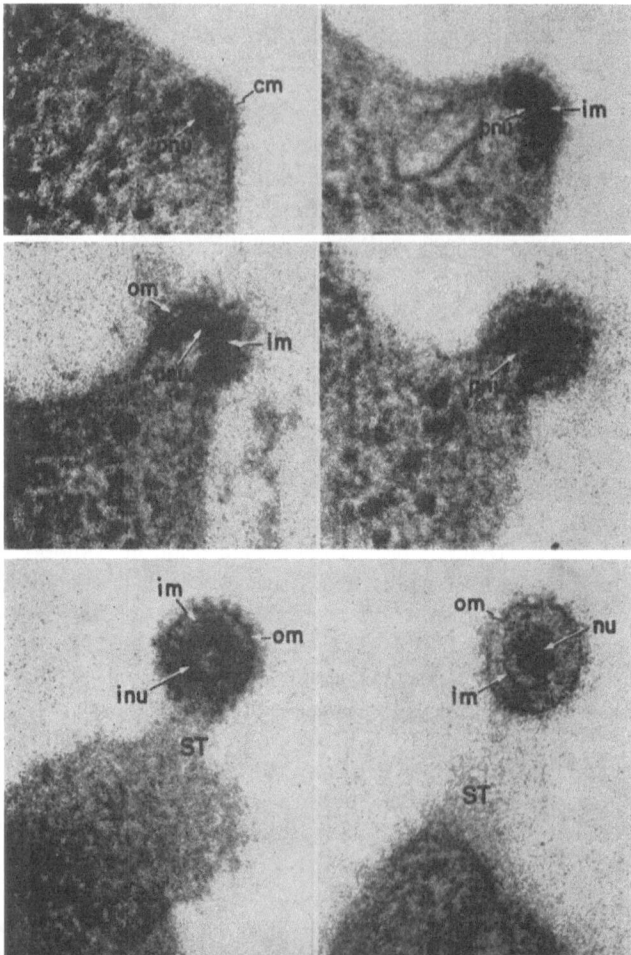

Fig. 25. Budding of AMV at cell membrane of different leukemic myeloblasts either from circulating blood or tissue culture after various periods *in vitro*. Micrographs show different stages in increased size of dense prenucleoid (pnu) first beneath cell membrane (cm) and subsequent involvement of cell membrane. Structures suggestive of outer (om) and inner particle membranes (im) are indicated. The lower left picture shows typical "immature" avian tumor virus particle before nucleoid (inu) condensation. Dense nucleoid (nu) characteristic of "free" particles is seen in the lower right picture. Surface of buds and particles peripheral to outer membrane is irregular and indistinct. ST = stalk. × 215,000. From DE THÉ, BECKER, and BEARD, 1964

a genuine viral component. Later studies utilizing an elegant method (HOLT and HICKS, 1961) for demonstrating ATP:ase at the ultramicroscopic level have shown that the viral ATP:ase is derived from the virus producing cell (NOVIKOFF, DE THÉ, D. BEARD, and J. BEARD, 1962). Myeloblasts rich in cell membrane associated enzyme produce ATP:ase positive particles, whereas nephroblastomas, lymphosarcomas, ovarian carcinomas and thymic blast cells, which only have traces of ATP:ase in the cell membrane produce enzyme-negative virions (THO-RELL, 1961; DE THÉ, HEINE, SOMMER, ARVY, D. BEARD, and J. BEARD, 1963; DE THÉ, ISHIGURO, HEINE, D. BEARD, and J. BEARD, 1963; DE THÉ, BECKER, and J. BEARD, 1964). AMV grown in chick fibroblasts also does not contain ATP:ase (BAUER, BAHNEMANN and SCHÄFER, 1965). The direct incorporation of an ATP:ase positive segment of the cell membrane into a viral envelope has been

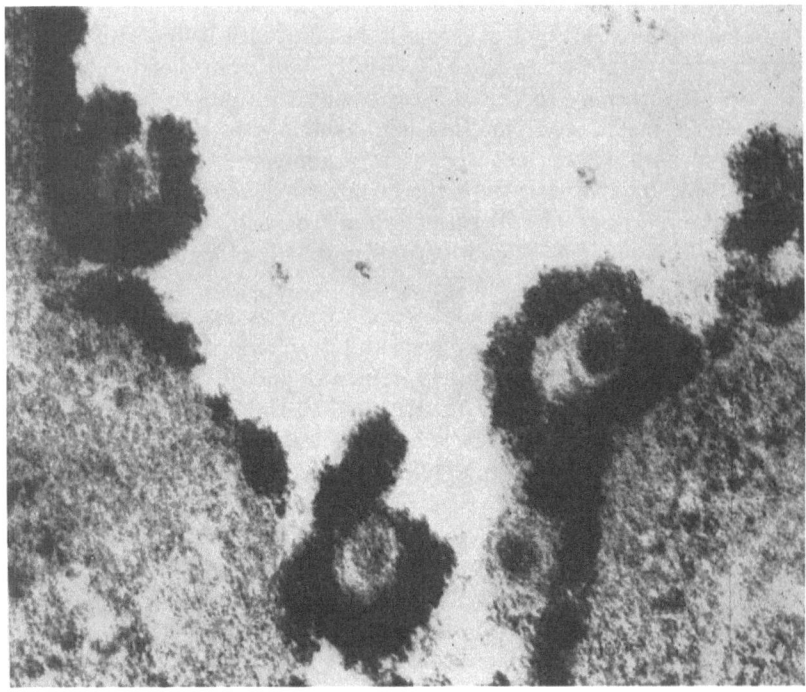

Fig. 26. Portion of surfaces of two blast-like cells. Four AMV particles are seen near the surface of the cells. ATPase reaction product is seen as dark material over the cell membrane and three of the virions × 70,000. From NOVIKOFF, DE THÉ, BEARD, and BEARD, 1962

visualized (Fig. 26) as have cartilage fibrils assimilated into the envelope of AMV produced by chondrocytes (HEINE, DE THÉ, ISHIGURO, SOMMER, D. BEARD, and J. BEARD, 1962). These findings provide clear evidence that AMV particles contain material specified by the genome of the host cell.

Other studies have employed normal chicken tissue antigen as a marker for host cell material in virus particles. It was early observed that antisera prepared against tissues from healthy chickens decreased the infectivity of avian leukemia or sarcoma viruses (GYE and PURDY, 1933; AMIES and CARR, 1939; KABATH and FURTH, 1941). Later on extensive experiments seemed to confirm that incubation of purified AMV or CEV with rabbit antiserum against normal chicken tissue or serum abolishes virus infectivity (ECKERT, SHARP, D. BEARD, GREEN, and J. BEARD, 1955; D. BEARD, BEAUDREAU, BONAR, SHARP, and

J. BEARD, 1957; D. BEARD and J. BEARD, 1957). The validity of these results is doubtful since no leukosis virus-free chickens were available at the time.

RUBIN (1956) challenged the accepted view of a direct neutralizing interaction between RSV and anti normal chick antibody by showing a requirement for complement. He also found that antiserum + complement was effective when added *after* the virus had penetrated the cells and concluded that the action by anti normal chicken serum was a pseudo-neutralization caused by an inhibition of transformed cells rather than a direct virus antibody reaction. BORSOS' (1958) and RYCHLÍKOVÁ and SVOBODA's (1960) findings that heterologous anti normal chicken serum failed to neutralize the necrotizing (in contrast to the transforming) effect by RSV in chick embryos or rats is in keeping with Rubin's interpretation.

Neutralization of infectivity of RSV (ROUS, ROBERTSON, and OLIVER, 1919) and perhaps also other avian tumor viruses may turn out to be more complex than originally thought. It is for instance possible that the anti-cell antibody may have prevented adsorption of virus in a non-specific manner (AXLER and CROWELL, 1968). In addition, the experiments have not employed pure virus strains and the presence of interfering virus was not critically excluded. As pointed out by P. VOGT (1965) the important problem of the effects of anti normal chicken antibody should be reinvestigated *in vitro*.

The viruses of the avian leukemia-sarcoma complex can be divided into four subgroups on the basis of surface antigens and genetic host resistance. Originally subgroups A, B and C were recognized (ISHIZAKI and P. VOGT, 1966; P. VOGT and ISHIZAKI, 1966a, b; PAYNE and BIGGS, 1966; M. HANAFUSA and T. HANAFUSA, 1966b; P. VOGT, ISHIZAKI and DUFF, 1966). Subgroup D has only recently been identified (BAUER and GRAF, 1969; DUFF and P. VOGT, 1969).

Localization of subgroup and type specific antigens has been inferred with FA staining. In infected tissue cultures, fluorescein-conjugated antibody stained extracellular material in areas with underlying virus particles, as visualized by electron microscopy. The antigens were accessible to antibody even without fixation indicating a location at the viral outer surface (P. VOGT and RUBIN, 1961; P. VOGT and LUYKX, 1963). The conclusion has been that the envelope of avian leukosis-sarcoma viruses contain a subgroup-specific and probably also a type-specific component (P. VOGT, ISHIZAKI, and DUFF, 1966; CHURCHILL, 1968; CHUBB and BIGGS, 1968) synthesized under the control of viral RNA. Table 11 lists members of the respective subgroups.

BAUER, TOZAWA, BOLOGNESI, GRAF, and GELDERBLOM (1969) succeeded in solubilizing virus specific envelope antigens by treating purified virus with a detergent. Subgroup A and B viruses yielded antigens specific for the respective groups with respect to induction of neutralizing rabbit antibody and capacity of autogenous interference (page 122). The antigenicity was protein- but not lipid-dependent.

Under special conditions, when cells are infected by a single virus type, the subgroup specificity is presumably controlled by the genome of the infecting particle. Many virus stocks have, however, been found to contain mixtures of related viruses where one of the partners is defective in the sense that it reproduces infectious particles slowly. Only if the cells are coinfected with another partner, the helper virus, will infectious progeny be formed with maximal efficiency

(H. HANAFUSA, T. HANAFUSA, and RUBIN, 1963). The helper virus genome supplies information for the synthesis of the subgroup specific antigens associated with infectivity (H. HANAFUSA, T. HANAFUSA, and RUBIN, 1964c). One important implication from this work is that an envelope component responsible for infectivity may be specified by a second viral genome (the helper virus). The concept of defective RSV is further discussed on page 124—127.

e) Viral Core Constituents

Apart from RNA, the viral core contains the group specific antigen for the avian leukemia-sarcoma complex. The reagent for the gs-antigen was discovered by HUEBNER, ARMSTRONG, OKUYAN, SARMA, and TURNER (1964) who found that sera from hamsters bearing large sarcomas induced by RSV-SR reacted in a CF test with RSV induced hamster and guinea-pig tumors. The antigen could be solubilized and was present in non-virus yielding tumor cells as well as viral suspensions. It was suggested that the antigen was a soluble subunit of the virus, whose synthesis did not require the formation of complete virus particles.

Subsequent studies have shown the gs-antigen in tumor cells induced by any member of the avian leukosis-sarcoma virus complex (SARMA, TURNER, and HUEBNER, 1964a, b; P. VOGT, 1967a). It gives rise to a specific antibody in hamsters, rabbits and monkeys (BAUER and JANDA, 1967). No antibody was detected in chicken sera with immunofluorescence methods (KELLOFF and P. VOGT, 1966; P. VOGT, 1967a), but ARMSTRONG (1969) was recently able to obtain reactions between gs-antigens and chicken serum antibody by immunoprecipitation. The gs-antigens are serologically identical regardless of the animal species infected with a leukemia-sarcoma virus (BAUER and JANDA, 1967).

The relationship between the gs-antigen and the viral particles has only been extensively studied with AMV and RSV. ECKERT, ROTT, and SCHÄFER (1964a, b) and BAUER and SCHÄFER (1965) showed that purified AMV was associated with a potent CF antigen reactive with rabbit antisera. The antigen was part of the viral core as evidenced by studies of disintegrated virions (BAUER and SCHÄFER, 1966). P. VOGT, SARMA, and HUEBNER (1965) employed immune sera from hamsters with a reactivity corresponding to the CF activity against the gs-antigen. Such sera sharply stained the cytoplasm of cells producing RSV(O) (page 126) and the membrane and the intercellular spaces of Rous cells, where virus production had been stimulated by helper virus. Weak staining was obtained in a few nuclei (PAYNE, SOLOMON, and PURCHASE, 1966) and all the cytoplasm (KELLOFF and P. VOGT, 1966; P. VOGT, 1967a) of the latter type of cells. The conclusion from these experiments has been that gs-antigen is synthesized in chicken and mammalian cells infected by members of the avian leukemia-sarcoma virus complex. The gs-antigen in extracellular virus could only be visualized after aceton treatment because this is necessary to expose the interior of the virions to effective contact with the fluorescent antibody (PAYNE, SOLOMON, and PURCHASE, 1966; KELLOFF and VOGT, 1966). The gs-antigen is clearly separated from the viral (V) antigen located at the viral surface, which elicits neutralizing antibody (cf. W. ROBINSON and DUESBERG, 1968).

Recent studies of the proteins from several purified members of the avian leukemia-sarcoma complex by polyacrylamide gel electrophoresis have suggested that the gs-antigen may consist of at least two polypeptides (DUESBERG, H. ROBINSON, W. ROBINSON, HUEBNER, and TURNER, 1968). Immunoprecipitation (ARMSTRONG, 1969; ROTH and DOUGHERTY, 1969) has produced up to 5 distinct bands. Chemical analysis of the gs-a component of the group specific antigen has shown alanine as the C-terminal (ALLEN, 1968) and proline as the N-ternal amino acid (ALLEN, 1969). This component is then probably a single polypeptide chain with a molecular weight of about 20,000 (ALLEN, 1969).

A new aspect of the origin of the gs-antigen was introduced by PAYNE and CHUBB (1968). An inbred chicken line (I) of phenotype C/O (see Table 12) contained gs-antigen which in appropriate crosses segregated as an autosomal dominant gene. Line I was virus-free in the electron microscope and extracts failed to show any helper activity. DOUGHERTY and DI STEFANO (1966) and

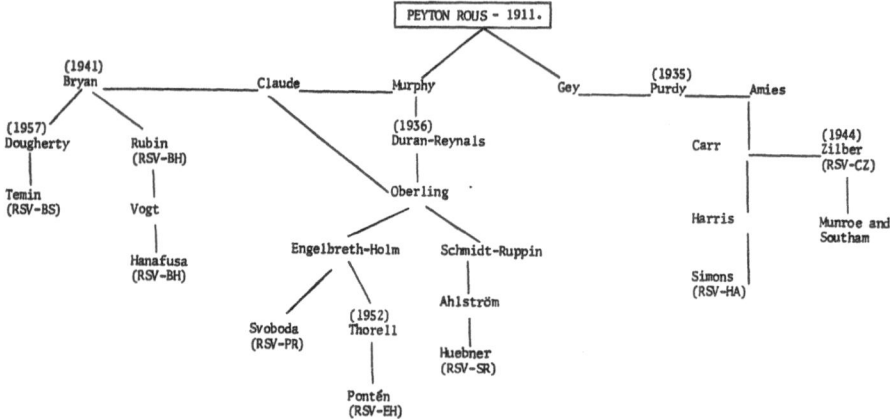

Fig. 27. Origin of the most widely employed strains of RSV. Modified after MORGAN and TRAUB, NCI Monograph, No. 17, 1964

DOUGHERTY, DI STEFANO, and ROTH (1967) also found an antigen in apparently virus-free chick embryo cells, which cross-reacted with viral gs-antigen in gel immunoprecipitation.

The surprising finding of cross-reacting antigens in virus particles and apparently virus-free cells has not yet been adequately explained. It is possible that the dominant cellular gene is a vestige of past virus infection with incorporation of viral genetic information in the cell's genome. This hypothesis requires the integration of RNA determined heredity in the cell genome. Such a phenomenon has never been observed, but is compatible with the hypothesis by TEMIN (1963), VIGIER and GOLDÉ (1964a, b) and PRINCE (1966) of a DNA Rous provirus synthesized on the RNA template of the RSV. An alternative would be that the gs-antigen contained in virus particles is of cellular origin. In gs-negative cells virus infection may lead to a derepression of the cell gene for gs-antigen synthesis. The antigen may then be incorporated in the viral nucleocapsid during maturation. A third possibility would be the actual pick up by virus particles of cellular genes directing synthesis of gs-antigen (PAYNE and CHUBB, 1968). The latter two explanations imply that repressed genes for gs-antigen synthesis occur even in distantly related animal species. The problem will be difficult to solve until it has been unequivocally shown that the serological identity between the antigen of PAYNE and CHUBB (1968) and the viral gs-antigen is not coincidental.

f) Origin of RSV Stocks

The origin of the most widely employed Rous sarcoma laboratory virus stocks is chartered in Fig. 27, which is modified after MORGAN and TRAUB (1964). The exact history of some of the stocks is unknown and most if not all probably represent mixtures of related viruses.

g) The Species Range of RSV

In his original description PEYTON ROUS stressed the narrow range of susceptible hosts. The first passage was done to "chickens from the same setting of eggs as the individual with the tumor" which was "a barred Plymouth Rock hen of light color and pure blood". Only after several transfers could unrelated chickens be infected but pigeons, ducks, rats, mice, guinea pigs and rabbits were resistant (ROUS, 1911). Many of the present RSV strains are known to have an unusually wide range of species susceptibility including mammals (ZILBER and KRYUKOVA, 1957a, b; SVET-MOLDAVSKY, 1957), birds, reptiles (SVET-MOLDAVSKY, TRUBCHENINOVA, and RAVKINA, 1967) and insects (BURDETTE and YOON, 1967) (Table 9). The RSV-SR and RSV-CZ were first noted for their effect on mammals but later studies have shown that RSV-PR, RSV-BH, RSV-BS and RSV-EH also have an extended species range. The reasons for the remarkable widening of the susceptibility spectrum which seems to have taken place after Rous' original discovery are unknown (ROUS, 1967), but may be related to extensive phenotypic mixing during laboratory passage (page 126).

The tumors induced *in vivo* have retained the original morphology of a spindle cell- or myxosarcoma regardless of the animal species.

h) Replication of Leukemia-Sarcoma Viruses in Avian Cells

Non-transforming virus strains (Table 11). *In vivo* these viruses either occur as non-disease producing passengers or they may give rise to leukemia. They do not alter cells morphologically or cause any transformation *in vitro*. All of them seem to be capable of acting as helper viruses. Three strains (RAV 1, RAV 2, RAV 3) have been recovered from stocks of RSV-BH (RUBIN and P. VOGT, 1962; H. HANAFUSA, 1964, 1965; P. VOGT and ISHIZAKI, 1965). RAV 4 and RAV 5 were isolated from a stock of RSV-BS (P. VOGT and ISHIZAKI, 1966) and RAV 50 from RSV-SR (H. HANAFUSA and T. HANAFUSA, 1966b). A Fujinami-associated virus (FAV 1) has also been found by P. VOGT and ISHIZAKI (1965). Stocks of avian myeloblastosis virus contain two viruses which both may function as helpers (P. VOGT, 1965). Recently these two agents have been named MAV 1 and MAV 2 because of their inability, alone, to induce myeloblastosis *in vivo* or stimulate myeloblast proliferation *in vitro* (MOSCOVICI and P. VOGT, 1968). An LV strain consisting of two non-defective potential helper viruses RIF 1 and RIF 2 has been described (see P. VOGT, 1967a).

Transforming virus strains (Table 11). These cause unrestrained and irregular growth transformation in cell culture and sarcomas and related solid mesenchymal

Table 11. *Subgroups of Avian Leukemia-Sarcoma Viruses. The Division is Based on Cross Reactivity between Antigens Eliciting Neutralizing Antibody in the Chicken, Host-range among Genetically Different Chickens and Patterns of Virus-virus Interference.* Mainly after P. VOGT and ISHIZAKI (1965, 1966a, b), P VOGT, ISHIZAKI, and DUFF (1966) and DUFF and P. VOGT (1969)

	Subgroup A	Subgroup B	Subgroup C	Subgroup D
Leukemogenic helper viruses No transformation *in vitro*	RAV-1 RAV-3 RAV-4 RAV-5 FAV-1 MAV-1 RIF-1	RAV-2 FAV-2 MAV-2 RIF-2	RAV-7 RAV-49	RAV-50
LV or AMV stock isolates	RPL-12, RPL-28 RPL-29, RPL-30 AMV-1	RPL-25 AMV-2		
Sarcomagenic pseudotype: transformation *in vitro*	RSV-PR (RAV-1) RSV-SR (RAV-1)	RSV-PR (RAV-2) RSV-SR (RAV-2)	"RSV-PR-C" = RSV-PR (RAV-49); ? "B 77" = RSV-PR (RAV-?)? RSV-SR (RAV-49) RSV-BH (RAV-7)	"RSV-SR-D" "RSV-SR-H" RSV-SR (RAV-50) "RSV-CZ-D"

neoplasms *in vivo*. The only extensively studied representatives are of RSV origin. Many strains have been artificially created in the laboratory by coinfection with a leukemia helper virus. The RSV particles which carry different coat proteins determined by the helper virus have been termed pseudotypes (RUBIN, 1965). The origin of the coat is then written in parenthesis, *e.g.* RSV (RAV 1).

AMV causes a characteristic myeloblast-like alteration in embryo cultures containing specific target cells (BEAUDREAU, BECKER, BONAR, WALLBANK, D. BEARD, and J. BEARD, 1960; BALUDA and GOETZ, 1961; BALUDA, MOSCOVICI, and GOETZ, 1964). This conversion does not strictly correspond to transformation as defined in Table 1 but seems entirely equivalent to the induction of myeloid leukemia *in vivo*. MCV causes a similar alteration (TODOROV and YAKIMOV, 1967).

Replication cycle. One tsep growth curves have been most easily studied with the transforming RSV, where the characteristic foci of altered cells can be used for titration of focus forming units (FFU) (TEMIN and RUBIN, 1958). Non-transforming strains can be quantitated on the basis of their reducticon of the number of foci produced by a challenge with a transforming virus strain (RUBIN, 1960 b). Most data are based on inpure virus stocks and some facts may have to be reinterpreted when pure strains will be more extensively studied.

Adsorption of RSV to susceptible chicken cells proceeds rapidly and within one hour 95% of cell-associated infectivity has disappeared. The eclipse lasts for 12—14 hours. It is followed by an exponential rise of FFU:s until a plateau is reached due to an equilibrium between synthesis and heat inactivation of the virus (RUBIN, 1956; TEMIN and RUBIN, 1959; VIGIER and GOLDÉ, 1959; PRINCE, 1960a; GOLDÉ and VIGIER, 1961; P. VOGT and RUBIN, 1961). Considerable asynchrony exists between individual cells.

Virus is released as a trickle from the cell membrane by budding and it acquires its infectious property during this final stage of maturation. The interval between completion of a fully infectious particle and its extracellular release is about 3 minutes. Viral antigen reacting with neutralizing FA sera becomes detectable shortly before infectious particles are formed. Antigen in cells grown *in vitro* seems to occur exclusively in the cytoplasm (MELLORS and MUNROE, 1960; MALMGREN, FINK, and MILLS, 1960; NOYES, 1960; P. VOGT and RUBIN, 1961).

The rather fast synthesis above was obtained when a rapidly multiplying, efficient helper virus was present in addition to transforming RSV. If viruses of different growth rates are employed as helpers, the rate of appearance of RSV focus forming capacity is directly correlated with the rate of helper virus synthesis (H. HANAFUSA and T. HANAFUSA, 1966a). It therefore seems likely that onestep growth curves of unpurified stock RSV reflect the rate of helper virus synthesis rather than the true rate of RSV synthesis.

When RSV-SR was studied in the apparent absence of helper virus, the final yield was about 1,000 times less than the maximal titer of about 5×10^7 FFU/ml obtained with RSV-BH (RAV 1) + RAV 1. The yield of focus forming RSV-SR could be increased by a factor of 50—100 by the addition of RAV 1 as a helper (H. HANAFUSA and T. HANAFUSA, 1966a). The final yield of RSV-SR possibly reflected the low true synthesis rate of an RSV particle in the absence of helper, but neither in this nor in other experiments has it been possible to isolate and test an RSV strain in guaranteed absence of helper virus.

Growth curves of leukemogenic helper viruses have also been difficult to obtain partly because of the purification problem but also because the titration methods lack the precision of the RSV focus assay (RUBIN, 1960b; P. VOGT and RUBIN, 1961). RAV1 and stock AMV have single step growth curves with an eclipse of about 18—24 hours and the virus titers rise exponentially during 2—4 days after which they level off (P. VOGT and RUBIN, 1963; P. VOGT, 1963b; H. HANAFUSA, T. HANAFUSA, and RUBIN, 1964b; P. VOGT, 1964; BAUER, BAHNE-MANN, and SCHÄFER, 1965). RIF1 and RIF2 have slower infectious cycles than for instance RAV1 (RUBIN and P. VOGT, 1962).

The cytoplasmic location of viral antigens, viral assembly at the cell membrane and other features of non-transforming virus strains seem to be identical with those observed in RSV + helper stocks (RUBIN and P. VOGT, 1962; P. VOGT and RUBIN, 1963; P. VOGT, 1963a). This may not be particularly significant since the RSV-stocks employed contained helper virus in excess and particles and antigens registered are therefore predominantly of helper origin also in that case.

The metabolic events after infection by RSV are not well known. The replicative form of the avian tumor virus genome has not been isolated. Viral replication is transiently dependent on DNA synthesis (BADER, 1965b). This aspect is further discussed on page 143—144.

i) Subgroups of Viruses of the Avian Leukemia-Sarcoma Complex

Neutralization tests. The serological evidence for the subgroups A, B and C is based on neutralization tests with chicken immune sera and has already been reviewed (page 116 and Table 11). The existence of a new subgroup D (DUFF, 1968; BAUER and GRAF, 1969) has recently been suggested.

Interference tests. The subgroup specificity can also be established in interference tests *in vitro*. Cells of a susceptible genotype are first infected by one virus and subsequently challenged by another. If both viruses belong to the same subgroup, multiplication of the challenge strain is curtailed. If the challenge virus belongs to a different subgroup, replication is unimpaired and resistance is even induced against a repeated challenge with members of the same subgroup (H. HANAFUSA, 1964; P. VOGT, 1965; P. VOGT and ISHIZAKI, 1966).

j) Genetically Determined Cell Susceptibility

It has long been known that individual birds differ in their sensitivity to natural infection by avian leukemia-sarcoma viruses. Breeding has produced flocks with a different susceptibility to various members of the avian leukemia-sarcoma complex (BURMESTER, FONTES, WATERS, BRYAN, and GROUPÉ, 1960; WATERS and FONTES, 1960; WATERS and BURMESTER, 1961).

The variable sensitivity rendered RSV pock titration on the chorioallantoic membrane of non-inbred embryos anomalous (KEOGH, 1938; RUBIN, 1955; PRINCE, 1958; VIGIER, 1959; DHALIWAL, 1963) and not until individual pretested embryos became available (TEMIN and RUBIN, 1958) could avian tumor viruses be titrated quantitatively.

Extensive analysis has shown that resistance to infection operates at the level of the individual cell and is manifest in mature birds, chorioallantoic membrane (PRINCE, 1958; BOWER, GYLES, and BROWN, 1964), mixed cell cultures (CRITTENDEN, OKAZAKI, and REAMER, 1963; PAYNE and BIGGS, 1964; PAYNE, CRITTENDEN, and OKAZAKI, 1968) and in cultures derived from separate tissues (R. WEISS, 1969a). Curiously the pattern follows the division of the viruses into serological subgroups. A single autosomal dominant gene controls cellular resistance to growth of viruses of subgroup A (PRINCE, 1958; WATERS and BURMESTER, 1961; BOWER, 1962; CRITTENDEN, OKAZAKI, and REAMER, 1964; CRITTENDEN and OKAZAKI, 1965, 1966; PAYNE and BIGGS, 1966). A second gene with similar characteristics controls the resistance against members of subgroup B (PURCHASE and OKAZAKI, 1964; RUBIN, 1965). The two genes are inhereted independently (CRITTENDEN, STONE, REAMER, and OKAZAKI, 1967).

Table 12. *Cellular Susceptibility to Subgroups of the Avian Leukemia-Sarcoma Virus Complex*

Subgroup	Susceptible chicken phenotypes[1]	Oncogenicity for hamsters
A	C/B C/BC C/C C/O	low or absent
B	C/A C/AC C/C C/O	low or absent
C	C/A C/AB C/B C/O	low or absent
D	C/A C/AB[2] C/B[2] C/O	high

[1] The nomenclature suggested by P. VOGT and ISHIZAKI (1965) is followed. The first letter(s) indicate the species. The letters behind the bars indicate the excluded subgroups. C/A for instance denotes chicken cells specifically resistant against virus members of subgroup A.

[2] The plating efficiency of subgroup D is reduced on cells resistant to subgroup B indicating a closer relatedness between these two subgroups than between any other subgroups (DUFF and P. VOGT, 1969; BAUER and GRAF, 1969).

The phenotypical expression of cellular resistance is believed to be either a barrier to penetration or uncoating, since virus adsorption is not impaired (P. VOGT, ISHIZAKI, and DUFF, 1966; STECK and RUBIN, 1966a, b; PIRAINO, 1967). The genome can be forced into susceptible cells by the use of irradiated cells as virus source. Once the penetration block had been overcome efficient synthesis took place showing that resistance does not directly affect virus multiplication (CRITTENDEN, 1968).

All cells cultivated from embryos with the dominant gene for susceptibility to infection by, for instance, viruses of subgroup B should be expected to be fully susceptible to infection by any member of this subgroup. No ceiling in transformation frequency similar to that encountered with PV or SV40 should be encountered (page 66). This prediction has been hard to prove mainly because of the difficulty in obtaining pure virus strains and avoiding interference between the individual members of the respective subgroups. TRAGER and RUBIN (1964) did, however, report that high multiplicities of a particular RSV stock (RSV8) transformed about 80—85% of the exposed cells. They concluded that "there is little or no restriction of the number of transformed clones" if genetically susceptible cells were used and viral interference circumvented. Close to 100% morphologically transformed cells has later been reported by HANAFUSA (1969).

The genetic resistance as well as the resistance induced by viral interference is not absolute but may be broken through with high doses of challenge virus (RUBIN, 1961; PAYNE and BIGGS, 1964).

The peculiar linkage within the subgroups between genetic susceptibility, interference and capacity to elicit neutralizing antibody has not been clarified.

It may be speculated that a cell (C/B) heterozygous for the dominant gene for susceptibility to subgroup A directs the synthesis of a specific membrane-bound receptor- like component essential for penetration of the virus particle. The cell membrane component is also "passively" incorporated during virus release in a manner akin to other cell membrane components. Interference can be achieved either directly by the addition of a large excess of interfering virus (STECK and RUBIN, 1966a, b) or purified subgroup-specific viral antigen (BAUER, TOZAWA, BOLOGNESI, GRAF, and GELDERBLOM, 1969) simultaneously with the challenge virus or indirectly by letting the exposed cells synthesize interfering virus before the challenge is carried out (STECK and RUBIN, 1966a, b). In the former case the receptor-like component may be blocked by the excess interfering particles, in the latter it may continuously be removed from the cell membrane during virus maturation thus preventing penetration of the challenge virus. Cells homozygous for susceptibility (C/O) probably produce two kinds of membrane components essential for virus penetration. Virus manufactured by such cells would be expected to contain both type A and B cell membrane-derived components. This prediction does not seem to have been critically tested, but may have bearing on the problem of the antiviral effect of sera against normal cells (page 115).

k) Viral Interactions, Helper Effects

The first indication of complex virus-virus interactions within the avian leukemia-sarcoma complex was provided by RUBIN (1960b). He found that a field strain of LV, which occurred in a large proportion of apparently healthy chick embryos conferred resistance to RSV. The virus isolate, termed RIF (Resistance Inducing Factor) could be titrated on the basis of its interference with challenge RSV.

A second virus with a similar resistance inducing capacity was isolated by limiting dilution by RUBIN and P. VOGT (1962). Because of its invariable presence in RSV-BH stocks, the agent was called RAV (Rous Associated Virus). RAV was serologically indistinguishable from RSV-BH in neutralization tests but distinct from RIF. After animal inoculation RAV gave rise to a high percentage of erythro-leukemias but not to sarcoma. It did not achieve any morphological alterations in cell cultures.

The nature of the intimate relationship between RSV and RAV was further studied first mainly by HANAFUSA and coworkers. The original approach was to infect chick embryo cells with a high dilution of stock RSV-BH (i.e. a mixture of RSV-BH and an excess of RAV) and cover the cultures with agar containing neutralizing serum against RAV to counteract virus spread. Foci of transformed cells only infected by a single infectious unit of RSV-BH were picked and assayed for virus production. More than 80% were found to release neither infectious focus-forming RSV-BH nor RAV using conventional susceptible chicken cells as test objects. Rous cells not producing easily detected infectious RSV-BH spontaneously were referred to as NP (non producing) cells. NP cells super-infected by RAV acquired the capacity to produce easily detectable infectious particles and synthesized both RSV-BH and RAV (H. HANAFUSA, T. HANAFUSA, and RUBIN, 1963).

The hypothesis was formulated that RSV-BH is a defective virus unable to complete a full infectious cycle. Only if a suitable helper virus such as RAV is added, is infectious focus forming RSV-BH synthesized. RAV is non-defective and fully capable of independent multiplication. The deficiency of RSV-BH involves a late step in the infectious cycle concerned with the formation of the viral envelope. RAV overcomes this deficiency by directing synthesis of coat proteins for both the RAV and the RSV-BH genome (H. HANAFUSA, T. HANA-FUSA, and RUBIN, 1963, 1964 a, b, c).

The original hypothesis of the obligate cooperation between a defective RSV and a related helper virus has been modified and extended in later experiments. All leukemia-sarcoma strains are now believed to be capable of replicating and produce extracellular C-particles but some of them are produced slowly and only have a narrow range of susceptible hosts. Their infectious character is therefore difficult to prove. These correspond to the originally "defective" strains, where the failure to detect infectiosity was caused by the employment of resistant host cells. The term NP (non-producing) cells has accordingly been changed to L⁻R (leukosis virus negative Rous cells) (H. HANAFUSA and T. HANAFUSA, 1968) L⁻R cells elaborate typical C particles by a membrane budding process (DOUGHERTY and DI STEFANO, 1965; COURINGTON and VOGT, 1967; HAGUENAU and H. HANA-FUSA, 1968) morphologically identical to that seen in cells infected with crude RSV stocks or leukemia virus. H. ROBINSON (1967) has also isolated particles from L⁻R cells with identical sedimentation behavior in gradient ultracentri-fugation as that of ordinary infectious RSV + RAV 1. The L⁻R RNA from the particles was indistinguishable (base ratio; sedimentation coefficient) from that of RSV + RAV 1 virus (Table 10). The particles contain the gs-antigen (DUESBERG, H. ROBINSON, W. ROBINSON, HUEBNER, and TURNER, 1968).

The helper virus provides essential coat proteins for its own genome and any other genome of the avian leukemia-sarcoma complex present in the same cell. If the latter is only replicating slowly a substantial proportion of it will be enveloped by a helper determined coat (H. HANAFUSA, T. HANAFUSA, and RUBIN, 1964 c). This can be established by the use of different helper viruses which impart their own serological specificity (H. HANAFUSA, T. HANAFUSA, and RUBIN, 1964 c) and other subgroup characteristics (host range, interference pattern) (H. HANAFUSA, 1965) to released RSV particles.

A criterion of an L⁻R cell is that its production of focus forming RSV is augmented by the addition of helper virus. It has been difficult to establish whether this is solely caused by the surplus of helper coats with an extended host range or if a true increase in the rate of RSV genome synthesis is also induced. H. ROBINSON (1967) calculated, on the basis of rate of viral RNA synthesis, that no increase of RSV RNA synthesis was induced by addition of RAV 1. HAGUENAU and H. HANAFUSA (1968), on the other hand, in a quantitative electron microscope study in L⁻R cells found only 1.5% of the number of particles encountered in cells infected with a non-transforming virus (RAV 1 was mainly studied). This is in agreement with the original contention of H. HANAFUSA and T. HANAFUSA (1966a) that the rate of helper virus synthesis determines the rate at which the "helped" virus is manufactured.

Particles released from L⁻R cells have been called RSV(0). Their biological activity was discovered by P. VOGT (1967c) and R. WEISS (1967). The titer of helper induced RSV from L⁻R cells is about 3 logs higher than that of RSV(0) alone (P. VOGT, 1967a).

L⁻R cells were later found to be heterogenous with respect to the RSV(0) particles released (H. HANAFUSA and T. HANAFUSA, 1968). Clones of L⁻R cells were isolated from C/0 chick embryo cells infected with RSV (RAV 1). In 33 out of 38 clones focus forming infectiosity could be established mainly by the use of Japanese quail cells. This virus was designated RSVβ(0). The remaining 5 clones synthesized particles with an identical density, but these did not transform any other cells. This entity for which so far no susceptible hosts have been found was termed RSVα(0).

R. WEISS (1969a) could in a thorough study confirm the existence of two RSV(0) viruses. RSVβ(0) could transform Japanese quail (phenotype Q/BC), European pheasant (Ph/B) but not bantam (B/0) or goose (G/AB) cells. About 40% of brown leghorn chickens (C/0) were also susceptible, but another chicken strain with phenotype C/0 was resistant. RSVα(0) was not infectious for any of the non-chicken hosts mentioned, nor did it infect for instance C/0 chickens. L⁻R clones releasing RSVβ(0) came from embryos which were spontaneously gs-antigen positive and belonged to the category described by PAYNE and CHUBB (1968) where the antigen was inherited as a dominant Mendelian gene product and was not associated with any evidence of infectious virus in contrast to L⁻R clones releasing RSVα(0) which all were from gs-negative embryos. If RSVα(0) is provided by a suitable coat from a helper virus (H. HANAFUSA and T. HANAFUSA, 1968) or if cell fusion is resorted to (R. WEISS, 1969a), the cell membrane barrier to infection by RSVα(0) is bypassed and replication becomes possible.

RSVβ(0) does not seem to fall into any of the known subgroups of avian leukemia-sarcoma viruses. It has no relationship to subgroup A and C. Anti-B sera do not neutralize RSVβ(0), although interference between subgroup B viruses and RSVβ(0) has been observed. One antiserum to subgroup D virus neutralized RSVβ(0), but subgroup D has another host range than RSVβ(0) (R. WEISS, 1969b).

The superinfection of L⁻R cells involves the presence within the same cell of two different but closely related genomes. The progeny virus particles have identical envelopes but different genomes. It has been pointed out that this may be regarded an extreme case of phenotypic mixing with one partner contributing zero to the viral phenotypes (H. HANAFUSA, T. HANAFUSA, and RUBIN, 1964c). True phenotypic mixing has been established by P. VOGT (1967b). Chick embryo fibroblasts of the multisusceptible C/0 phenotype were simultaneously infected by RSV-BH coated with RAV 1 plus an excess of RAV 1 and RAV 2. The progeny particles had antigens both of RAV 1 and RAV 2 type indicating phenotypic mixing with envelope components synthesized under the direction of both helper genomes. If the virus mixture was passaged at high multiplicity, pheno-

typic mixing was maintained; at low multiplicities, when most cells were singly
infected by helper virus, pure RSV coated with RAV 1 or RAV 2 tended to outgrow particles with mixed envelope properties. Since the capacity to infect
cells is a function of envelope antigens, phenotypic mixing led to a broadened
host susceptibility. It may be suggested that phenotypic mixing is responsible
for the spectular widening of host susceptibility which high titer laboratory strains
of RSV have evolved during their numerous passages as high titered stocks.

l) Infection of Other Bird Cells than Chicken Cells

Already MURPHY and ROUS (1912) were able to produce tumors in duck and
pigeon embryos thus proving that RSV infection is not limited to its species of
origin. FUJINAMI and SUZUE (1928) and FUJINAMI and HATANO (1929) managed
to infect ducks with their chicken derived FSV. Later on FUJINAMI (1930) reported

Table 13. *Susceptibility and Resistance of Bird Species Other than Chickens to Subgroups
of Avian Leukemia-Sarcoma Viruses*
After DUFF and P. VOGT (1969) and R. WEISS (1969a)

| Species | Subgroup | | | | | RSV_β (O) | RSV_α (O) |
| | A | B | C | | D | | |
			RSV (RAV-7)	RSV (RAV-49)			
Japanese quail	S	R	R	(S)	R[1]	S	R
Turkey	S	R	R	S	R[1]		
Ringneck pheasant	S	R	R	S	R[1]		
European pheasant						S	R
Duck	R	R	R	S			
Bantam						R	R
Goose						R	

[1] Resistant to RSV-SR-D and RSV-OZ-D but comparatively susceptible to RSV
(RAV-50) (Table 11) (DUFF and P. VOGT, 1969).

infection of quails, pigeons and paddy-birds with the same virus. DES LIGNERIS
(1932) produced tumors in guinea-fowls and turkeys by minced Rous chicken
sarcoma tissue and ANDREWES (1932) managed to induce sarcomas in pheasants
(review by FOULDS, 1934a, b).

With the advent of tissue culture, cells from most of the species mentioned
as well as a few additional ones have been found susceptible to infection by
members of the avian leukemia-sarcoma complex. Susceptible target cells respond
with morphologically transformed foci essentially similar to those of formed by
chicken cells (FREEMAN, 1963). Turkey cells synthesize RSV-BH and RSV-BS
(PRINCE, 1960a, b) as efficiently as chick cells.

As a recent development the different susceptibility of various bird species
has been used for the analysis of RSV subgroups and RSV(O) (Table 13). A rather
surprising finding is the high and possibly general susceptibility of Japanese
quail cells to RSVβ(O), in contrast to chicken cells where only a proportion of
embryos belonging to special strains respond (P. VOGT, 1967c; R. WEISS, 1967,
1969a; H. HANAFUSA and T. HANAFUSA, 1968). This may explain why stock

RSV-BS lost infectivity for chickens after passage through quail embryos (RAU-SCHER, REYNIERS, and SACKSTEDER, 1964).

These studies make it clear that species belonging to the order *Galliformes* are fully susceptible to RSV. Cells respond with virus synthesis and transformation apparently in the same principal manner as chicken cells do. The genetic basis for differences between and within species with respect to subgroups of RSV has not yet been worked out.

Duck cells (order *Anseriformes*) also seem susceptible in the same principal manner as in the order *Galliformes* but members of the order *Columbiformes* (*e.g.* pigeon) may be resistant unless high doses of virus are used (SARMA, LOG, HUEBNER, and TURNER, 1969). Other orders within the subclass *Carinatae*, to which *gallus domesticus* belongs, do not seem to have been tested. It may be more adequate to regard RSV and related viruses not as members of an *avian* complex but rather of a *galliform-anseriform* virus complex.

m) Infection of Mammalian Cells by RSV

Early observations of survival or growth of xenografted Rous of Fujinami chicken sarcoma tissue in mice (ROSKIN, 1927; GREENE, 1951; AGEENKO, 1957), rats (MILONE, 1928; FUJINAMI and HATANO, 1929; SCHMIDT-RUPPIN, 1959; SIGEL, SCOTTI, SCHULZ, and BURNSTEIN, 1960), guinea-pigs (FUJINAMI and HATANO, 1929; SHRIGLEY, GREENE, and DURAN-REYNALS, 1945), rabbits (GREENE, 1951), hamster (KUWATA, YASUMURA, and KANISAWA, 1958; KUWATA, 1960) were generally interpreted as growth of chicken sarcoma cells in the foreign host.

Later studies employed crude RSV preparations apparently rendered cell-free by filtration. Such virus stocks induced hemorrhagic cysts (ZILBER and KRYU-KOVA, 1957a, b; SVET-MOLDAVSKY and SKORIKOVA, 1957) or sarcomas (SVET-MOLDAVSKY, 1958) in rats. ISING-IVERSEN (1960) showed by the use of a chromosome marker that a transplantable mouse sarcoma line induced by implantation of Rous chicken sarcoma was of murine origin. AHLSTRÖM and JONSSON (1962) implanted minced chicken sarcoma tissue induced by RSV-SR into new-born rats. The resulting tumors were composed of cells of rat origin — an unequivocal demonstration of tumor induction in the host. These studies have later been extended to a number of mammalian hosts (Table 9).

Mammalian cells differ from bird cells by their almost universal incapacity to produce virus infectious for chickens or the respective mammalian host. The two exceptions are a particular hamster tumor induced by RSV-PR (SVOBODA and KLEMENT, 1963) and rat tumor cells infected by B 77 sarcoma virus (ALTANER and SVEC, 1966; ŠIMKOVIČ, POPOVIČ, ŠVEC, GRÓFOVÁ, and VALENTOVÁ, 1969), which release infectious RSV even many years after the original infection.

The virology of the interaction between RSV and mammalian cells is, in contrast to the well explored virus-bird cell interactions, underdeveloped. Virus-like C particles budding from the cell membrane can, however, be found (DOUGHERTY and DI STEFANO, 1965; RABOTTI, BUCCIARELLI, and DALTON, 1966; COURINGTON and P. VOGT, 1967; BUCCIARELLI, RABOTTI, and DALTON, 1967; VALENTINE and BADER, 1968). Such particles yield RNA physico-chemically similar to virus RNA (W. ROBINSON, H. ROBINSON, and DUESBERG, 1967; VALENTINE and

BADER, 1968). The cells also contain the gs-antigen. The failure to synthesize infectious virions may be related to a "defect" in the cellular machinery which precludes manufacture of infectious viral envelopes perhaps because helper virus synthesis cannot be sustained.

2. Transformation in vitro by Members of the Avian Leukemia-Sarcoma Complex

a) Predominantly Sarcomagenic Virus Strains

Morphological aspects. The multitude of effects caused by RSV and other sarcomagenic strains may be arbitrarily separated into two categories a) morphological changes in individual cells and b) changes affecting cell population behavior. Only the latter have direct bearing on cell transformation as defined in this monograph.

The morphological and growth control disturbances of the established infection may be directly determined by the introduced viral genome and is thus independent of the viral coat. The RSV-BH genomes are said to give rise to compact, extensively multilayered foci of round or spindle-shaped basophilic cells regardless of the helper virus used to provide it with an infectious coat (TEMIN, 1960; RUBIN, 1964). RSV-BS, RSV-SR and FSV genomes cause more diffuse foci. RSV-HA, RSV-PR and RSV-CZ only cause a small degree of multilayering (P. VOGT, 1967a). These markers hold some promise in the elucidation of the complex cellular changes induced by RSV. Their stability and reliability have, however, not been critically tested and it is not excluded that the observed differences are spurious and due to for instance variations in the environment or the employment of heterogenous embryonic cell cultures.

Cytological Alterations

CARREL (1924a, b, 1926), CARREL and EBELING (1926), LUDFORD (1937), LO, GEY, and SHAPRAS (1955) and others designed experiments to determine the type of susceptible cells. Some of these authors also described morphological alterations *in vitro* but did not achieve any clear and systematic understanding. For an extensive discussion see LEVINE (1939a, b, c, d).

HALBERSTAEDTER, DOLJANSKI, and TENENBAUM (1941), DOLJANSKI and TENENBAUM (1942) and TENENBAUM and DOLJANSKI (1943) investigated normal chick embryonic muscle exposed to RSV (in the form of irradiated sarcoma) and explanted Rous tumors. Three types of cells were described which now constitute the classical cytological response to RSV. No other oncogenic virus induces this spectrum and spontaneous transformations have other cytological traits (page 27). Tumor explants show the same pathognomonic morphology as cells primarily exposed to RSV *in vitro*. All cell types identified *in vitro* have counterparts *in vivo* in the light and electron microscope (cf. HAGUENAU and BEARD, 1962).

(1) Round, basophilic cells (Fig. 28). These ameboid elements have a strongly basophilic, homogenous or finely granular, moderately abundant cytoplasm sometimes extended into pseudopodia. The nucleus is round or oval, eccentrically placed with a prominent outline. A few large nucleoli are typical. The

chromatin occurs as dense clumps and mitosis is frequent. The cell size varies between 6—18 μ (DOLJANSKI and TENENBAUM, 1942). The cells tend to grow as grape-like easily detached clusters.

Electron microscopy of cells in culture shows particulate aggregates of unknown significance in the nuclei. The free cytoplasmic ribosomes are considerably increased, mitochondria were numerous and the Golgi apparatus is hyperplastic (HAGUENAU and J. BEARD, 1962; DI STEFANO and DOUGHERTY, 1965). The cell surface showed an increased layer of acid mucopolysaccharides which was thicker after FSV than RSV. It was also seen after infection with non-transforming RAV (MORGAN, 1968).

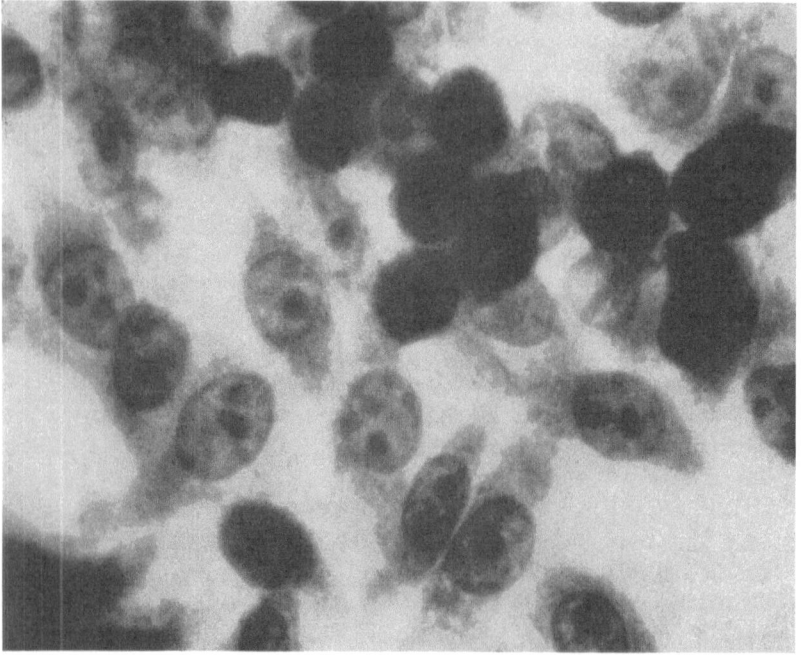

Fig. 28. Round basophilic cells from a culture of bovine fibroblasts transformed by RSV. The basophilic cells have such an affinity for the dye that their nuclei are difficult to distinguish. The tendency to grow as loose clusters is evident. May-Grünwald-Giemsa. Appr. magn. × 2,500

The cytoplasmic basophilia may be due to the increased concentration of ribosomes (RUBIN, 1960b; GOLDÉ, 1962; SUNDELIN, 1967) and perhaps also to acid mucopolysaccharides accumulated along the membrane (ERICHSEN, ENG, and MORGAN, 1961; TEMIN, 1965).

The round, basophilic cells have been observed in all species susceptible to alteration by RSV *in vitro*. Their function is unknown but they have been compared with macrophages, monocytes and lymphocytes. A few cells of identical morphology have been noted in explants of embryos not exposed to RSV (DOLJANSKI and TENENBAUM, 1942).

All standard RSV preparations induce the formation of basophilic round cells. Certain strains produce a pure population whereas others give rise to mixtures with other cell types (PURCHASE and OKAZAKI, 1964; SMIDOVÁ, VALENTOVÁ, SMIDA and MEDZIHRADSKÝ, 1965). The characteristic response does not depend on the selection of a preexistent basophilic round cell since it can be induced with almost 100% efficiency in embryonic chick fibroblastic cells plated for colony formation (TRAGER and RUBIN, 1964). The alteration has also been described in epithelial-like chick iris cultures (EPHRUSSI and TEMIN, 1960) unfortunately without any photographic documentation, in human glioma cells (PONTÉN and MACINTYRE, 1968; MACINTYRE, GRIMES, and VATTER, 1969) and in chicken myoblasts (KAIGHN, EBERT, and STOTT, 1966; RABIN, 1967). TEMIN (1960, 1961) has isolated a stable mutant, morph^r, from RSV-BS, which induces a round cell change partially independent of the type of target cells.

HANAFUSA (1969) showed that RSV-SR induced typical basophilic round cells in 90% of chick embryo cells within 24 hours after exposure to RSV-SR. This was accompanied by capacity to grow in soft agar. It is the most rapid gross morphological alteration reported for any oncogenic virus *in vitro*.

The round basophilic cell *in vitro* probably correspond to the compact macrophage-like cell often seen in histologic sections of sarcoma tissue (TEUTSCHLAENDER, 1923; AHLSTRÖM and JONSSON, 1962).

(2) The basophilic polygonal (spindle) cell has a nucleus indistinguishable from that of the previous cell type. The cytoplasm has the same tinctorial properties but the cell shape is polygonal, oval or elongated with pointed ends. These cells are firmly attached to their substrate and tend to grow on top of normal cells. DOLJANSKI and TENENBAUM (1942) observed transitional forms between round and spindle-shaped basophilic cells. Clones from RSV-EH altered bovine fibroblasts always contained a mixture of basophilic round and spindle forms supporting their interconvertibility (STENKVIST, 1966c). This view is contradicted by the findings of TEMIN (1960) who isolated an RSV-BS mutant, morph^f, which induces a stable alteration of fibroblasts and iris epithelium into basophilic spindle cells (EPHRUSSI and TEMIN, 1960). The discrepancy can be explained if cells giving rise to daughter cells of varying morphology are originally infected both by morph^r- and morph^f-like genomes which then segregate independently among the progeny cells. Experiments with cloned virus and cells are necessary to resolve how the characteristic "Rous cell morphology" is controlled.

In vivo, the bulk of the sarcomas is made up of fibroblasts (ROUS, 1910) corresponding to the basophilic polygonal cell.

(3) The vacuolated (giant) cell (Fig. 29). This element is extremely variable in size (10—600 μ). It contains from one to a dozen or more nuclei with a peculiar tendency to arrange themselves symmetrically ("mirror image") (STENKVIST and PONTÉN, 1964). The nuclei contain one or two large nucleoli and basophilic inclusion bodies have been described. The cytoplasm is more or less distinctly divided into a central and a peripheral zone. The former is strongly basophilic and corresponds to a hypertrophic Golgi zone. The latter is vacuolated and may contain small eosinophilic granules (TENENBAUM and DOLJANSKI, 1943). It is firmly attached to its substrate and has a peculiar ability to prevent other cells from adhering to its surface. Even in a superdense culture with extensive multilayering, the giant cells manage to repel other cells so that their upper surface always is in direct contact with the fluid medium.

The vacuolated cell was first observed in aging cultures and is regarded a degenerate form. This is supported by the absence of mitosis and inability to undergo serial passage. It probably arises by fusion of cells rather than nuclear division within a common cytoplasm. Multinucleated cells also occur *in vivo* (ROUS, 1910).

In addition to these classical types the following morphologies have been recorded.

(4) Polykaryocytes. MOSES and KOHN (1963) described multinucleated syncytia ("polykaryocytes") 4 days after exposure of chick fibroblast cultures to

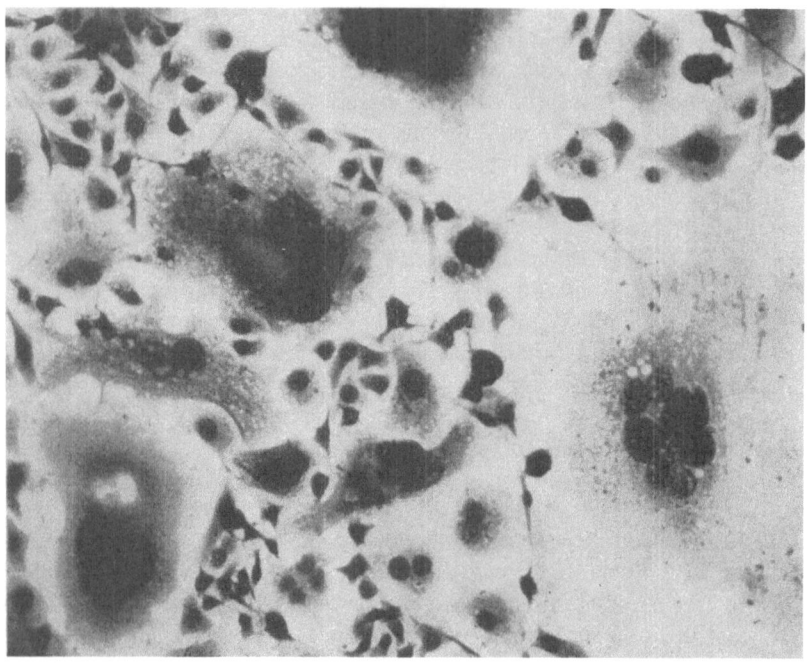

Fig. 29. Vacuolated multinuclear giant cells from cultures of chick embryo cells exposed to RSV. Note several cells with many nuclei often arranged in the center of cell. Multinucleated cells are only very rarely overgrown by other cells and do not wander of top of their neighbors. They probably represent fused cells and will eventually degenerate. Most of the remaining elements are basophilic polygonal (spindle) cells. May-Grünwald-Giemsa. Appr. magn. ×1000

RSV-BS. The large cells contained 10 or more nuclei and resulted from cell fusion. Polykaryocytes do not divide and eventually perish (KOHN and MOSES, 1965). It is doubtful if they represent a unique response or an exaggeration of the formation of the vacuolated giant cells.

(5) Non-basophilic polygonal ("epithelial-like") cells. This type was observed among clones isolated from bovine lung fibroblast cultures exposed to RSV-EH (STENKVIST, 1966 c). The monomorphic elements have a weakly stained (Giemsa) cytoplasm and oval nuclei with inconspicuous nucleoli. They resemble the cells observed by CALNEK (1964) after infection by LV, but also variants seen in non-infected cultures. Their relation to RSV has not been clarified.

(6) Cells with fragmented nuclei. LEVINSON (1967) described foci of chicken cells with fragmented nuclei by 36 hours after exposure to RSV-EH. The alteration could be used for titration purposes. The relationship between this striking effect and the classical Rous changes has not been determined.

Cell Population Changes

(1) Unrestrained growth transformation (Fig. 30). The proliferation of basophilic round or spindle cells is not controlled by the local cell density as efficiently as that of normal cells. This important observation was first clearly expressed

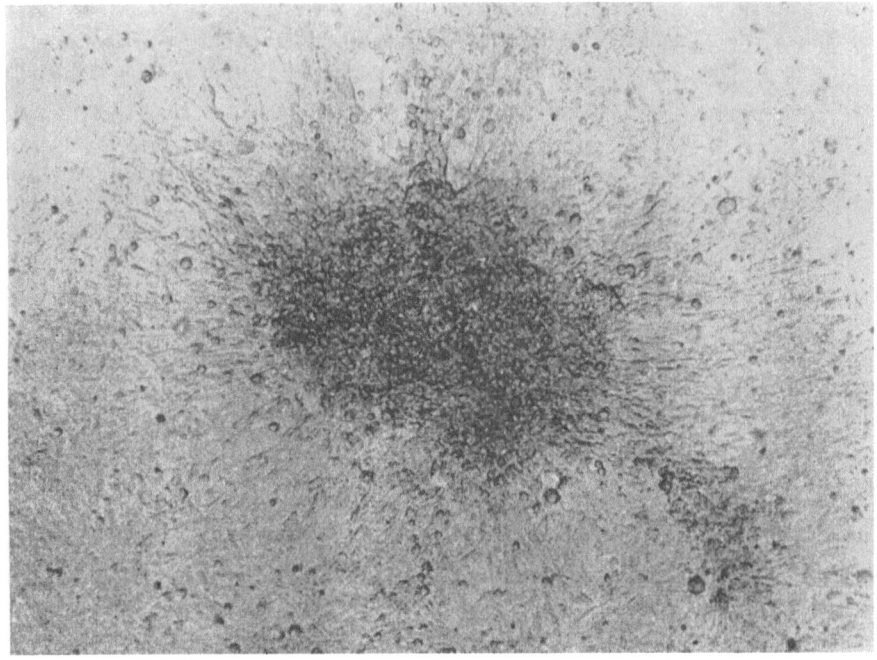

Fig. 30. Typical RSV focus viewed under agar in an unstained condition. The focus consists of rounded cells surrounded by a corona of radiating spindle cells. Unrestrained growth transformation is indicated by the capacity of cells to form a tightly packed three-dimensional cluster. Picture from RUBIN (1960c)

by TEMIN and RUBIN (1958), who introduced a quantitative titration of RSV based on the occurrence of discrete foci of multiplying altered cells (MANAKER and GROUPÉ, 1956) in stationary chicken cells kept under agar. Under these conditions embryonic fibroblasts remain essentially confined to a monolayer except in virus infected areas where they acquire the ability to overcome the growth restraint and multiply to form clusters of basophilic round or spindle cells.

Unrestrained growth transformation of basophilic round or spindle cells has been seen in all species susceptible to RSV transformation *in vitro* as listed in Table 9. Its quantitative aspects in mammalian cells have been studied by STENKVIST (1966b) and MACIEIRA-COELHO (1967a, b). Bovine cells transformed by RSV-EH grew to a cell density of at least twice the normal level. At this increased density DNA synthesis proceeded at a virtually undiminished rate as a sign of pronounced decrease of cell cycle inhibition (Fig. 3).

Chicken cells transformed in bulk by RSV have been reported to grow slower than control cells and not to exceed the normal terminal cell density (PRINCE, 1960a; GOLDÉ, 1962; RUBIN, 1966a, b). This failure to obtain signs of unrestrained growth in altered avian cells may have been due to depletion of the medium because these cells consume a growth factor rapidly. Under well defined conditions with frequent subcultivation and medium changes, chicken cells transformed by RSV have shown an increased terminal density and maintenance of proliferation also in dense cultures (TEMIN, 1966; TEMIN, 1967c). Under optimal conditions, RSV infected cells show signs of unrestrained growth transformation in the form of unabated DNA synthesis by 24—36 hours post infection (RUBIN and COLBY, 1968) even in a dense culture. RAV has no capacity to release chicken cells from a density-dependent proliferation restraint (COLBY and RUBIN, 1969).

(2) Irregular growth transformation. Microscopic observation of Rous foci in chicken cultures strongly suggests a lack of inhibition of cell movement. The basophilic spindle cells show a haphazard heaped-up arrangement as if they had moved on top of each other without restraint (P. VOGT, 1963a). Unfortunately no cinematographic confirmation of this subjective impression has been published. It is still possible that the cells, due to decreased adhesiveness, are passively squeezed into piles. BHK-21 hamster cells transformed by RSV-SR form colonies of cells growing in disarray strongly suggestive of irregular growth transformation (MACPHERSON, 1965). Rous cells are anchorage-independent as shown by a capacity to grow in free suspension (BADER, 1968) or soft agar (MACPHERSON, 1965; HANAFUSA, 1969).

MACIEIRA-COELHO (1967a) has claimed *heightened* contact inhibition of cell movement after transformation of human fibroblasts by RSV-SR. Basophilic spindle cells attained an increased local cell density indicating some loss of the control of cell division but in spite of this the degree of nuclear overlapping decreased. His interpretation that unrestrained and irregular growth transformation were dissociated is, however, open to question. It was not excluded that the tighter packing of the transformed human cells was caused by an altered cell shape with a diminished area of attachment to the substrate. The results underline the necessity of quantitative microcinematographic observation.

(3) Infinite growth transformation. As discussed in connection with the DNA viruses, the capacity of any agent to induce infinite growth transformation can only be evaluated against the background tendency of spontaneous infinite growth transformation.

Comparatively few studies are on record with stable chicken, bovine or human cells. In the avian system it has been claimed (H. HANAFUSA, T. HANAFUSA, and RUBIN, 1963) that L⁻R cells induced by RSV-BH have a capacity for infinite proliferation. These experiments involved addition of normal viable cells as "feeders" and it was not excluded that such elements were constantly transformed. PONTÉN (1970) studied cells transformed by RSV-SR in serial passage under a strict regimen. In the beginning Rous cells grew faster but in the long run growth was better among normal cells. A normal cell population had a potential of about 20 doublings whereas a Rous population had only a potential of 12 doublings. No indefinitely propagable lines were formed. Artificial mixtures of cells of different sex showed that there was a constant recruitement of transformed cells from a pool of untransformed elements and confirmed the limited division potential of morphologically altered fibroblasts. BADER (1968) reported that RSV infected chicken cells could grow in suspension in contrast to normal cells apparently, however, without forming established lines.

STENKVIST (1966a) did not obtain any established lines in several attempts involving a large number of passages of bovine and human cells. The transformed populations had a shorter total life span than uninfected cells. Infinite growth transformation thus does not occur in cells not prone to spontaneous transformation.

Unstable rodent species have not been systematically investigated but established lines have been produced by infection of rat cells *in vitro* (SVOBODA and CHÝLE, 1963; VESELÝ and SVOBODA, 1965). Since uninfected controls died out, it is possible that RSV infection increases the probability of infinite growth transformation.

The formation of established lines after RSV exposure has been correlated with chromosome damage. PONTÉN and LITHNER (1966) observed that SV 40 which did cause heteroploidy in bovine cells also induced infinite growth transformation whereas RSV infected lines which showed no virus related chromosome damage failed to become established. SCHENDLER and HARRIS (1967) could not confirm the findings of NICHOLS, LEVAN, CORIELL, GOLDNER, and AHLSTRÖM (1964) and NICHOLS, LEVAN, HENEEN, and PELUSE (1965) of RSV-SR induced chromosome breakage in human lymphocytes, but attributed the lesions to a probable contamination with Mycoplasma. KATO (1967) on the other hand, claimed an increased frequency of breaks in Chinese hamster cells exposed to RSV-SR and RSV-MH. The pattern of breaks was the same in controls and virus exposed cultures and it could not be excluded that RSV acted indirectly enhancing an inherent (or PPLO dependent) tendency to chromosome damage.

In summary, all strains of RSV induce at least a temporary unrestrained growth transformation. The process is accompanied by a specific cell alteration into basophilic round or polygonal (spindle) cells rich in RNA and acid mucopolysaccharides. Infinite growth transformation occurs only against an unstable background and cannot be regarded a specific RSV effect. Irregular growth transformation has not been sufficiently well investigated. In the section on oncogenic avian RNA viruses transformation will therefore be synonymous with a characteristic cytological alteration into basophilic round and/or spindle cells combined with unrestrained growth transformation.

b) Predominantly Leukemogenic Strains

Avian myeloblastosis virus. BEAUDREAU, BECKER, BONAR, WALLBANK, D. BEARD, and J. BEARD (1960) described the successful infection of adult chicken bone marrow cells by strain BAI-A of AMV. Uninfected cultures could only be maintained for a short period and failed to grow as suspensions. Infected cultures were after 2—3 weeks converted to uniform populations of myeloblasts which readily grew in suspension and liberated large amounts of virus. The cultures showed considerable individual variations but could often be subcultivated

60 days or more. In all respects the cultures resembled those obtained from explanted leukemic bone marrow from diseased birds (BEAUDREAU, BECKER, SHARP, PAINTER, and J. BEARD, 1958; BONAR, WEINSTEIN, SOMMER, D. BEARD, and J. BEARD, 1960; BECKER, BEAUDREAU, CASTLE, GIBSON, D. BEARD, and J. BEARD, 1962). Released virus particles were ATP:ase positive signifying that the cell membranes of the myeloblasts were as rich in enzyme *in vitro* as *in vivo*. The claim that indefinitely propagable cell lines were formed was not clearly documented and MOSCOVICI and P. VOGT (1968) obtained results compatible with a finite life span of AMV altered cells.

BALUDA and GOETZ (1961) described the development of foci of myeloblasts in Petri dish cultures from a variety of embryonic or adult chicken organs. Tissues such as spleen which contain abundant granulopoietic stem cells showed numerous foci of myeloblasts in contrast to, for instance, muscle which contains few target cells. Myeloblasts can assume a spindle shape simulating fibroblasts under certain conditions (BUBENIK, DONNER, CHÝLE, KOLDOVSKÝ, MALASKOVÁ, LIBÁNSKÝ, VEPŘEK, and ŘIMAN, 1968).

The conversion to myeloblasts was not correlated with virus production; unconverted cultures produced as much infectious virus as myeloblast-converted. The converting effect of AMV resembled the effects *in vivo* and it was concluded that the culture alterations were analogous to the tumorigenic effect in the living bird. It has not been established if the conversion involves selection of preexisting cell types or an induced cell change. A quantitative assay based on myeloblast foci in yolk sac cells kept under agar has been developed (MOSCOVICI, 1967).

A recent analysis by MOSCOVICI and P. VOGT (1968) suggests that converting AMV may be analogous to RSV(O) by not belonging to any of the known avian virus subgroups (Table 11). It may owe its infectivity and broad host range to formation of pseudotypes such as AMV(MAV-1) and AMV(MAV-2). As an isolated particle AMV may either be non-infectious or have a very narrow host range.

Aleukemic myelocytomatosis virus. Two strains (Mc-29 and Mc-31) of MCV induced foci of basophilic myeloblast-like cells in chicken bone marrow cultures (TODOROV and YAKIMOV, 1967). Strain Mc-29 further caused rapid proliferation and cytological alterations in the form of small, basophilic, triangular cells in chick embryo cultures within 72 hours after infection (LANGLOIS, SANKARAN, HSIUNG, and J. BEARD, 1967). At a low virus dilution enumeration of foci can be used for titration (LANGLOIS and J. BEARD, 1967). Mc-29 contains the group-specific antigen of the avian leukemia-sarcoma complex and crude stocks are serologically related to subgroup A and B (FRITZ, LANGLOIS, D. BEARD, and J. BEARD, 1968). The ultrastructure of the infected cells with a high content of free ribosomes, a poorly developed rough endoplasmic reticulum and smooth cell membranes is characteristic of primitive cells (HEINE, LANGLOIS, ŘÍMAN, and J. BEARD, 1969). The cell population doubling time is 24 hours (LANGLOIS, BOLOGNESI, FRITZ, and J. BEARD, 1969). The nature of the interesting alterations caused by aleukemic myelocytomatosis virus in terms of cell growth control has not yet been determined and its precise virological relationship to members of the avian leukemia-sarcoma virus complex remains to be elucidated.

Chicken erythroblastosis virus. CEV multiplies in fibroblastic chicken cells *in vitro* (RUFFILI, 1938; DOLJANSKI and PIKOVSKI, 1942). Bone marrow cultures also supported virus synthesis (LAGERLÖF, 1960a), but did not change morphologically.

Lymphomatosis virus. Multiplication of field strains (RUBIN, 1960a; RUBIN, CORNELIUS, and FANSHEIR, 1961) as well as laboratory "helper" strains (RUBIN and P. VOGT, 1962; H. HANAFUSA, T. HANAFUSA, and RUBIN, 1964a; P. VOGT, 1964) in embryonic chick fibroblast cultures has been clearly established. The cells maintain their normal cytology at least during comparatively short observation periods (cf. P. VOGT, 1965). CALNEK (1964) has, however, described conversion of cultures exposed to LV to an epithelioid appearance 3—6 weeks after infection. This change does not seem comparable to transformation as defined here. It did not lead to formation of established lines.

In summary, avian leukemia viruses can infect most or perhaps all chicken tissues *in vitro*. The cells invariably respond by virus synthesis. CEV leaves cultures morphologically unaffected while AMV causes selective growth of myeloblasts particularly in explants rich in granulopoietic precursor cells. The process is similar to virus induced myeloblast leukemia *in vivo*. LV and MCV have caused cytological alterations of unknown significance.

No infinite growth transformation has been clearly documented. The cultures alterations induced by AMV, LV or MCV do not conform with criteria for unrestrained or irregular growth transformation given in Table 1. For AMV and MCV this may well be due to the difficulty in defining a "leukemic cell transformation" *in vitro* as discussed on page 15.

c) Tumorigenicity

RSV transformed cells. Avian cells are the only elements of a stable species which have been extensively studied. Microscopic examination of chicken tumors will show that Rous sarcoma cells have one important indication of malignancy — the capacity to infiltrate surrounding normal tissue (ROUS, 1910). From histological evidence ROUS (1911) also concluded that metastases were formed by descendents of tumor cells from the primary site, rather than circulating virus infecting normal cells.

Infiltrative capacity is not sufficient for clinical malignancy, particularly not with mesenchymal cells which can have this property normally; cells also have to multiply without the restraint imposed on non-neoplastic cells.

MORGAN and ANDRESE (1961, 1962) used sex chromatin as a marker to determine the fate of RSV transformed chicken cells inoculated into allogeneic hosts. They concluded that the transformed cells multiplied extensively after implantation and could thus be regarded as malignant. These studies are not in accord with those of PONTÉN (1962a, 1964), who noted a rapid disappearance of injected cells in allogeneic as well as syngeneic (determined by skin histocompatibility) chickens. He concluded that Rous sarcoma cells even when implanted *in vivo* have a finite life span and that progressive tumor growth is maintained by temporarily unrestrained proliferation of lethally affected cells combined with continuous conversion of normal elements by released virus or intercellular transfer of subviral material with transforming capacity. The latter route may be the most important one since neutralizing antibody has no effect on the growth rate of established sarcomas (DOUGHERTY, STEWART, and MORGAN, 1960).

The reason for the discrepancy is probably the unreliability of the sex chromatin markers in birds (MILES and STOREY, 1962) compared to the sex chromosome marker used by PONTÉN. From results with immunized animals BERGS and GROUPÉ (1962, 1963) have also concluded that RSV sarcomas in the turkey spread by virus infection rather than unlimited tumor cell division.

H. HANAFUSA, T. HANAFUSA, and RUBIN (1964c) inoculated allogeneic L⁻R cells and found progressively growing tumors until about day 12 when most regressed. A few tumors, however, recurred and killed their hosts. It was not determined if the tumors were composed of implanted cells or induced from host cells. L⁻R cells can only be propagated for any length of time if normal viable "feeder cells" are added at intervals. It would be interesting to find out if this reflects a continuous recruitment of newly transformed cells or depends on a peculiar metabolic insufficiency in transformed cells. P. VOGT's (1967d) observation of a transfer of the RSV genome, presumably affected by RSV(O), from transformed L⁻R C/A cells to normal C/B feeder cells supports the hypothesis of a continuous recruitment analogous to that suggested *in vivo*.

ZYLBERBAUM and FEBVRE (1965) reported lung tumor formation in new-born rats inoculated intravenously by human cells transformed by RSV-BS. It was not determined if the tumor cells could be serially transplanted. Their cytology did not conform to the classical pattern and the finding of a large proportion of heteroploid cells makes it difficult to exclude contamination for instance by HeLa cells. These interesting findings should therefore be confirmed using appropriate cell markers.

In summary, avian Rous cells are invasive and proliferate excessively *in vivo*. It remains to be shown that infection *in vitro* or *in vivo* by RSV of avian cells or cells from any other stable species not prone to spontaneous transformation will lead to infinite growth transformation or serially transplantable tumor lines. Their classification as neoplastic or malignant will therefore depend on the definitions adopted.

Cultures from unstable rodent species which show, after RSV exposure, infinite growth transformation without infectious virus synthesis, form tumor lines which are only transmissible *in vitro* by viable cells (SVOBODA and CHÝLE, 1963; FEBVRE, ROTHSCHILD, ARNOULT, and HAGUEMAN, 1964; VESELÝ and SVOBODA, 1965; MACPHERSON, 1965; VESELÝ, SVOBODA, and DONNER, 1966; MACPHERSON, 1966). Rhesus monkey leptomeningeal cells converted to typical morphology by RSV-SR gave rise to metastasizing poorly differentiated sarcomas after inoculation into allogeneic new-born monkeys (RABOTTI, LANDON, PRY, BEADLE, DOLL, FABRIZIO, and DALTON, 1967).

In these studies cell-free preparations of the tumor tissue failed to induce tumors and the evidence is strong that the tumors represent growth of the inoculated cells. But in view of the frequent observation of RSV-like particles in transformed mammalian cells (page 128) and the possibility of subviral cell to cell transfer of RSV genome (page 142), the alternative remains that tumors may have been induced in the host, since no cell markers were employed.

Cells infected by leukemogenic viruses. Myeloblast populations created by exposure to AMV have only a limited life span after inoculation *in vivo* (BALUDA, 1962) and thus resemble RSV transformed chicken cells. Erythroblasts induced by CEV *in vivo* likewise seemed to have only a limited life span in serial allogeneic transplantation where they will be overgrown by cells induced in the recipients (LAGERLÖF, 1962; PONTÉN, 1962a, b).

LAGERLÖF (1960b) studied the results of implantation of chicken bone marrow cultures exposed to CEV. The cells were morphologically indistinguishable from uninfected controls and gave rise to rapidly progressing erythroleukemia after allogeneic implantation. The author's conclusion that a true transplantation took place was not confirmed by the use of any cell markers.

Cells exposed to LV *in vitro* do not seem to have been tested for their tumorigenicity *in vivo*. Lymphosarcoma cells induced *in vivo* have, however, formed transplantable lines (PONTÉN, 1963; PONTÉN and BURMESTER, 1967) in apparent contrast to cells transformed by other members of the avian leukemia-sarcoma complex.

d) Relation between Virus Synthesis and Transformation

RSV infected mass cultures of avian cells show an eclipse phase of about 15 hours followed by rapid exponential increase in infectious (focus forming) virus titer until a plateau reflecting a balance between virus synthesis and thermal inactivation is reached after 4—6 days (TEMIN and RUBIN, 1959; VIGIER and GOLDÉ, 1959; PRINCE, 1960b; P. VOGT and RUBIN, 1961; GOLDÉ and VIGIER, 1961). Closer analysis, including the observation of viral antigen in individual cells (P. VOGT and RUBIN, 1961; P. VOGT and LUYKX, 1963), and assay for the number of cells releasing virus when plated on top of susceptible indicator cells (RUBIN, 1960b), has revealed a considerable asynchrony even after infection by a large excess of RSV. The fraction of cells which produced virus at 24 hours was only about 1% and it took 3 days until practically every cell was engaged in virus synthesis. The interval between assembly of a complete particle and its extracellular release has been estimated to be 3 minutes (RUBIN, BALUDA, and HOTCHIN, 1955).

The accumulation of cytoplasmic and nuclear RNA (SUNDELIN, 1967) which coincides with the morphologic alteration into basophilic spindle or round cells (TEMIN and RUBIN, 1958) is first seen about 48 hours post infection suggesting a short delay between onset of virus synthesis and morphologic alteration. The cytologic changes seem to occur simultaneously with unrestrained growth transformation because the first multiplying Rous foci are also seen after 48 hours (TEMIN and RUBIN, 1958). This fits the findings of RUBIN and COLBY (1968) that infected cells reach higher than control densities already during their first or second cell cycle.

The delay between virus synthesis and transformation explains why fibroblasts with normal morphology may act as virus yielders (TRAGER and RUBIN, 1964; SUNDELIN, 1967). TRAGER and RUBIN (1966a, b) determined the fate of such virus producers in extensive cloning experiments. When virus yielding apparently non-transformed populations were cloned, the clones were either transformed or normal. The former type inevitably yielded virus in contrast to the latter which generally was negative. When an occasional normal clone produced virus, subcloning traced this down to the presence of a few transformed cells. In no case could a non-transformed RSV releasing clone be serially carried without eventually acquiring a transformed character. The reason for these delayed transformations is unknown but may be related to environmental factors. Transformation can selectively be suppressed by fetal calf serum (RUBIN, 1960c), the absence of tryptose phosphate broth in the medium (PRINCE, 1962) or contamination with mycoplasma (PONTÉN and MACPHERSON, 1966) and it seems likely that other undiscovered modifying factors exist.

L⁻R cells differ from the above pattern by not showing any abundant synthesis of virus belonging to any of the conventional subgroups (Table 11). Instead RSVβ(O) with its restricted host range or particles for which no susceptible cells have been found are manufactured. This does not, however, affect the cytological response or the lack of growth control in the transformed cells.

Mammalian cell cultures exposed to RSV show a longer latent time until characteristic foci appear and the number of virus particles required to produce one transformed focus is increased over that effective in chicken cells by a factor of 10^2 in bovine and 10^4 in human cells (STENKVIST, 1966c). Observations of mass cultures also strongly suggest that the majority of the cells fail to respond cytologically. In the established BHK-21 hamster line the transformation frequency was below 1% when scored as capacity for growth in soft agar (MACPHERSON, 1965).

Synthesis of virus of the same character as the bulk of the virus stock used for infection is absent in mammalian cells (SVOBODA and CHÝLE, 1963; JENSEN, GIRARDI, GILDEN, and KOPROWSKI, 1964; STENKVIST, 1966d). This parallels the non-infective character of Rous sarcomas induced in mammals *in vivo* (SVOBODA, CHÝLE, SIMKOVIC, and HILGERT, 1963; KLEMENT, CHÝLE, and SVOBODA, 1963; SIMKOVIC, SVOBODA, and VALENTOVÁ, 1963; AHLSTRÖM, BERGMAN, FORSBY, and JONSSON, 1963; HUEBNER, ARMSTRONG, OKUYAN, SARMA, and TURNER, 1964).

These facts make it abundantly clear that extensive virus replication is not required for transformation in avian or mammalian cells. The long held thought that L⁻R cells and transformed mammalian cells do not produce any virus particles has been modified and the increasing number of cases where virus-like particles are observed may mean that all transformed cells are in fact virus producers but that the rate of production of transforming virions depends on the rate of helper virus synthesis. The original hypothesis (P. VOGT and RUBIN, 1961) that virus production *per se* is responsible for the membrane change which underlies irregular growth transformation and invasive growth may still turn out to be correct, but is contradicted by the finding of mammalian cells with no production of virions or viral RNA (VALENTINE and BADER, 1968). It seems more likely that synthesis of mature virions bears no direct relationship to transformation.

e) Reversion and Loss of RSV Genome

The studies by TRAGER and RUBIN referred to in the previous section suggested that the RSV genome is not stably associated with the transformed cell genome, since it could be lost if rapid cell division was stimulated early after infection. These experiments were complicated by the continuous synthesis of infectious virus. MACPHERSON (1965) — using cloning in soft agar — investigated the persistence of the RSV genome in BHK-21 cells, which had undergone irregular growth transformation under the influence of RSV-SR and did not yield any directly demonstrable infectious virus. Transformed clones were characterized by a) the capacity to produce infectious avian sarcomas after inoculation of viable cells into chicks b) a disorganized, multilayered growth pattern after plating on a solid substrate c) enhanced tumorigenicity after transplantation into hamsters d) the presence of group specific CF antigen and e) a colony forming efficiency in agar of about 5%. A proportion of single cell isolated from transformed clones was found to produce cell populations which had lost all of the characters acquired at transformation. They could be considered examples of transformed cells which had reverted to normal with the possible exception that they appeared resistant to reinfection by RSV-SR (MACPHERSON, 1966). These events were compatible with a "non-integrated" state of the RSV-genome which could be diluted out during multiplication of transformed cells. Using less rigorous criteria STENKVIST (1966c) suggested a similar reversion in transformed bovine cells.

Taken together the studies of the last two sections suggest 1. That the presence of one or more RSV genomes will sooner or later lead to unrestrained growth transformation of fibroblasts (but perhaps not of all other cell types). 2. That no strict correlation exists between the replication of the genome of the cell and the virus. 3. That the viral genome may be lost through unequal distribution at cell division and 4. That the loss of the RSV genome leads to a return of physiologic cell growth controls *i.e.* a reversion from a transformed to a normal state. However, in most instances the RSV genome remains permanently associated with the infected and transformed cell.

f) Transmission of the RSV Genome

Animal to animal transfer. Under crowded laboratory conditions (BURMESTER, FONTES, and WALTER, 1960; BURMESTER and FREDRICKSON, 1966) and presumably also in nature, avian sarcoma virus spreads by horizontal transmission like any conventional contagion. This also holds true for the numerous field strains of LV which have been isolated. Such strains persist as endemic agents through a complicated interplay involving genetic susceptibility, acquired tolerance and immunity. Chronically infected dams transmit LV by laying congenitally infected eggs (BURMESTER, 1957; RUBIN, CORNELIUS, and FANSHIER, 1961; RUBIN, FANSHIER, CORNELIUS, and HUGHES, 1962; RUBIN, 1962) and the virus is thus transmitted vertically as well.

The species range by any virus of the avian sarcoma-leukemia complex is largely if not entirely determined by its envelope. H. HANAFUSA, and T. HANAFUSA (1966b) studied the importance of the coat protein of RSV-BH and RSV-SR in detail using different helper viruses. They consistently failed to induce tumors in new-born hamsters if the viral envelope was provided by RAV1 regardless of whether the viral genome was derived from an RSV-BH or RSV-SR strain. On the other hand, if a newly isolated helper virus — RAV50 — was used to coat the genomes both were tumorigenic. Later studies (DUFF and P. VOGT, 1969) makes it likely that the subgroup D coat confers high infectiosity and tumorigenicity for mammalian cells to the RSV genome (Table 13). Since LV and related leukemia viruses capable of acting as helpers are widespread in most chicken flocks, it seems likely that the spread particularly of defective RSV genomes is governed by accompanying helper viruses and that infection of mammals could occur provided appropriate helpers are available.

Cell to cell transfer. Mixed cultures of C/A L⁻R cells (induced by RSV-BH) and C/B cells (Table 12) were tested for the possible transfer of the RSV genome from C/A to C/B cells by isolating foci of transformed cells. Transformed L⁻R foci were challenged by RAV 1 (which only replicates in C/B cells) and RAV 2 (which only replicates in C/A cells) to distinguish between the two phenotypes. Surprisingly the majority of activable transformed foci were of the C/B type indicating efficient intercellular transfer of RSV genome in the absence of helper virus. Such transfer is presumed to be affected by the virus-like RSV(O) particles observed in L⁻R cells in the electron microscope but may also have been carried out by "subviral" agents. Close cell contacts were required for the exchange (P. VOGT, 1967 d).

Transformed mammalian cells do not, with only a few exceptions (SVOBODA and KLEMENT, 1963; ALTANER and SVEC, 1966), produce directly demonstrable virus. Despite this apparent non-infectiosity viable cells readily produce infectious RSV when brought in contact with chicken cells either *in vivo* or *in vitro* (cf. P. VOGT, 1965). Some insight into this phenomenon has been gained by experiments in cell culture.

SVOBODA (1960, 1961) induced a transplantable rat sarcoma, designated XC, by RSV-PR. Neither in this line or a cell line, MR_5, transformed *in vitro* by SVOBODA and CHÝLE (1963) could virus synthesis be induced by X- or UV-irradiation, manipulation of the temperature or altering the composition of the growth medium (CHÝLE, KLEMENT, and SVOBODA, 1963). SIMKOVIC, VALENTOVÁ, and THURZO (1962), SVOBODA (1964) and VIGIER and SVOBODA (1965) detected RSV in XC cells which were cocultivated with chick embryo fibroblasts. CHÝLE (1964) showed that direct cell-cell contact was necessary since chicken cells separated from XC or MR_5 cells by a collodium membrane (pore size $0.3-0.5~\mu$) failed to produce RSV. The production of RSV is markedly enhanced by fusion of transformed and normal chicken cells by Sendai virus (SVOBODA, MACHALA, and HLOZÁNEK, 1967; VIGIER, 1967; YAMAGUCHI, TAKEUCHI, and YAMAMOTO, 1967; SVOBODA, HLOZÁNEK, and MACHALA, 1968).

Hamster cells transformed by RSV-BH cannot be activated by superinfection with a helper virus such as RAV 50 and no virus synthesis is induced after inoculation into chicks or cocultivation with avian cells. Only if RAV is added to a mixed culture will infectious RSV-BH be produced (SARMA, VASS, and HUEBNER, 1966; VIGIER, 1966; H. HANAFUSA and T. HANAFUSA, 1966 b). An apparent analogy was found when L⁻R chick cells were cocultivated with normal chick cells — synthesis of large amounts of infectious virus required superinfection by a helper virus (P. VOGT, 1967 a). The most straightforward interpretation is that the RSV genome in the form of RSV(O) is transferred from the transformed cells to the normal indicator cells where synthesis of infectious particles may be augmented by a suitable helper virus. Cell to cell transfer of RSV and entrance of virus into cells is facilitated by close cell to cell contacts and/or treatment with Sendai virus.

g) Metabolism of RSV Infected Cells

The metabolic requirements for the establishment of chronic virus synthesis and transformation has been investigated by TEMIN and others using various inhibitors of nucleic acid and protein synthesis.

Actinomycin D inhibits DNA-dependent RNA synthesis. Many RNA viruses are not suppressed by this antibiotic (REICH, FRANKLIN, SHATKIN, and TATUM, 1961). RSV and RAV were, however, strongly inhibited at concentrations of actinomycin where the cells were morphologically intact (TEMIN, 1963b, 1964a; BADER, 1964b). This effect is also found with the physico-chemically similar myxovirus influenzae (BARRY, IVES, and CRUICKSHANK, 1962).

Inhibitors of DNA synthesis including amethopterin, 5-bromodeoxyuridine, 5-iododeoxyuridine, mitomycin C, cytosine arabinoside, excess thymidine, 5-fluorouracil and 5-fluorodeoxyuridine have shown a more or less clearcut inhibition of RSV and RAV synthesis in treated chicken cells (VIGIER and GOLDÉ, 1964b; BADER, 1964a, b; TEMIN, 1964a, b; FORCE and STEWART, 1964; BADER, 1965a, b; BADER, 1966a, b, c; TEMIN, 1967b; NAKATA and BADER, 1968). These findings indicate that DNA synthesis is required for the production of infectious virus. An analogous situation has not been found with any other RNA virus. This requirement seems to be transitory because DNA inhibitors added later than 12—16 hours after exposure to RSV were ineffective. Actinomycin D, on the other hand, was capable of interrupting already established synthesis but larger concentrations were required than for prevention of initiation of RSV synthesis (BADER, 1965b, 1966b). The morphological alteration into typical Rous cells was effected in the same way as virus synthesis (BADER, 1965a). The interpretation has been that initiation of virus synthesis and genotypic fixation of transformation require DNA synthesis, but that the phenotypic expression of transformation (in this case the morphological change) can develop without DNA synthesis or cell division (NAKATA and BADER, 1968). When the early requirement for DNA synthesis has been fulfilled virus synthesis continues even if DNA synthesis is interrupted.

It has been speculated that the requirement for DNA synthesis reflects the making of replicating DNA "provirus", which serves as a template for the RNA of RSV (VIGIER and GOLDÉ, 1964b; TEMIN, 1964b). This would explain why inhibitors of DNA synthesis are ineffective on already established infections in contrast to actinomycin which acts on DNA-dependent RNA synthesis. TEMIN (1964c) claimed to have shown a new Rous cell DNA sequence homologous to RSV RNA and therefore the chemical equivalent of the assumed provirus. Virus was labeled with H^3-uridine and the binding of partially purified RNA to DNA from normal and transformed chicken cells was compared using either the Nygaard-Hall filter method or the Bolton-McCarthy technique where the DNA is affixed in an agar gel. Temin's interpretation was based on a slightly increased amount of bound radioactivity but can hardly be accepted as valid in view of the low activity of the imperfectly purified RSV-RNA a few tenths of a per cent of which hybridized at the most. In two brief reports homology was found between RNA from RSV (J. HAREL, L. HAREL, GOLDÉ, and VIGIER, 1966) or AMV (L. HAREL, J. HAREL, LACOUR, and HUPPERT, 1966) and chicken cell DNA. In contrast to Temin's results, DNA from uninfected or virus infected cells showed the same affinity for viral RNA, a result also obtained by BADER (1966a) and YOSHIKAWA-FUKADA and EBERT (1969).

TEMIN (1967b) exposed partially synchronized chick embryo cultures to RSV and watched the effect of inhibitors of DNA synthesis added at different times

RSV synthesis could be depressed by the DNA inhibitors, even if these were added in the G2 period of the cell cycle suggesting that the newly synthesized DNA required for RSV production may not be the ordinary DNA of the cell's chromosomes. This result is consistent with the "DNA-provirus" hypothesis but may also reflect a requirement for synthesis of, for instance, mitochondrial DNA. A corollary finding was that prevention of the first mitosis after infection prevented virus synthesis while interference with later cell divisions had no such effect.

No proof of any DNA-provirus has yet been given and the apparent requirement for DNA synthesis may reflect some other process related to events in the normal cell cycle. Its likelihood has been considerably increased by the recent finding of an RNA-dependent DNA polymerase in members of the avian and murine leukemia-sarcoma virus complexes (BALTIMORE, 1970; TEMIN and MIZUTANI, 1970). See page 183.

GOLDÉ (1962) and TEMIN (1965) found no significant increase in the average amount of DNA in Rous cells over that of normal cells. Single cells infected by RSV-SR and analysed by microspectrophotometry (SUNDELIN, 1967, 1968) showed, however, increased amounts of DNA in a proportion of virus-yielding basophilic round or spindle cells. Post-mitotic G1 cells showed up to 5 times the normal diploid G1 amount indicating that the altered cells divided after having accumulated supernormal amounts of DNA. These interesting findings may be the cytochemical corollary to the DNA synthesis required for initiation of RSV synthesis and transformation. It should be observed that chromosome analysis has shown avian RSV transformed cells to be diploid *in vivo* and *in vitro* (PONTÉN, 1963 and unpublished). Polyploidization does, therefore, not seem likely as an explanation for the increased amount of DNA. This increase in DNA ought to be demonstrable by chemical analysis of mass populations and the reason for the discordant results is unexplained.

Recent studies have shown an increased thymidine incorporation into DNA of RSV infected chick embryo fibroblast cultures as early as 8 hours post infection, *i.e.* well before morphological alterations appear (KÁRA, 1968a). Stimulation of DNA synthesis was accompanied by induction of deoxycytidylate deaminase and uridine kinase (KÁRA, 1968b), but it has not been established that the effect on DNA synthesis concerns the transforming cells directly.

Other metabolic changes in RSV transformed cells have attracted less attention but an increased RNA content has been found with macrochemical methods (GOLDÉ, 1962) and microspectrophotometry (SUNDELIN, 1967). RNA is increased in the nucleus and the cytoplasm. An elevated incorporation rate in nucleolar RNA has been noted by HAMPTON and EIDINOFF (1962).

Increased glycolysis in RSV infected avian and mammalian cell cultures was reported by MORGAN and GANAPATHY (1963) and TEMIN (1968a). STECK, KAUFMAN, and BADER (1968) have suggested that this may be a direct reflection of unrestrained growth transformation and correlated to a capacity to proliferate under crowded condition and in low serum concentration. Infection by RSV leads to a 5-fold increase in cellular hyaluronic acid synthetase activity (ISHIMOTO, TEMIN, and STROMINGER, 1966) and increase in production of hyaluronic acid (TEMIN,

1965). Large amounts of this polysaccharide have been found in Rous cells and extracellular fluid (WARREN, WILLIAMS, ALBURN, and SEIFTER, 1949; HARRIS, MALMGREN, and SYLVÉN, 1954; ERICHSEN, ENG, and MORGAN, 1961; GROSSFELD, 1962). Converted rat cells do not show an increased production of acid mucopolysaccharides, which is therefore not an obligate event in transformation (TEMIN, 1968b).

Chicken and duck cells transformed with RSV or FSV show a reduced requirement for serum, i.e. a concentration of serum can be found at which only transformed cells are capable of multiplication (TEMIN, 1966, 1967a). This finding parallels results with most transformed cells and does not seem to be specifically concerned with virus induced transformation. Increased multiplication of RSV-SR transformed chick embryo cells in limiting amounts of serum was apparently caused by more efficient utilization of factor(s) necessary for multiplication rather than increased binding or uptake of the factor(s) (TEMIN, 1969).

Insulin can at least partially replace the unidentified serum component necessary for multiplication of transformed chicken and duck cells and has a growth stimulating action on the transformed cells (TEMIN, 1967b, 1968a, b).

Untransformed chicken cells produce an unidentified, non-dialyzable "conditioning factor" which enhances the growth of sparsely seeded chick embryo cell. If cultures are started from different numbers of cells, a critical low density will be reached below which no proliferation will take place. Addition of "conditioning factor" will make cell growth possible even below this threshold (RUBIN, 1966a). Avian cells transformed by RSV cease to produce "conditioning factor" and instead produce a growth inhibitory factor which restrains proliferation of other Rous cells (RUBIN, 1966b). RUBIN (1966c) in a review of the cell surface in carcinogenesis suggested that the inhibitory substance disrupts the surface structure of Rous cells causing them to commit suicide at least *in vitro*.

h) Avian Tumor Viruses and Differentiation

The absence of definite evidence for the creation of new cellular genotypes suggests that the avian RNA viruses may operate at an epigenetic or phenotypic level. The possibility that members of the avian leukemia-sarcoma complex act by interfering with normal cell differentiation has received some experimental support.

AMV has been studied from this point of view by BALUDA and GOETZ (1961), BALUDA (1962) and BALUDA, GOETZ, and OHNO (1963). Their findings have indicated a strong correlation between the number of myeloblasts in different organs during embryogenesis and susceptibility to the action of AMV. The myeloblasts obtained after virus infection are — except for the presence of virus particles — indistinguishable from normal immature myeloblasts suggesting that AMV may act by preventing differentiation of myeloblasts into granulocytes. When yolk sac cultures, which normally only yield fibroblast-like cells and macrophages, were exposed to AMV an abundance of myeloblasts appeared, which has been interpreted as a virus induction of specific differentiation of yolk sac cells to myeloblasts (BALUDA, GOETZ, and OHNO, 1963). This interpretation cannot, however, be accepted until it is shown that the target cultures were completely devoid of myeloblasts. The possibility remains that AMV selectively favored growth of a few myeloblasts present below the level of detectability. Even explants from muscle have given rise to myeloblast-like cells after exposure to AMV suggesting that non-hematopoietic tissue can undergo this type of conversion (C. Moscovici, M. Moscovici, and Zanetti, 1969). This interesting observation needs morphologic documentation that the rounded cells indeed are myeloblasts and for instance not identical to the "basophilic round cells" described on page 130. Studies *in vivo* do not contradict an effect on cell differentiation. Solid tumors such as nephroblastomas caused by AMV are characterized by varying degrees of differentiation and resemble embryonic tissue (Thorell, 1958; Baluda and Jamison, 1961; Ishiguro, D. Beard, Sommer, Heine, de Thé, and J. Beard, 1962).

CEV has failed to produce morphologic effects *in vitro* but certain observations *in vivo* have bearing on its possible specific interference with normal cell differentiation. Pontén and Thorell (1957) observed that erythroleukemia induced by CEV starts as a focal accumulation of erythroblasts along the borders of the bone marrow sinusoids, *i.e.* the same areas where immature red cell precursors are found normally (cf. Lucas and Jamroz, 1961). Some erythroleukemia cells, analysed microspectrophotographically, have a small amount of the two types of hemoglobin which occur physiologically. The cytoplasmic RNA concentration is much higher than that of normal erythropoietic cells with the same content of hemoglobin (Cowles, Saikkonen, and Thorell, 1958; Ambs and Thorell, 1960). Whether this RNA increase should be interpreted as a sign of abnormal differentiation (Ambs and Thorell, 1959) or viral RNA synthesis has not been determined. The results with CEV can well be explained on the basis of a "maturation arrest" in virus infected cells.

Because of the uncertainty of the direction of differentiation among fibroblastic cells it has not been possible to interpret RSV effects in terms of fibroblast differentiation.

Cloned myoblasts responded to infection by RSV-BH by conversion to basophilic round cells indistinguishable from transformed fibroblasts (Kaighn, Ebert, and Stott, 1966; Lee, Kaighn, and Ebert, 1966). Uninfected myoblasts will fuse to form multinucleated myotubes — differentiated structures with no DNA synthesis. Myotubes infected by RSV-BH started to synthesize DNA within 24 hours. This nucleic acid synthesis was not followed by transformation or cell division but cytoplasmic vacuolization and death of the myotubes (Lee, Kaighn, and Ebert, 1968). The results show that DNA synthesis can be induced in well differentiated cells which normally have no capacity for cell division, but are otherwise difficult to interpret in terms of viral effects on differentiation.

There is thus — particularly with the leukemogenic viruses — some evidence that these agents may act as specific inhibitors of normal differentiation. Nothing contradicts that the leukemogenic action of AMV and CEV can be explained by such an effect. Infected blast cells will not mature and instead remain at a stage where rapid proliferation prevails. Apart from this maturation arrest the cells can be regarded normal in the sense that analogous blast cells are found in normal embryonic or adult blood cell formation. This hypothesis would be strengthened if "reversions" akin to those seen in RSV transformed fibroblasts (page 140) occurred in "leukemic" erythro- or myeloblasts and if they would involve resumption of normal differentiation and function.

The Murine Leukemia-Sarcoma Complex

GROSS' (1951 a, b) fundamental discovery of a virus which induces lymphatic leukemia in mice led to the isolation of a large number of leukemogenic agents in rodents. HUEBNER (1967) stated that some 200 isolates from mice had been made in his laboratory alone. The viruses have been difficult to characterize by immunological and other methods and many isolates may represent identical viruses or mutations of one parent virus. Their classification is imperfect and Table 14 includes only viruses which have been more intensively investigated in cell culture. The table shows that the responses *in vivo* are complex covering a broad spectrum from necrosis to obvious malignant neoplasia exclusively involving hematopoietic or non-hematopoietic mesenchyme. The separation between predominantly sarcomagenic and leukemogenic strains is less sharp than in the avian leukemia-sarcoma complex.

The murine leukemia viruses as well as the analogous avian agents seem to be very widely distributed in the respective hosts. They are usually innocent and have been isolated from normal mice, X-irradiated animals and solid tumors where they occurred as passengers. Only rarely and late in life do they cause leukemias in control animals. They are almost exclusively transmitted "vertically", *i.e.* embryos become infected before birth from their virus-carrying mothers (GROSS, 1951 b, 1952, 1961 a, b; LAW and MOLONEY, 1961). Many (or all?) cell lines of rodent origin including the well known L-strain (page 21) (SCHÄFER, FRANK, and PISTER, 1968) harbor leukemia viruses which cause no overt cytopathic or other effects.

The biological and virological aspects have been reviewed repeatedly (GROSS, 1958a, 1961 a, b; DMOCHOWSKI, 1960; MILLER, 1961; SINKOVICS, 1962; MOLONEY, 1962, 1964; LÉVY and PÉRIÈS, 1964; BOIRON, LÉVY, and PÉRIÈS, 1967).

1. Pathology and Classification

The pathological effects may be described in the same categories as in the avian leukemia-sarcoma complex. Malignant lymphoma (usually lymphosarcoma) starts in the thymus and later involves the spleen, liver, bone marrow and lymph nodes (DUNN, MOLONEY, GREEN, and ARNOLD, 1961). Although often referred to as a "leukemia", this disease is not ordinarily accompanied by circulating abnormal cells. This type of neoplasia dominates after infection by Gross (GLV), Moloney (MLV), Kaplan (KLV) and many other strains. Its latent period is usually measured in months. It is indistinguishable from most lymphomas which arise "spontaneously" or after irradiation (DAWE, LAW, MORGAN, and SHAW, 1962). Variants may occur including reticulum cell sarcoma and Hodgkin-like lesions. Sometimes thymic lymphosarcoma is accompanied by myeloid, stem cell, lymphoid or even erythroic leukemia (GROSS, 1966). A recent extensive discussion of the pathology of murine malignant lymphoma has been published by DUNN and DERINGER (1968).

The second variety is represented by myeloid leukemia, which may be seen in a pure form after inoculation of Graffi virus (GRAFFI, 1957). In thymectomized mice inoculated with GLV (GROSS, 1960) or MLV (DUNN, 1961) pure myeloid leukemia may also occur after a long latent time (SIEGLER and RICH, 1966). A

Table 14. *Effects of Selected Members of the Murine Leukemia-Sarcoma Complex in vivo and in vitro. Many of the Entries must be Regarded as Preliminary because of Uncertainty of the Exact Composition of the Virus Stock Employed etc.*

Virus	Susceptible species	Princip. path. response in vivo	Virus synthesis		Transformation in vitro		
			Parent type[1]	Host range variants	Foci[a]	Irreg. unrest. growth	Infin. growth
GLV (Gross)	mouse rat hamster	lymphosarcoma lymphosarcoma	+	—			
MLV (Moloney)	mouse rat hamster	lymphosarcoma lymphosarcoma	+	—			
RLV (Rauscher)	mouse rat hamster	lymph.sarc.+eryth.leuk. lymphosarcoma	+ +	— —	— — +	— —(+) +	— —(+) +
FLV (Friend)	mouse rat hamster	eryth. leukemia eryth. leukemia	+	—			
KLV (RadLV) (Kaplan)	mouse rat hamster	lymphosarcoma	+	—			
GiLV (Graffi)	mouse rat hamster	myel. leukemia myel. leukemia n.t.	+ n.t.				

Murine leukemia viruses

Murine sarcoma viruses[3]	Host species	Effect in animal					
MuSV-M (Moloney)	mouse	atyp. granuloma (sarc.)	+ +	+ +	+ +	+	—
	rat	sarcoma	+ +	+	+	+	+
	hamster	n.t.	— +				●?
	calf	n.t.		+			
	cat						
MuSV-H (Harvey)	mouse	atyp. granuloma (sarc.)	+		+	+	—
	rat	sarcoma	(+)		+	+ ●?	
	hamster	sarcoma	— ●?				
MuSV-K (Kirsten)	mouse	hem. cysts + sarcoma (no effects on mice)	+		+	+	—
	rat			+[4]	+	+	—
	hamster				+	●?	

[1] Reproduction of virus with identical serology and host range as input virus

[2] Foci of morphologically altered cells as depicted in Fig. 33. These may or may not be grown as established lines

[3] Preparations always contaminated by MuLV

[4] The effects listed were produced by a hamster adapted variant "Ki-MSV (O-H)" described by KLEMENT, HARTLEY, ROWE and HUEBNER (1969)

n.t. = not tested.

stock of Rauscher virus (RLV) has given late myeloid leukemias in mice (BOIRON, LÉVY, LASNERET, OPPENHEIM, and BERNARD, 1965).

The third response which may be compared to erythroleukemia in chickens has never been seen spontaneously. It is induced by the Friend (FLV) and Rauscher (RLV) leukemia virus. Within 2—3 weeks the spleen and liver become greatly enlarged due to massive accumulation of red cell precursors of different maturity. At this stage splenic rupture often kills the animal. The neoplastic nature of this peculiar erythroblastosis has never been clearly established (FRIEND, 1957b; RAUSCHER, 1962; SIEGEL, WEAVER, and KOLER, 1964; BOIRON, LÉVY, LASNERET, OPPENHEIM, and BERNARD, 1965) but the establishment of continuous culture lines of proerythroblasts from FLV inoculated mice suggests that the erythroid response is at least potentially neoplastic (FRIEND, PATULEIA, and DE HARVEN, 1966; PATULEIA and FRIEND, 1967; ROSSI and FRIEND, 1967). FLV and RLV do not give rise to pure erythroleukemia, because unquestionably malignant histological reticulum cell sarcomas may be observed at a late stage after FLV infection (FRIEND, 1957b). Cell clones from such "reticulum cell sarcomas" carried *in vitro* or *in vivo* constantly differentiated into proerythroblasts (PATULEIA and FRIEND, 1967) suggesting that the tumors could be regarded solid "proerythroblastomas" rather than true reticulum cell sarcoma. The thymic lymphosarcoma which develops in mice surviving the early RLV induced erythroblastosis (RAUSCHER, 1962) is histologically indistinguishable from the lymphosarcomas seen after GLV infection.

The fourth type of pathology is a sarcoma or "atypical granuloma", which has not been observed as a spontaneous disease. HARVEY (1964) inoculated plasma from a leukemic rat infected with stock MLV into newborn BALB/C mice and obtained anaplastic sarcomas at the site of inoculation in addition to malignant lymphoma and leukemia. MOLONEY (1966) likewise obtained locally growing sarcoma in mice injected with stock MLV. By continued passage of cell-free extracts from the sarcomas, preparations were obtained which showed a reproducible capacity to induce sarcoma at the site of inoculation and a reduced leukemogenic activity. It is not known if the Harvey and Moloney isolates represent different viruses or variants of the same agent. Their species range and principal effects *in vitro* are found in Table 14. They will be referred to as MuSV-H and MuSV-M, respectively.

MuSV-H induces splenomegaly within 2—3 weeks with infiltration of cells interpreted as malignant reticulum cells, erythroblasts (HARVEY, 1964; CHESTERMAN, HARVEY, DOURMASHKIN, and SALAMAN, 1966) or lymphoblasts (MITCHINER, 1967). The solid mesenchymal tumors at the site of inoculation are undifferentiated spindle-cell sarcomas (HARVEY, 1964). MuSV-M induces pleomorphic sarcomas some of which are angiosarcomas (STANTON, LAW, and TING, 1968) and others poorly differentiated rhabdomyosarcomas (MOLONEY, 1966; PERK and MOLONEY, 1966; PERK, MOLONEY, and JENKINS, 1967). In addition lymphangiectatic cysts were noted (CHESTERMAN, HARVEY, DOURMASHKIN, and SALAMAN, 1966). STANTON, LAW, and TING (1968) in a thorough study of mice inoculated with MuSV-M examined 49 rapidly enlarging tumors grossly resembling malignant neoplasms. Out of these only 6 were histologically verified as undifferentiated sarcomas or hemangiosarcomas. The remaining were "atypical granulomas" composed of granulation tissue rich in capillaries, polymorphnuclear leukocytes, lymphocytes and histiocytes. Such lesions only contained scattered atypical mesenchymal cells with plump nuclei which resembled sarcoma cells (Fig. 31). Atypical granulomas released MuSV but were not transplantable (LAW, TING, and STANTON, 1968). In accordance with established but unfortunate practise "sarcoma" will be used for any solid rodent tumor induced by MuSV even if many would probably be "atypical granulomas" on histological examination.

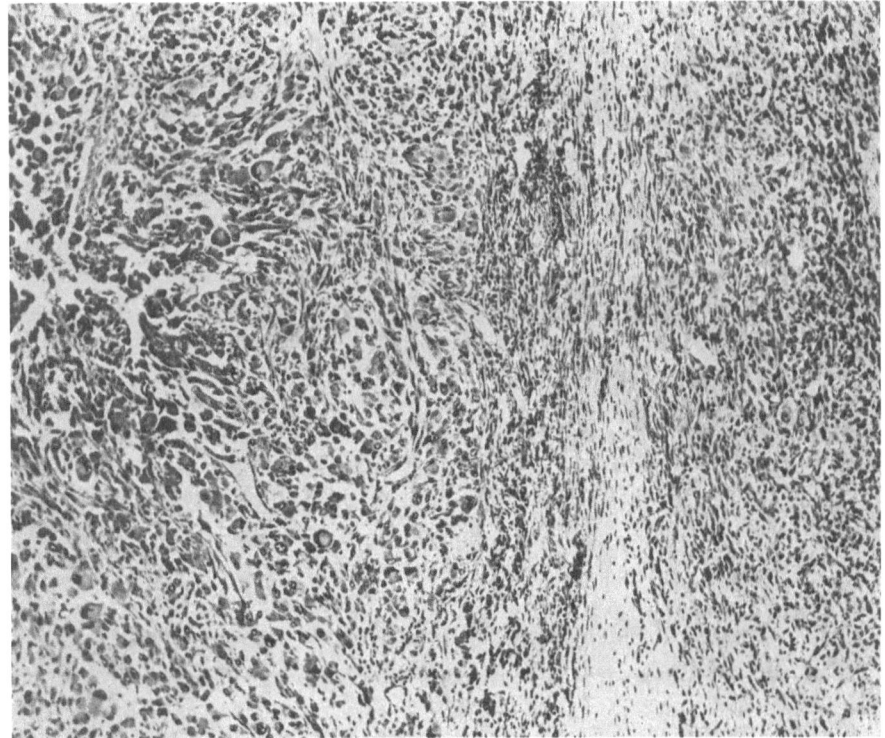

Fig. 31. Anaplastic mesenchymal cell sarcoma adjacent to atypical granulomatous reaction. Note similarity between neoplastic cells and scattered cells in the granulomatous area. Neoplastic cells show prominent perinuclear zones. Hematoxylin and eosin. ×128. From STANTON, LAW, and TING, 1968

BERMAN and ALLISON (1969) compared the effects of MuSV-H and MuSV-M in mice. Both strains induced sarcomas, but the former in addition also caused an erythroblastic response similar to that induced by FLV. These authors concluded that the sarcomatous lesions were neoplastic, because the tumors ultimately became transplantable.

A possible variant of MuSV-M has been described, which induces splenomegaly and lymph node enlargement in addition to locally invasive pleomorphic undifferentiated sarcomas (CHIRIGOS, SCOTT, TURNER, and PERK, 1968).

The complicated response to MuSV is best summarized as a variable mixture of necrosis, inflammation, lymphangiectasies and atypical cell proliferation, which may progress into transplantable sarcomas. Dominance of the necrotizing or proliferative effects may be determined by such factors as the age of the recipient, the virulence of the virus stock and the immunological status of the recipient — factor known to be of importance also in connection with virus induced avian sarcomas.

The pathological picture after infection by MuLV or MuSV can only be used for a coarse classification. This may partly be due to the impurity of the virus stocks employed. The pathology is very similar to that of the avian leukemia-sarcoma complex and all lesions seen in virus inoculated mice have histological counterparts in the virus infected chicken or chick embryo (cf. DURAN-REYNALS, 1953).

GROSS (1966) has proposed that all leukemia-lymphoma inducing isolates are identical or least closely related variants of a single virus, which he has named "mouse leukemia virus type A", not to be confused with the "A-particles" described below. Differences in the pathological responses were ascribed to genetic differences in the host and variations in viral virulence. Type A is extremely common in mice and accounts for spontaneous leukemias and lymphomas in high leukemia strains such as AK. It is also important for the development of leukemia/lymphoma after X-irradiation and other physical carcinogens. The disease induced by FLV is regarded a laboratory artifact caused by an agent different from "type A". The special response caused by RLV can be explained if this virus is a mixture of "type A" and FLV (GROSS, 1964a, b). The former component is responsible for the lymphosarcomatosis and transmissibility to rats, whereas the latter induces the splenomegalic component which can only be elicited in mice. This view was supported by the apparent removal of the splenomegaly inducing capacity of RLV after passage through rats, which only develop lymphosarcoma (GROSS, 1966). CHAMORRO (1967) has separated one component inducing erythroblastosis from another causing reticulum cell sarcoma in a stock of FLV.

2. Physico-chemical Properties of Virions

Members of the murine leukemia-sarcoma complex are fragile (DALTON, HAGUENAU, and MOLONEY, 1964) and easily inactivated but have been concentrated by differential centrifugation and partially purified by density gradient techniques most often in sucrose or polymeres of carbohydrates with no loss of infectivity (review by ROBINSON and DUESBERG, 1968). Cells and tumor tissue are unsuitable as starting material because of their high content of particulate material derived from cell membranes with properties similar to virus particles. Viremic sera and tissue culture fluid have been used and their products analysed, although no perfectly pure preparations have been obtained. No consistent differences have been found between various MuLV and MuSV preparations.

Ultrastructural studies have shown two types of particles in association with infected cells, termed A and C according to BERNHARD (1960). The C particles occur as immature or mature forms. The mature C particle has a diameter of about 100 mμ and contains a dense central nucleoid of about 65—70 mμ bordered by an envelope. An inner membrane, surrounding the nucleoid, may sometimes be observed, particularly when the nucleoid has shrunk to give the particle a "signet-ring" like appearance. The immature C particle envelope is of the same size and shape as that of the mature virion. Two concentric rounded shells with a diameter of 75—80 and 60—65 mµ, respectively, are found inside. The center is electron translucent (Fig. 32). Both types of C particles are formed by budding from the cell membrane and may be seen in the culture fluid, extracellularly in close contact with the cell membrane or inside vacuoles but never intracellularly. The interior of RLV C particles has been analysed after treatment with digitonin (DE THÉ, 1967) or Tween-ether (DE THÉ and O'CONNOR, 1966). The latter technique showed the outer shell of immature C particles as a hollow coiled barrel-like structure. The inner electron dense shell and the center of the particle became translucent after treatment by RNase indicating that RNA is located in these structures. Mature C particles were believed to arise by the fusion of the two internal shells of immature precursors to give a dense nucleoid (Fig. 32).

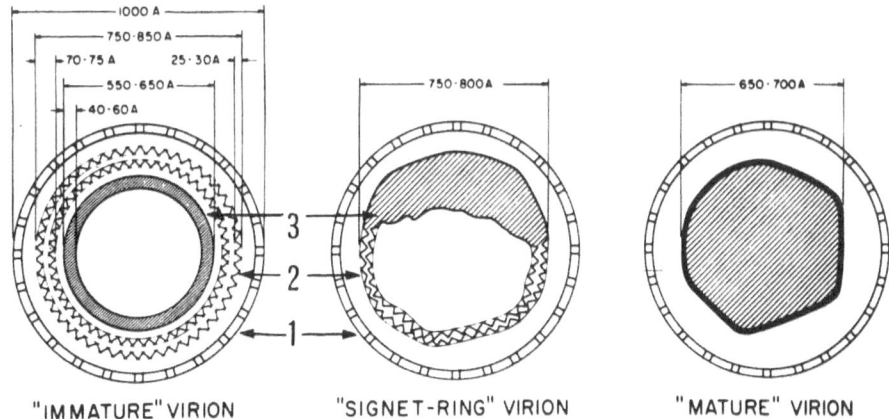

"IMMATURE" VIRION   "SIGNET-RING" VIRION   "MATURE" VIRION

Fig. 32. Diagram giving the sizes of the different types of virions seen in RLV preparations. The "immature" virion (at left) presents: 1, the envelope; 2, the outer shell, 75 Å thick, corresponding to a hollow, coiled cylinder; 3, the inner shell possibly containing the RNA. In the "signet-ring" type of virion (middle), the outer shell (2) is somewhat collapsed over the dense inner nucleoid (3). In the mature virion (at right) the two internal shells are fused together to give a dense nucleoid, smaller in the size than that of immature or signet-ring virions. From DE THÉ and O'CONNOR, 1966

The C particles are indistinguishable from the virions of the avian leukemia-sarcoma complex and in all likelihood they represent the etiological agents of mouse leukemia. C particles have also been observed budding from rat or mouse sarcoma cells induced by MuSV *in vivo* or *in vitro* (DALTON, 1966; LECLERC, SILVESTRE, LÉVY, OPPENHEIM, and VARET, 1967; MITCHINER, 1967). These have been assumed to represent MuSV, but this question cannot be settled until MuSV has been isolated from its associated MuLV (see below!).

Type A particles have been a source of confusion. Different definitions have been used and a subdivision into A1, A2 and A3 particles was introduced, where the A1 particles seemed to be identical to the C type (cf. ANONYMOUS, 1966 and HALL, HARTLEY, and SANFORD, 1968). In this review the A particle is defined (ANONYMOUS, 1966) as an intracytoplasmatic particle usually found in the expanded cisternae of the endoplasmic reticulum but sometimes also in the cytoplasmic matrix. The round particles have a diameter of about 70 mµ and an electron lucent center of about 30 mµ. Sometimes, especially in particles budding from endoplasmic reticulum, a unit membrane can be seen along the outer surface.

The role of the A particle, which has been found in many leukemias (HEINE, GRAFFI, HELMCKE, and RANDT, 1957; DE HARVEN and FRIEND, 1958, 1960; DALTON, LAW, MOLONEY, and MANAKER, 1961), has not been understood. HALL, HARTLEY, and SANFORD (1968) made clones of established spontaneously transformed murine cell lines carrying C and A particles probably as passengers. Only C particles could be passed onto Swiss mouse embryo cell cultures and it was concluded that the A particle is neither infective nor developmentally associated with the C particle. It remains to be proved that the A particle is indeed a virus and its relation to mouse neoplasia must be regarded questionable.

The chemical composition of MuLV purified C particles is probably very similar to that of avian leukemia-sarcoma viruses. The quantitate overall chemical composition of MuLV does not seem to have been reported. The particles have a buoyant density of about 1.16 g/ml (O'CONNOR, RAUSCHER, and ZEIGEL, 1964; OROZLAN, JOHNS, and RICH, 1965; DUESBERG and ROBINSON, 1966).

The envelope contains lipid as attested by its sensitivity to ether; 75—80% of the lipid of RLV or MLV is phospholipid (largely lecithin). Minor differences in the proportion of different fatty acids of RLV and MLV preparations are probably attributable to differences in the lipids of the cell membranes incorporated in the viral envelope during maturation (JOHNSON and MORA, 1967).

The outer coats of RLV and MLV contain ATPase when the viruses are isolated from mouse plasma. The amount did not seem to be as high as in avian myeloblastosis virus and a proportion of the particles were negative (DE THÉ, 1966a, b). The origin of the viral enzymatic activity is unclear since megakaryocytes, generally believed to be a major site for virus synthesis (DALTON, 1962), do not show any ATPase in the cytoplasmic channels where virus budding takes place. Virus releasing leukemic thymus and lymph node cells were likewise free of membrane-associated ATPase demonstrable by ultramicroscopic histochemistry (DE THÉ, 1966b). It is possible that the enzyme positive particles had been formed by organs or cells not studied. DE THÉ (1966a) also described alkaline phosphatase in the envelope of MLV a finding ascribed to the incorporation of enzyme during budding from the leukemic lymphoblasts, whose membranes are positive for the enzyme.

The RNA of RLV and GLV contains heavy ($\sim$70S, mol. weight 10×10^6) and light ($\sim$4S) fractions after density gradient centrifugation (GALIBERT, BERNARD, CHENAILLE, and BOIRON, 1965; MORA, McFARLAND, and LUBORSKY, 1966; BALUDA, 1966; ROBINSON and DUESBERG, 1966; BASES and KING, 1967). The RNA is easily degraded during preparation (MORA and McFARLAND, 1965) and the light fraction has been interpreted as a degradation artifact. This interpretation has been questioned by BLAIR and DUESBERG (1968), who also obtained a 36S component after melting of the fast sedimenting RNA. In analogy with similar findings on RSV RNA (DUESBERG, 1968) this was interpreted as an indication that the 70S RNA is not a single large polynucleotide but rather a specific aggregate of smaller subunits held together by weak physical bonds. Melting at various ionic strengths and the response to RNase indicate that at least large segments of the parent molecule are single stranded (cf. ROBINSON and

DUESBERG, 1968). No biological activity has been found associated with MuLV RNA. The nucleotide composition of MuLV RNA does not differ from that of oncogenic avian RNA viruses (Table 10).

Little is known about the proteins of MuLV, but one report (DUESBERG and ROBINSON, 1966) has shown that RLV contains at least two major protein fractions by polyacrylamide gel electrophoresis.

3. Immunological Properties of Mouse Leukemia Viruses

A detailed account is beyond the scope of this monograph. Many virus strains (*e.g.* GLV) are at most only weakly antigenic and the almost ubiquitous occurrence in mice of GLV and related strains with frequent vertical spread (GROSS, 1951 b) and consequent tolerance has made it hard or impossible to produce murine antisera (cf. BOIRON, LÉVY, and PÉRIÈS, 1967). Heterologous antisera, particularly from rats and rabbits, have been used but have often had low titers. Another difficulty shared by all viruses which incorporate cell membrane in their envelope is the presence of antigens of cellular origin in viral antigen preparations.

The viral antigens may be divided into a) group specific (gs) and b) type or strain specific. A similar localization of antigens in the virions of the murine and avian leukemia-sarcoma complexes with a group specific complement fixing antigen in the interior and strain specific antigens reactive with neutralizing antibody in the envelope has been conjectured.

a) The group-specific (gs) antigen was demonstrated by GEERING, OLD, and BOYSE (1966). These authors produced hyperimmune sera directed against GLV induced rat lymphomas transplanted to adult syngeneic hosts, which resisted growth of the tumor. The serum reacted with ether disrupted GLV, FLV, MLV, RLV or Gross lymphoma extracts from which intact GLV had been removed by centrifugation. A reaction of identity was obtained by immunodiffusion against these 5 antigens. It was concluded that there was a common determinant located in the virus particles. These results were confirmed in complement fixation tests by HUEBNER and associates (cf. HUEBNER, 1967) with MuSV rat antiserum. Previously, HARTLEY, ROWE, CAPPS, and HUEBNER (1965) had obtained sera from rats carrying a transplanted (virus-yielding) RLV induced lymphoma which reacted in complement-fixation tests with mouse embryo tissue culture cells infected with RLV, MLV, FLV (but not GLV) and with extracts of tumors produced in mice or rats with these viruses and GLV. The CF test was applied as the first *in vitro* assay for the growth of the non-cytopathic MuLV but the nature of the antigen(s) was not clarified. A potent CF serum against gs-antigen has been produced in rabbits by injection of degraded RLV (SCHÄFER and SEIFERT, 1968).

Purified gs-antigen from FLV had properties of a basic protein physicochemically (but not immunologically) very similar to the avian gs-antigen (SCHÄFER, ANDERER, BAUER, and PISTER, 1969).

b) The type-specific envelope antigens are still very imperfectly known, mainly because of methodological difficulties.

Tests employing neutralization, specific cytotoxicity and the immunofluorescence of untreated virions, which theoretically should measure antibodies reacting with surface antigens, have been hard to interpret. Most antisera have been produced in rodents which are susceptible to the oncogenic effect of the virus. The antibodies are therefore not only directed against viral surface antigens but also against the tumor cell specific antigens which must be suspected to arise in the course of the immunization (G. KLEIN and E. KLEIN, 1964).

Mouse and rat antisera obtained after infection by live virus or virus releasing lymphoma cells react with homologous lymphoma cells in cytotoxic and FA tests (SLETTENMARK and KLEIN, 1962; FINK and MALMGREN, 1963; E. KLEIN and G. KLEIN, 1964; review by OLD and BOYSE, 1964).

Cytotoxic sera against the Friend, Moloney or Rauscher agents (the FMR group) reacted against lymphoma cells induced by any of the three viruses but not against Gross lymphoma cells. Cytotoxic activity could be absorbed from a rat antiserum against GLV by Gross lymphoma cells but not by FMR cells. From these observations it was concluded that the antigens associated with leukemias induced by the FMR group are closely related and distinct from those associated with GLV (OLD, BOYSE, and LILLY, 1963; GEERING, OLD, and BOYSE, 1966). The distinction between FMR and Gross' lymphoma cells has also been established by transplant rejection (G. KLEIN, SJÖGREN, and E. KLEIN, 1962; GLYNN, McCoy, and FEFER, 1968) and other techniques as reviewed by WAHREN (1966). Since all lymphoma cells release virus, it has not been possible to ascertain if the FMR cross reactivity reflects a common induced cell antigen or a common viral surface antigen. The absence of a correlation between degree of virus release and immunogenicity of lymphoma cells has suggested the former alternative (G. KLEIN and E. KLEIN, 1964).

Sera from rats immunized with different MuLV contain cross-reacting neutralizing antibodies — in addition to group-specific CF antibody. The degree and kind of cross-reactivity has not been consistent between different laboratories (cf. MOLONEY, 1962; OLD and BOYSE, 1964; GROSS, 1966) but the titer is usually highest against the strain used for immunization. Selected sera can be employed in the serological typing of MuLV grown in cell culture on the basis of differences in the neutralization titers against various strains (HUEBNER, HARTLEY, ROWE, LANE, and CAPPS, 1966; HUEBNER, 1967; IGEL, HUEBNER, DEPPA, and BUMGARNER, 1967).

4. Genetic Susceptibility to Murine Leukemia Viruses

Friend leukemia virus differs in its capacity to induce disease depending on the genetic host background. C57B1 mice are for instance relatively resistant (FRIEND, 1957a; FIELDSTEEL, DAWSON, and BOSTICK, 1961). Susceptibility of other types of MuLV also seems to be a hereditary character (MOLONEY, 1960; RAUSCHER, 1962; LILLY, BOYSE, and OLD, 1964). ODAKA (1969) managed to introduce genetic susceptibility to FLV into resistant C57B1 mice and produced a pair of congenic lines which only differed with respect to single autosomal susceptibility gene. Splenomegalic erythroblastosis was used as an indicator of susceptibility and it was not determined whether virus synthesis *per se* was affected. MEIER, MYERS, and HUEBNER (1969) carried this type of experiment a step further by demonstrating that tumorigenesis rather than virus synthesis is genetically controlled. They compared the incidence of "spontaneous" MuLV associated leukemia and the frequency of "spontaneous" virus infection and found that genetically resistant mice only developed neoplasms in low frequency in spite of sustaining viral synthesis apparently as efficiently as genetically susceptible congenic animals which developed leukemia in high frequency.

These results resemble those obtained in the avian system by demonstrating segregating single autosomal susceptibility genes. In the avian system viral synthesis seems to be effected in parallel with transformation or tumor formation (page 122).

5. Murine Leukemia Viruses in Cell Culture

Studies on the multiplication of MuLV are greatly hampered by the absence of any easy way of titrating virus. Time- and animal consuming bioassays with large statistical errors have been employed and several conflicting reports have appeared. Assays based on FA (YOSHIDA, SMITH, and PINKEL, 1966), CF (HART-LEY, ROWE, CAPPS, and HUEBNER, 1965), viral interference (SARMA, CHEONG, HARTLEY, and HUEBNER, 1967), syncytium formation in the RSV rat tumor XC line after cocultivation with mouse embryo cells carrying MuLV (KLEMENT, ROWE, HARTLEY, and PUGH, 1969), induction of specific surface antigens (NOR-DENSKJÖLD, E. KLEIN, TACHIBANA, and FENYÖ, 1970) or the facilitation of focus formation by defective MuSV (FISCHINGER and O'CONNOR, 1968) have begun to alter this picture, but no quantitative routine method has been worked out yet.

Gross leukemia virus. Survival and possibly also synthesis of GLV *in vitro* was reported by GROSS, DREYFUSS, and MOORE (1961) in mouse embryo cells observed during 32 days. IOACHIM, BERWICK, and FURTH (1966) established continuous lines from explanted Gross rat lymphomas which released lymphoma-genic virus during at least 2—3 years. One line was composed of lymphoblasts growing in symbiosis with reticulum- or fibroblast-like cells; the other almost entirely consisted of lymphoblasts. These two growth patterns resemble the two types seen in spontaneous lymphoblastoid transformation of human cells (page 31). IOACHIM and BERWICK (1968) infected rat embryo thymus, spleen and kidney cultures with a cell-free filtrate of GLV. When first tested after 65 days infected thymus released GLV detected in a mouse bioassay. Spleen and kidney remained negative throughout the entire experiment, whereas thymus synthesized virus continuously. Slowly after 2—3 months the infected cultures changed in morphology to a wavy growth pattern with tight packing and piling up of cells. The proportion of epithelial-like cells increased and atypical nuclei were noted. These changes represented an accentuation of alterations also seen in the con-trols. Infected syngeneic thymus cells inoculated intraperitoneally or under the kidney capsule of adult immunodepressed hosts gave intraabdominal reticulum cell sarcomas with a latent time of about 5—6 months. The authors presumed that this neoplastic transformation was induced by GLV. In view of the high frequency of spontaneous infinite growth transformation and the correlation of this phenomenon with tumorigenicity in rat thymus (IOACHIM and FURTH, 1964) as well as other tissues (page 19), it is, however, equally likely that the *in vitro* transformation was spontaneous and that GLV acted as a selective agent favoring growth of cells with an enhanced tumorigenicity.

Moloney leukemia virus. Early reports described persistence or synthesis of MLV in mouse spleen (MANAKER, STROTHER, MILLER, and PICZAK, 1960), mixed embryo or kidney cultures (GINSBURG and SACHS, 1961a, b, 1962; SALAMAN, ROWSON, and HARVEY, 1962) for periods up to a year. Later MANAKER, JENSEN, and KOROL (1964) infected established BALB/C mouse spleen cells with MLV and obtained a line — MT77 — which released leukemogenic virus for an observation period of 4 years. The MT77 line has been grown in suspension with release of virus (TOPLIN, RICCARDO, and JENSEN, 1965). Mixed primary or secondary mouse embryo cells synthesize MLV as indicated by FA tests and appearance of CF antigen (HARTLEY, ROWE, CAPPS, and HUEBNER, 1965). The process is not accompanied by any transformation or CPE *in vitro*. The tumorigenicity of infected cells does not seem to have been tested.

Rauscher leukemia virus. An established line (JLS-V6) from thymus and spleen of normal BALB/C mice could support growth of RLV during long periods without any cytopathic effects (WRIGHT and LASFARGUES, 1965), as did lines of Swiss mouse embryo (SINKOVICS, BERTIN, and HOWE, 1966) or rat kidney origin (DUC-NGUYEN, ROSENBLUM, and ZEIGEL, 1966).

TYNDALL, VIDRINE, TEETER, UPTON, HARRIS, and FINK (1965) and TYNDALL, TEETER, OTTEN, BOWLES, VIDRINE, UPTON, and WALBURG (1966) observed CPE ascribed to RLV in an established line — JLS-V5 — originated from spleen and thymus of normal weanling BALB/C mice. Exposed cultures showed clumped cells with cytoplasmic eosinophilia and irregular nuclear staining. Later fibroblast-like cells appeared in a certain criss-cross pattern. The sequence passed over into overt cytopathogenicity with rounded, degenerating cells. The stage with criss-cross growth was referred to as a "transformation", but without convincing documentation. The alterations resemble effects by PPLO and not until the possibility of accidental contamination has been rigorously excluded can any CPE or transformation by RLV be accepted. Infected "transformed" cultures were somewhat more tumorigenic than the uninfected control line, which had undergone spontaneous infinite growth transformation. Increased tumorigenicity is compatible with a PPLO infection which may select for tumorigenic cell variants (RUSSEL, NIVEN, and BERMAN, 1968). Other studies with the JLS-V5 line have not confirmed the "transformation" or CPE (BERNARD, SILVESTRE, TANZER, PÉRIÈS, and BOIRON, 1965).

BARBIERI and BARSKI (1968) infected a tumorigenic established murine line of C57B1 origin with RLV. A chronic infection ensued demonstrated by immunofluorescence or release of C particles. The RLV exposed cultures showed a striking decrease in tumorigenicity.

RLV synthesis in established carrier cultures was inhibited by actinomycin D but unaffected by the DNA synthesis inhibitors cytosine arabinoside and 5-fluorodeoxyuridine (BASES and KING, 1967).

FREEMAN, PRICE, IGEL, YOUNG, MARYAK, and HUEBNER (1970) could transform (irreg. + unrestrained growth) rat embryo cultures with RLV and a chemical carcinogen; virus or carcinogen alone was, however, ineffective. Hamster embryo cultures were morphologically altered after exposure to RLV (RHIM, HUEBNER, and TING, 1969).

Friend virus. A continuous virus shedding line has been obtained from histo-

logical reticulum-cell sarcomas explanted in culture (FRIEND, PATULEIA, and DE
HARVEN, 1966).

Early attempts to cultivate FLV in mouse cells not already infected *in vivo*
were only met with temporary success and no consistent serial passage of the
virus was obtained by MOORE and FRIEND (1958) and MOORE (1963) in spite of
extensive trials. Short term virus release possibly reflecting virus synthesis was
reported by VIGIER and GOLDÉ (1962), OSATO, MIRAND, and GRACE (1964) and
CHAMORRO, LATARJET, VIGIER, and ZAJDELA (1962) without any cytopathic or
other morphological cell alterations. Subsequently YOSHIKURA, HIROKAWA,
YAMADA, and SUGANO (1967), reported an established mouse line which could
be infected *in vitro* and support continuous synthesis of FLV.

Kaplan virus. This agent, originally isolated from a radiation induced C57B1
lymphoma (LIEBERMAN and KAPLAN, 1959), has not been grown effectively as
an independent agent *in vitro* (FISCHINGER and O'CONNOR, 1969b). Only in the
presence of MuSV has a restricted synthesis been reported (IGEL, HUEBNER, DEPPA,
and BUMGARNER, 1967).

Other murine leukemia viruses. The Graffi agent and a few other less well
defined MuLV have been propagated *in vitro* as reviewed by BOIRON, LÉVY, and
PÉRIÈS, 1967).

In summary. All MuLV, except KLV, including a number of fresh isolates
multiply in mouse embryo cultures, when assayed by the CF technique (HUEBNER,
1967).

Long term carrier cultures have been established from spontaneously transformed murine cells infected with GLV, MLV, RLV and FLV. Prolonged passage
of RLV has given attenuated virus strains with reduced leukemogenic potency
but unaltered immunogenicity (WRIGHT and LASFARGUES, 1966; BARSKI and
YOUN, 1966; SINKOVICS, BERTIN, and HOWE, 1966). The other viruses have
retained their leukemogenic activity in bioassays over a number of years. No
conclusive evidence of any specific CPE or transformation has been presented.

6. Murine Sarcoma Viruses in Cell Culture

A new field in the *in vitro* study of murine leukemia-sarcoma viruses was
opened by the demonstration of HARTLEY and ROWE (1966) that extracts from
tumors induced by MuSV-M induced focal changes in mouse cell monolayers. The
altered areas seen already after 3—5 days were composed of a mixture of basophilic round and polygonal cells of an appearance resembling the corresponding
Rous elements (page 130). The cells piled up on top of normal cells (Fig. 33). This
alteration of mouse cells by MuSV-M was subsequently confirmed (SIMONS,
DOURMASHKIN, TURANO, PHILLIPS, and CHESTERMAN, 1967). Rat cells were found
to be susceptible to conversion by MuSV-H (TING, 1966) and MuSV-M (BERNARD,
BOIRON, and LASNERET, 1967). The former virus also caused focus formation in
hamster embryo cells (SIMONS, BASSIN, and HARVEY, 1967) with a predominance
of basophilic spindle cells.

If virus spread is prevented by addition of agar or neutralizing serum, no
or only few foci appear indicating that foci are caused by virus spread with
subsequent cytological alteration rather than by cell proliferation (HARTLEY and

Rowe, 1966). The inability of mouse cells infected by MuSV-M to form foci if virus spread was prevented was also noted by O'Connor and Fischinger (1968), Yoshikura, Hirokawa, Ikawa, and Sugano (1968) and Bernard, Guillemain, Périès, and Boiron (1968). Bather, Leonard, and Yang (1968) plated MuSV-M altered cells onto irradiated feeder cells and obtained no foci indicating that virus spread to healthy cells is essential for focus formation. The proliferation of altered cells was probably not inhibited by the irradiated normal cells because

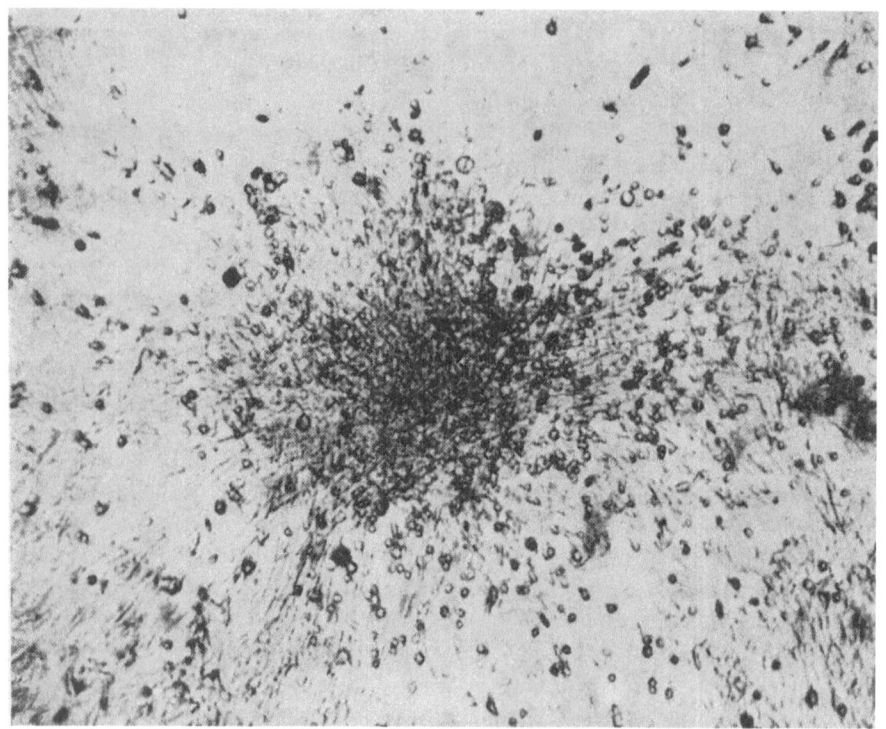

Fig. 33. Focus of altered cells induced in Swiss mouse embryo tissue culture by Moloney sarcoma virus (MSV) 10 days post infection. Appr. magn. × 100. From Hartley and Rowe, 1966

no foci were found when altered cells were seeded onto non-susceptible living chick embryo cells. This results points to a possible difference between mouse cells infected by MuSV and chicken cells transformed by RSV, since the latter will grow on top of normal chick embryo cells (R. Weiss, 1969 c). Focus formation by MuSV-H in hamster cells was also suggested to require virus synthesis (Simons, Bassin, and Harvey, 1967).

This focus formation by MuSV has often been termed "transformation". This practise is not followed here, since it has never been convincingly shown that MuSV induces any of the growth control disturbances defined as transformations in Table 1. This may not only be a matter of semantics. It should be recalled that a majority of the "sarcomas" induced by MuSV-M in mice were "atypical granulomas" often with prominent necrosis. It is possible that the focus formation by MuSV *in vitro* essentially reflects a cytopathic effect rather than a growth control disturbance. PARKMAN, LEVY, and TING (1970) have pointed out that the interaction with rat cells may be different. Neutralizing anti-viral serum has no effect on the enlargement of rat cell foci and continuous lines of MuSV altered rat cells are easy to obtain. Probably, therefore, MuSV stocks are capable of inducing irregular, unrestrained and infinite growth transformation (Table 1) in rat but not in mouse cells. An obvious explanation could be the excessive synthesis of the murine leukemia (helper) component of stock MuSV, which is typical for the former but not the latter species.

A recent preliminary report (THOMAS, BOIRON, STOYTCHKOV, and LASNERET, 1968) has demonstrated replication of MuSV-M in bovine cells. The synthesized virus induced foci in mouse embryo cells.

7. Interactions between Murine Leukemia and Sarcoma Viruses

The interrelations between MuLV and MuSV have not been as thoroughly investigated as the corresponding phenomena in the avian leukemia-sarcoma complex. The prevailing hypothesis is, however, that the two complexes are principally similar in that sarcomagenic members may be "defective" and leukemogenic viruses act as helpers. As with RSV the term "defective" does not mean absence of virus synthesis but rather that production of infectious particles containing sarcomagenic genomes may be "helped" by concomitant infection with a leukemia virus. The latter directs the synthesis of a surplus of an essential coat protein which increases the titer of MuSV assayed by focus formation. The above picture is similar to that of the avian system but it may be oversimplified because there is now evidence of even more complex interactions.

Evidence for the existence of "defective" MuSV-M particles was first presented by HARTLEY and ROWE (1966), who studied the relation between virus dilution and number of foci either in an established BALB/C mouse line or secondary cultures of Swiss mouse embryos. With most MuSV preparations the number of foci was inversely proportional to the dilution at low dilution. At high dilution the number of foci decreased with the square of the dilution as a "two hit" response. A direct "one hit" proportionality could be obtained over the entire dilution range if the MuSV-M preparation was mixed with a large excess of MLV or RLV. The original MuSV-M stock was shown to contain non-focus forming MLV in 1000-fold excess over focus-forming MuSV by terminal dilution. It was postulated that focus formation required dual infection by MuSV-M and MLV and that at a low dilution this is spontaneously achieved by the presence of MLV in the stock. At high dilution addition of extra MLV can provide effective dual infection. Since focus formation required virus synthesis it cannot be determined if MuSV is "defective" with respect to virus synthesis, the morphological alteration underlying focus formation or both. The difference between a RSV and MuSV focus may be fundamental. The former does not require virus synthesis with reinfection of cells and can be solely caused by cell proliferation, whereas the latter is only observed in connection with synthesis and spread of virus. It is built up of cells apparently incapable of multiplication.

Dose-response curves compatible with Hartley and Rowe's hypothesis were subsequently obtained by BASSIN, SIMONS, CHESTERMAN, and HARVEY (1968),

BERNARD, GUILLEMAIN, PÉRIÈS, and BOIRON (1968) and YOSHIKURA, HIROKAWA, IKAWA, and SUGANO (1968). The latter authors obtained the helper effect with FLV.

O'CONNOR and FISCHINGER (1968) and FISCHINGER and O'CONNOR (1968) analysed the interaction between MuSV and helper MuLV. The theoretical relationships between stock virus dilution and the numbers of foci were calculated for virus preparations with different ratios of infective MuLV/MuSV units. The shape of such curves depends a) on the MuLV/MuSV ratio and b) whether all or only part of the MuSV particles are defective with respect to ability to incite focus formation. If there is a very high MuLV/MuSV ratio and if all MuSV virions are defective, focus formation follows a one hit curve over the entire range of dilutions. A titration curve of a virus stock with a lower MuLV/ MuSV ratio will be linear until a dilution point where a proportion of the cells infected by defective MuSV are not any longer also infected by MuLV. At this point the plot will abruptly change to a two hit curve where the number of foci is inversely proportional to the square of the virus dilution; the location of this point along the dose axis depends on the original MuLV/MuSV ratio. O'CONNOR and FISCHINGER (1968) found that many virus harvests deviated significantly from the pattern predicted for 100% defective MuSV virions. The dose-response curves did not show any abrupt change in slope and were concave upwards countrary to the theoretical predictions. This behavior could be explained if it was assumed that a proportion of the MuSV particles were non-defective, *i.e.* capable of focus induction in the absence of dual infection with MuLV. Virus harvested daily from 3 T 3 cells 2—8 days post-infection showed titration patterns indicating a progressive increase in the proportion of non-defective MuSV. On day 2 almost all MuSV seemed to be defective in contrast to the harvest from day 8 which — according to computed model curves — contained at least 70% non-defective MuSV. A virus stock from mouse cells infected by a MuSV preparation may thus contain at least three different particles — helper MuLV, competent MuSV and defective MuSV. The existence of competent MuSV has also been inferred from the shape of dose-response curves by BERNARD, GUILLEMAIN, PÉRIÈS, and BOIRON (1968). MuSV-M synthesized by bovine (THOMAS, BOIRON, STOYTCHKOV, and LASNERET, 1968) or rat (PARKMAN, LEVY, and TING, 1970) cells may be non-defective according to titration patterns.

Competent and defective MuSV-M can be physically separated by differential centrifugation, sucrose gradient centrifugation or differential ultrafiltration. Competent MuSV then behaves as an interviral aggregate between MuSV and MuLV. The adherence is not accomplished by simple mixing but sedimentation at high speed or freezing in low molar citrate solution is effective. These results strongly suggest that genuinely competent single MuSV-M virions may not be formed, but that competence is a reflection of aggregation with suitable helper viruses (O'CONNOR and FISCHINGER, 1969).

The aggregating tendency of RNA tumor viruses was elegantly utilized by FISCHINGER and O'CONNOR (1969a) who rendered MuSV infectious for cat embryo cells. MuSV and feline leukemia virus (FelLV) were spun down together under conditions similar to those giving "competent" MuSV-MuLV aggregates. These preparations gave foci on cat embryo cells morphologically resembling those induced by MuSV on mouse cells. On passage the virus stock became exclusively infective for cat cells. Preinfection by FelLV induced resistance against challenge with aggregate MuSV-FelLV. The authors suggest that modifications of this technique may provide a delivery system for various genetic messages into receptive cells.

MuLV can be assayed by its helper effect with defective MuSV if the MuLV/MuSV ratio and the proportion of defective MuSV are known in a standard MuSV preparation (FISCHINGER and O'CONNOR, 1968). This *in vitro* assay is currently the fastest, most sensitive and accurate titration method for MuLV and it seems to be generally applicable to all strains. Its main drawback is that it requires a very careful selection of the reagents and that the quantitation of MuSV foci is fraught with difficulty (BATHER, LEONARD, and YANG, 1968).

HARVEY and EAST (1969) separated a leukemogenic entity, probably MLV, from their stock of MuSV-M by vertical transmission from three MuSV-M infected female BALB/C mice. Six successive generations of mice developed lymphatic leukemia of the type commonly associated with MLV but showed no evidence of the "early" sarcomatous lesions and the erythroblastotic splenomegaly caused by inoculation of crude MuSV-M (CHESTERMAN, HARVEY, DOURMASHKIN, and SALAMAN, 1966; HARVEY, 1968; EAST and HARVEY, 1968).

FISCHINGER, MOORE, and O'CONNOR (1969) encountered some anomalous and still not completely understood facts in their attempts to separate "helper" MLV from stock MuSV-M *in vitro*. 3T3 cells were infected by stock MuSV-MuLV and a cell clone "the 1C line" was isolated. It showed no evidence of a sarcomagenic genome and maintained the high degree of density dependent cell cycle inhibition characteristic of the 3T3 line. The 1C line produced "C particles" with a high degree of helper activity with respect to focus formation by incompetent MuSV. The virus was specifically neutralized by anti-MLV serum and induced lymphatic leukemia and other lesions in NIH Swiss mice of the same type as those seen after infection by Moloney leukemia virus. All evidence pointed to the separation of MLV from the MuSV stock. MLV seems to grow slowly *in vivo* and cannot alone be maintained by rapid serial passage. Only if it is part of the MuSV-MuLV complex is rapid propagation possible, which suggests that the leukemogenic virus is "helped" by the presence of sarcomagenic genomes. This is the reverse of the commonly observed effect *in vitro* where MLV acts as a helper virus. The mechanism and importance of this functional and apparently very complicated synergism between two murine tumor RNA viruses have not yet been elucidated.

Further support for the hypothesis of defective MuSV was presented by HUEBNER, HARTLEY, ROWE, LANE, and CAPPS (1966). They explanted a hamster tumor induced by MuSV-M *in vitro* and obtained a line, which did not yield any virus infective for mouse cells *in vitro*, newborn hamsters or mice. The cells were also negative for CF ability when tested against broadly reactive antisera from rats immunized with MLV or MuSV-M. Cocultivation of these cells with mouse embryo cells in the presence of MLV, RLV or FLV yielded large amounts of focus forming MuSV. Preliminary tests with selected neutralizing rat antisera suggested that the MuSV focus forming particles had the same serological properties as the MuLV used for the superinfection. The data suggest that the MuSV genome was wrapped up in a envelope provided by the MuLV. Hamster sarcoma cells induced by MuSV-H released virus infective for hamsters but not for mice. Superinfection with MLV caused synthesis of virus infective for mice suggesting that MuSV-H genomes were provided with coats synthesized under the direction of MLV acting as a helper (BASSIN, SIMONS, CHESTERMAN, and HARVEY, 1968).

IGEL, HUEBNER, DEPPA, and BUMGARNER (1967) studied the apparent helper effect of KLV, a virus originally isolated from radiation-induced lymphomas in C57Bl mice (LIEBERMAN and KAPLAN, 1959). KLV does not grow in mouse embryo cells when assayed by the CF technique of HARTLEY, ROWE, CAPPS, and HUEBNER (1965). Nonvirus producing hamster cells from a tumor induced by MuSV-M were mixed with C57Bl cells infected with KLV and incubated *in vitro* for 20 days. The cells were minced and then passaged four times with the addition of more KLV at the first passage. Cell-free extracts were finally obtained, which were highly oncogenic in mice and also produced foci in mouse embryo cultures. These results are compatible with rescue of defective MuSV genome by addition of helper KLV.

FISCHINGER and O'CONNOR (1969b) confirmed the inability of KLV to replicate in mouse embryo cells, but it could nevertheless be quantitatively assayed by its capacity to enhance focus formation by defective MuSV-M. In the latter case KLV was apparently synthesized — a finding supporting the hypothesis that non-replicating leukemia viruses may exist, which need a synergistic interaction with a sarcoma virus in order to undergo replication as infectious units. KLV is probably closer to a natural carcinogen than MLV and the hypothesis may therefore be of considerable biological significance for the understanding of "spontaneous" carcinogenesis.

The results reviewed are highly suggestive of similar interactions in the murine and the avian leukemia-sarcoma complexes including "defectiveness" of sarcomagenic viruses and helper effects of leukemogenic ones. It should be emphasized that this analogy has not been rigorously proven and alternative explanations are possible.

It is important to remember that no experiments have used mice checked as rigorously as chickens for being free of leukosis virus. The virus stocks have never been plaque-purified and there are thus numerous possibilities of inadvertent introduction of extraneous interfering viruses. Focus formation in mouse embryo cells, which is used to titrate stocks of MuSV, is possibly basically a cytopathic response and may not reflect sarcomagenic potency. It is influenced by the presence of MuLV, which in certain instances may be covertly present in the assay cultures. Other factors such as the age of the embryo at the time of preparation of cell cultures, the time at which the virus is added (BATHER, LEONARD, and YANG, 1968) and the cellular genotype (HUEBNER, 1967) also have a pronounced influence.

The only stringent test of an infection analogous to the L⁻R situation in the chicken — the production of transformed cells *in vitro* which still carry the gs-antigen and the subsequent rescue of this marker by a defined helper virus has not been possible to perform in the murine system. Infection of mouse cells *in vitro* has invariably led to foci of apparently lytically affected cells which release infectious virus. The only possible L⁻R equivalents have been derived from tumors induced *in vivo* in rats (TING, 1967) or hamsters (HUEBNER, HARTLEY, ROWE, LANE, and CAPPS, 1966; BASSIN, SIMONS, CHESTERMAN, and HARVEY, 1968). In the hamster tumors of HUEBNER *et al.* there was no proof of persisting MuSV genome in the form of gs-antigen and the hamster tumors of BASSIN *et al.* released virus infective for hamsters. The rat tumors had some virus-specific antigenicity demonstrated in transplantation resistance test. Their epithelial histology was, however, clearly deviant and a pick-up of MuSV in a neoplasm not directly caused by MuSV cannot be excluded. Recent analysis has shown that the rat tumor cells — MSB-1 line — release type C particles detected in the electron microscope (VALENTINE and BADER, 1968) and also a virus infective for rat kidney cells. Several examples (Table 14) are on record where the virus released by rat or hamster cells transformed by MuSV has altered host range specificity. The virus is no longer infective for mouse cells but sarcoma-genic, infective and transforming for rat or hamster cells, respectively. No satis-factory explanation for this phenomenon has been provided.

At least 3 different patterns of virus release by tumors or altered cells may be discerned a) synthesis of parental type of virus *i.e.* the mixture of "helper" and sarcomagenic virus types characteristic of the particular MuSV used b) release of host range variants as exemplified in Table 14 and c) no synthesis of infective virus but persistence of a viral genome which can be rescued by manipulations involving addition of a helper leukemia virus (KLEMENT, HARTLEY, ROWE, and HUEBNER, 1969).

In conclusion, the evidence is strong that various members of the murine leukemia-sarcoma complex interact with each other in a complex pattern. The details of this are unknown but present data strongly favor a model similar to the avian leukemia-sarcoma complex. Future research will probably show that focus formation is related to true transformation and also to induction of neo-plasia *in vivo*. The general conditions and metabolic requirements for focus formation by MuSV including the necessity for DNA synthesis early after infection conform with those established for RSV (BATHER, LEONARD, and YANG, 1968; NAKATA and BADER, 1968). Infected cultures also reach higher densities and have a decreased requirement for serum in analogy with avian sarcoma virus effects (TEMIN, 1968 b, c).

Synopsis of Transformation by RNA Viruses

RNA tumor viruses are — in contrast to transforming DNA viruses — natural carcinogens. They cause mesenchymal neoplasms *i.e.* various types of sarcomas and leukemias but not carcinomas. They occur in birds, mice, cats and possibly also in cattle (cf. DUTCHER, 1965), humans (cf. DMOCHOWSKI, YUMOTO, GREY, HALES, LANGFORD, TAYLOR, FREIREICH, SHULLENBERGER, SHIVELY, and HOWE, 1967; MORTON, HALL, and MALMGREN, 1969) and other species (cf. TODARO and HUEBNER, 1969).

Their characteristic C-particle morphology with maturation by budding from the cell membrane is identical regardless of species or cell origin and type of neoplasia induced.

Extensive data are only at hand for the oncogenic avian and murine RNA viruses. For each species a "complex" of many closely related individual viruses exists, defined by common group-specific antigens located in the interior of the virions. This synopsis is primarily based on data from the avian sarcoma-leukemia virus complex but the evidence from the other "complexes" seem to fit well into the same picture.

The virus-cell interaction can be divided into *infection i.e.* delivery of the viral genome to the cell's interior, *viral synthesis i.e.* the reproduction of the genome and the formation of infective virions capable of undergoing new cycles of replication and *transformation*, which in the case of fibroblasts means a loss of controlled locomotion and replication.

The susceptibility to *infection* is primarily governed by cellular genes which determine the presence of cell membrane receptors which interact with specific viral coat proteins. Neither the physiological state nor the degree or type of cell differentiation seem to be important. The serologically distinguishable viral coat proteins determine the host range and the avian virus complex has been divided into four subgroups on the basis of neutralization tests, host range and interference patterns.

The capacity to induce and sustain *synthesis of infectious progeny* varies both with different viruses and target cells. Target cells of the species of origin or closely related species are always superior with regard to virus synthesis; cells of foreign species cannot usually reproduce infectious viral particles in spite of replicating viral genome. The cell membrane barrier to infection can be bypassed by cell fusion and other procedures. In such cases virus synthesis will be possible even in avian cells genetically resistant to infection from the outside by avian RNA tumor viruses. Large differences exist between individual members of both the avian and murine virus complex with respect to rate of production of infectious particles in their respective species of origin ranging from no definite evidence for production of infectious particles to highly efficient virus synthesis. Sarcomagenic particles often belong to the former category whereas leukemogenic viruses belong to the latter class.

Multiple infection of the same cell by different viruses is common. This often gives rise to synergistic relationships between the infecting particles. The classical example is provided by avian leukemia viruses which act as helpers for RSV by furnishing RSV genomes with coat proteins essential for broad range infectivity. A similar situation is found in the murine field. Most virus stocks and field isolates are mixtures of related virus strains whose coexistence is explained by synergistic relationships.

Infection and/or virus synthesis does not automatically lead to *transformation*. The viral genome has to match the histiotypic differentiation of the cell *i.e.* a particular type of target cells requires specific genomes for transformation. This requirement is unrelated to permissiveness with respect to virus synthesis. Fibroblasts of any species can be regarded examples of specific target cells for sarcomagenic RNA viruses. The viral genome can either be introduced into the susceptible cell by natural infection or by an artificial procedure circumventing the cell membrane barrier to infection.

The presence of a sarcomagenic genome in the appropriate target cell will presumably inevitably lead to a loss of normal growth control. The genome may in some way direct an alteration of the structure of the cell periphery to render it less susceptible to growth and locomotion control exerted by neighboring cells. A leukemogenic genome such as that of AMV introduced into a fibroblast does not visibly alter the cell morphology or interfere with its growth and locomotion control. If AMV, on the other hand, is introduced to cultures containing its specific target cells, myeloblasts, a rapid conversion takes place by which myeloblasts become completely predominant.

The viral genome may be present in different forms in transformed cells. Certain facts suggest that viral RNA or a DNA copy of it may be integrated in the cellular genome, but no rigorous proof exists. Other experiments are on record where apparently non-integrated viral genomes have been lost presumably after unequal distribution to daughter cells. The cells which had got rid of the viral genome regained normal proliferation and locomotion control. These results strongly indicate that integration into the cellular genome is not a requirement for transformation but for persistence of the viral genome. This question will probably not be settled until a method has been found which unequivocally identifies viral RNA chemically linked to cellular DNA or a specific integrated DNA base sequence which can be transcribed to viral RNA sequences.

Transformed fibroblasts usually contain the entire sarcomagenic genome. In most instances transforming and sarcomagenic C-particles are formed by a slow trickle but the particles may be non-infectious. Release of C-particles during long times or perhaps indefinitely is compatible with cell survival.

Unrestrained and irregular transformation by sarcomagenic RNA viruses are intimately coupled and accompanied by typical morphological cell alterations. They occur readily in stable and unstable species. In the former infinite growth transformation is not superimposed and the cells are therefore mortal. After a period of rapid proliferation with impaired control of locomotion and growth, the division rate declines in RSV transformed cells which eventually undergo a premature death, compared to uninfected controls. Established RSV transformed lines of for instance chicken or human origin have not been possible to obtain.

RSV transformed cells from unstable species readily form established lines probably reflecting their inherent tendency to spontaneous infinite growth transformation.

It seems likely that RNA tumor viruses are carcinogens which in a rather direct way influence growth control and cell differentiation but have no mutation-like effect on the chromosomal DNA. There is no convincing proof of any significant chromosome damage, genetic imbalance or unequal distribution of DNA caused by mitotic errors; three features associated with transformation by DNA viruses.

V. General Survey and Discussion

A. Characteristics of Transformed Cells

One of crucial steps in biological evolution must have been the transition from unicellular to multicellular life. Whereas protozoa compete to survive, metazoan cells will have to cooperate. Cell division is to be geared to the needs of the entire organism and kept in check even in the presence of an abundant food supply. In higher organisms part of this control is mediated by hormones and probably also by other diffusible substances such as tissue specific "chalones" (cf. BULLOUGH, 1969). These types seem, however, to be superimposed on a more direct local control dependent on cell proximity, which presumably is phylogenetically older and thus more primitive.

The proximity mediated control was greatly clarified when *in vitro* studies were initiated. Metazoan cells were first considered completely devoid of any growth control once they had been removed from the intact organism and put in culture. They were accordingly expected to behave as unicellular organisms *i.e.* to have an infinite potential to divide and even more importantly not to exert any restrictive influence on each others growth or locomotion. Failure to divide was simply interpreted as due to inproper medium conditions.

Further studies have shown that this is not the whole truth. It has gradually been realized that cells from higher organisms retain some form of growth control in culture and the following three types have been singled out in this review:

a) a finite potential for cell proliferation,

b) a proliferation restraint correlated with local cell density and

c) a paralysis of plasma membrane movement triggered by cell to cell contact.

Permanent malfunction of any of these is defined as a transformation regardless of whether it involves a genetic or epigenetic change.

The following terms were introduced:

a) infinite growth transformation,

b) unrestrained growth transformation, and

c) irregular growth transformation, for lack of these three control mechanisms, respectively. Unrestrained growth transformation (syn. lack of contact inhibition of mitosis, lack of cell cycle inhibition, lack of density dependent inhibition of proliferation) has been most extensively studied, no doubt because it lends itself to quantitative experiments. Its presence is established by continued high DNA synthesis and cell division at a cell concentration where confluent control cells would not enter the cell cycle to any significant degree. The saturation (terminal) cell density has often been used as a quantitative expression of the degree of cell cycle inhibition.

The establishment by Todaro's group of allogeneic and syngeneic murine cell lines with a high degree of contact inhibition of mitosis (AARONSON and TODARO, 1968a) and the induction in such lines of unrestrained growth transformation by PV, SV40 and MuSV (AARONSON and TODARO, 1968a, b; JAINCHILL and TODARO, 1970; AARONSON, HARTLEY, and TODARO, 1969) has permitted a correlation between saturation density and tumorigenicity. Figure 34 shows that populations with high saturation densities are strongly tumorigenic whereas cells which stop growing at low concentrations are only weakly or not all tumorigenic. It has been concluded that the tissue culture property that correlates best with tumorigenicity is the loss of contact inhibition of cell division (AARONSON and TODARO, 1968b).

Density- dependent proliferation control is present when cells stop multiplying at a certain local concentration. The triggering mechanism for the proliferation block has not been elucidated. Three hypotheses may be formulated:

a) cells secrete a locally active inhibitory factor,

b) membrane contact is responsible *per se*, or

c) the medium is exhausted with respect to specific growth factor(s).

The assumption of a *secretion of inhibitory molecules* is theoretically unattractive. It calls for a mechanism which prevents autoinhibition of the cell by its own secretion but still permits inhibition when the same substance is produced by a neighboring cell. Except for a few unconfirmed articles (BÜRK, 1966, 1967), most investigators have also reported negative results when extracts of inhibited cells or "spent" medium have been added to growing populations (cf. SCHUTZ and MORA, 1968b). Another type of negative evidence is provided by experiments with so called microexudates. It has long been known that cells in culture shed macromolecules, which precipitate on the glass or plastic cell support (MOSCONA

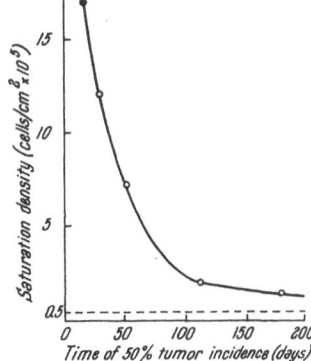

Fig. 34. Relationship between saturation density and tumor-forming ability. Unfilled symbols denote established mouse lines selected for different saturation densities; the filled symbol indicates an SV 40 transformed line. From AARONSON and TODARO, 1968b

and MOSCONA, 1952). The microexudate greatly facilitates growth and lack of it is probably partly responsible for poor growth of sparsely seeded cells (STOKER and SUSSMAN, 1965; REIN and RUBIN, 1968). No inhibitory effect of a microexudate has been described, which would be expected if cells secreted inhibitory substances. If extracellular inhibitors exist they must then be extremely labile and destroyed in the preparation of cell extracts and "conditioned" surfaces.

Can contact per se induce a block of cell division? One difficulty of this theory is of geometrical nature. Cultivated cells grow as thin disks. The bottom and the upper cell surface together comprise close to 100% of the entire plasma envelope and the contact making lateral borders only contribute a small part of the total surface. Even if the cell is surrounded by other cells laterally, its upper surface is still "free" and it is almost impossible to conceive of a mechanism which could sense the small increment in "occupied" surface provided by lateral contact if all parts of the cell surface were equally sensitive to contact. To make the hypothesis tenable it has to be assumed that the lateral part of the plasma membrane is specifically provided with the sensing mechanism and that both the bottom and upper surfaces are inert in this respect. Anatomical evidence indicates that different parts of the surface of attached cells in culture indeed perform different functions. Extensive time-lapse filming of normal glia-like cells (PONTÉN, WESTERMARK, and HUGOSSON, 1969) and fibroblasts (ABERCROMBIE and AMBROSE, 1958)

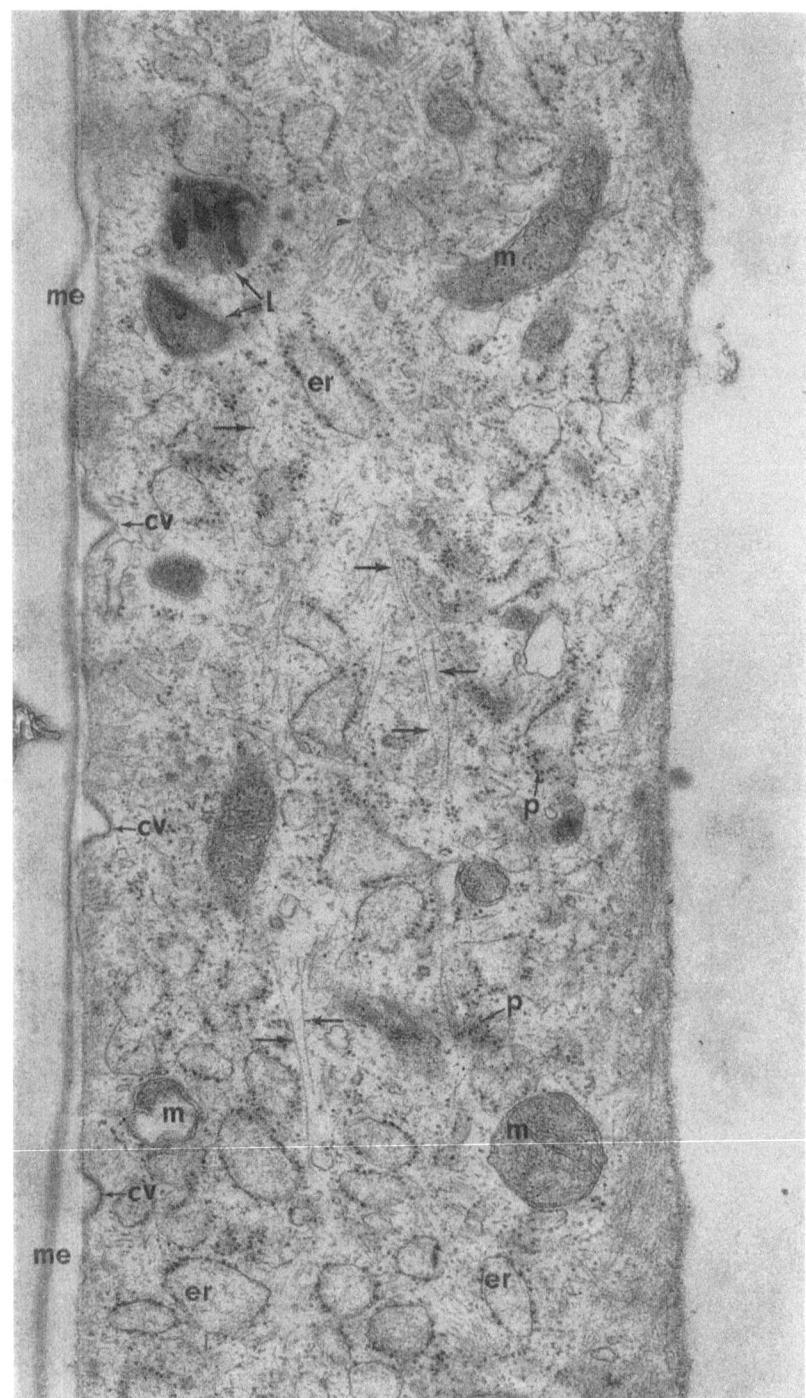

Fig. 35

has shown that ruffled membranes are exclusively confined to lateral cell borders and protoplasmic extrusions (INGRAM, 1969), where large pinocytotic vesicles are also formed (ABERCROMBIE and AMBROSE, 1958). The top surface of glia-like cells accumulates microtubules simulating a brushborder but has no endocytosis in contrast to the lower smooth surface which displays active endocytosis (Fig. 35) (ERIKSSON, BRUNK, WESTERMARK, and PONTÉN, 1970). These observations are compatible with the specific differentiation of the lateral part of the cell membrane required by the theory of contact mediated cell cycle inhibition in a monolayer.

The activity of ruffled membranes along the lateral border of glia cells continues up to the very point when the critical terminal cell density is reached where proliferation stops (PONTÉN, WESTERMARK, and HUGOSSON, 1969). Detailed autoradiographic analysis of another system indicated that monolayer cells in lateral contact with other cells but not yet completely surrounded by them still showed DNA synthesis. This did not cease until the entire lateral border had made contact with adjacent cells (GURNEY, 1969; VASILIEV, GELFAND, DOMNINA, and RAPPOPORT, 1969). It is, however, difficult to define what contact is in this sense, since the mere fact that two cells are closely situated does not necessarily imply a junctional interaction and *vice versa*.

The indication that an individual cell is able to distinguish between partial and complete lateral contact with other cells (VASILIEV, GELFAND, DOMNINA, and RAPPOPORT, 1969) is difficult to reconcile with the action of diffusing extracellular molecules. It is also not easy to formulate a theory based on a specific membrane-membrane interaction ("negative signal") which triggers inhibition of the cell cycle, since one would then expect inhibition to occur as soon as any contact between neighboring cells had ensued. The apparent requirement for total paralysis of the ruffling instead suggests that the ruffled membrane is the site for a "positive signal" which tells the cell to maintain proliferation. Partial inhibition of membrane movement does then not block this signal completely. To switch off the division cycle all "positive signals" have to be curtailed. In normal cells this is achieved when cells come in close mutual proximity along their entire lateral circumference. Transformed cells on the other hand fail to immobilize each other by contact and "positive signals" will always persist. As a consequence entry of new cells into the cell cycle is never turned off which is registered as unrestrained growth transformation with its increased terminal density and continuing DNA synthesis.

The number of "positive signals" per unit time may be inversely correlated with the duration of the cell cycle, particularly the G1 period. Certain transformed cells receive only few "positive signals" and have a very slow growth

---

Fig. 35. Electron microscopic picture of portion of a cultured human glia cell. The cell rests along its contact surface on a thin "microexudate" (me). The free surface (directed toward the culture medium) is different from the lower surface in that there are dense accumulations of cytofilaments along the former, while such elements are lacking along the contact surface. Furthermore, "coated vesicles" (cv) suggestive of endocytosis are exclusively present along the surface in contact with the microexudate. Note abundance of haphazardly arranged cytofilaments (arrows) in the central portions of the cytoplasm.

er = rough surfaced endoplasmatic reticulum,
L = lysosomes,
m = mitochondria,
p = "polysomes" on rough surfaced endoplasmatic reticulum.

Cells fixed in 3 % cacodylate-buffered glutaraldehyde (pH 7.2) and postfixed in 2 % s-collidine-buffered OsO₄ and embedded in an Epon-Araldite mixture. ×40,000 rate at high densities, whereas others are more insensitive to contact and receive an undiminished number of "positive signals" even under crowded conditions with consequent rapid, unabated multiplication. The essential, presumably qualitative defect is the failure among cells with unrestrained growth transformation ever to turn the cell cycle off completely. Such a model fits tumor cell behavior *in vivo*. Spontaneous neoplasms vary greatly with respect to rate of proliferation and the population doubling time is often of the order of months. Their common feature is a failure to stop growing regardless of growth rate *per se*.

The nature of the membrane-membrane interaction which abolishes the hypothetical "positive signal" remains elusive. The ultrastructural correlate may be a peculiar "tight junction" where parts of the cellular envelopes have merged (review by FURSPAN and POTTER, 1968).

STOKER (1964) proposed that cell membranes are equipped with an emittor/ receptor system for density dependent proliferation control. Extensive contacts between emittors and receptors along the lateral borders would inhibit the "positive signal" for cell division. Unrestrained growth transformation could depend on the loss of emittor or receptor or both. STOKER and collaborators (STOKER, SHEARER, and O'NEILL, 1966; STOKER, 1967) could show that PV transformed BHK hamster cells could be inhibited by normal mouse cells. A prerequisite was that the latter were non-dividing. Sparse mouse cells did not retard the growth of added PV cells. PV cells did not influence each others' growth. This can be explained if it is assumed that the emittor but not the receptor function is destroyed after PV transformation. PONTÉN and MACINTYRE (1968) obtained similar results with bovine cells.

SV40 and RSV transformed cells are, in contrast, not inhibited by non-dividing cell cycle inhibited cells according to most investigators (TODARO and GREEN, 1966a; MACINTYRE and PONTÉN, 1967; R. WEISS, 1969c) and should therefore lack the receptor (and emittor?) of STOKER. BOREK and SACHS (1966a) and EAGLE, LEVINE, and KOPROWSKI (1968) obtained results partially at variance with those discussed above. This may have been due to imperfections in the techniques for distinguishing transformed from normal cells. It is also clear that the capacity of PV transformed cells to respond with inhibition after contact with non-dividing normal cells is not an obligatory feature. R. WEISS (1969c) noted that PV transformed cells could change from sensitivity to insensitivity and PONTÉN (unpublished) grew cells from PV mouse tumors which were unresponsive to contact with cell cycle inhibited cells.

Synthesis of DNA in bacteria is initiated from a specific attachment point at the cell membrane (review by PARDEE, 1968). In eukaryotic cells one or several attachment points with a similar function may exist at the nuclear membrane (COMINGS and KAKEFUDA, 1968; ALFERT and DAS, 1969). This may be the site where the "positive signal" initiated from the ruffled membrane is received after it has traversed the cytoplasm.

A further recent development implicates a change in the intracellular communication system as responsible for uncontrolled growth. BENDICH, VIZOSO, and HARRIS (1967) described intercellular cytoplasmatic bridges in transformed hamster cells and inferred from autoradiographic evidence that they may contain DNA. CONE (1968) claimed that a mitotic stimulus may spread through a syncytium formed by intercellular processes among transformed cells and speculated that this may be a reason for the uncontrolled growth in cancer. More work with untransformed cells is needed before it can be accepted that a physical difference exists between normal and transformed cells with respect to the presence of intercellular bridges carrying vital information for triggering the division cycle. It should be pointed out that these studies stand in a certain contrast to those of LOEWENSTEIN et al. (cf. LOEWENSTEIN, 1968a, b for reviews), where a break-down of intercellular communication in cancerous tissue has been established. This decreased junctional communication was, however, not recapitulated by transformed cells in vitro (LOEWENSTEIN, 1968b).

The hypothesis that density- dependent inhibition of mitosis is caused by exhaustion of essential factor(s) present in the serum of the medium is supported by HOLLEY and KIERNAN (1968). They observed a direct correlation between terminal cell density and serum concentration in growth curves of 3T3 cells and obtained only very slow growth of sparsely seeded cells in medium removed from confluent 3T3 cultures as if the stationary state of the latter was caused by medium depletion. KRUSE and MIEDEMA (1965) by using perfusion at very high rates obtained a progressive increase in cell density among normal embryonic human lung fibroblasts also explicable on the basis of continued supply of an essential growth factor.

These and other experiments clearly establish that a non-dialyzable factor in fresh serum influences the division rate. The major issue is whether this factor is non-specific in the sense that lack of it is the only explanation for the diminished growth rate which crowded normal cells show. If this were true, the in vitro proliferation of normal cells from higher organisms would be principally similar to that of unicellular competing organisms. Several facts argue against this interpretation.

Dense and sparse cells immersed in the same medium do not stop multiplying simultaneously but a period exists during which the sparse cells grow in spite of the stationary state of adjacent confluent cells. The terminal cell density has never been shown to be lower than that corresponding to confluency, i.e. the stage at which no empty solid surface is left. Both these observations strongly suggest that contact per se modifies the response to the serum growth factor. This modification does not seem to operate through a change in the rate of uptake since dividing and non-dividing 3T3 cells removed the factor(s) from the medium at the same rate (JAINCHILL and TODARO, 1970).

At present it seems most likely that contact per se and one or several growth factors contained in fresh serum interact to produce density-dependent inhibition of mitosis. The nature of this interaction can only be a matter of speculation. It is possible that intimate cell to cell contacts prevent the cells from taking up the growth factor and that the serum growth factor is identical to the "positive signal" discussed above. This is, however, contradicted by the finding that stationary cells remove the factor from the medium as effectively as exponentially growing cells. A somewhat more likely suggestion may be that the growth factor modifies the cell membrane response to intercellular contact. It may — up to a certain limit — stimulate the release of "positive signals" and thus counteract the effect of establishment of contacts.

The biochemical events in the suppression of DNA synthesis under conditions of density-dependent control are still rather obscure. KILLANDER and RIGLER (1965) have made the important discovery that the functional state of the DNA-protein complex contained in the nuclear chromatin can be assessed by its capacity to bind acridine orange (AO). Resting lymphocytes have a comparatively small capacity, while stimulation by phytohemagglutinin leads to a rapid increase in AO-binding capacity interpreted as due to weakening of nucleic acid-protein bonds (KILLANDER and RIGLER, 1969). BOLUND, RINGERTZ, and HARRIS (1969) observed a similar increased AO-binding capacity of hen erythrocyte nuclei when these were incorporated into HeLa cell cytoplasm. These changes in the DNA-protein complex appear to be due to a transfer of arginine-rich histones from DNA to other acidic groups in the chromatin of cell nucleus (ZETTERBERG and AUER, 1968; AUER, ZETTERBERG, and KILLANDER, 1970). The biological significance of these chromatin changes is still unclear but it may represent a necessary preliminary step in the activation of genes (RIGLER and KILLANDER, 1969). ZETTERBERG and AUER (1970) correlated the state of the chromatin with cell density in primary mouse kidney cultures. The epithelial component of the cultures showed DNA synthesis and division up to a certain terminal cell density. Subsequently the local density of the epithelium increased further due to a denser packing or a local retraction of the epithelial sheet. The chromatin was in an "active state", i.e. showed a high AO-binding during the DNA synthesis period and the first phase of denser packing. Eventually, however, when a high local cell concentration was reached the chromatin became "inactivated". It is not known if the chromatin "inactivation" primarily reflects a late step in density-dependent suppression of cell division or is correlated with other events such as organization of epithelium into defined islands.

The fragmentary state of our knowledge about cell cycle inhibition in vitro precludes any deeper insight into the phenomenon. The following simple model seems to be compatible with existing data:

a) all nuclei in cells with ruffling membranes (or more exactly membrane segments not "engaged" by contact with other cells, a solid surface or the fluid medium) receive stimuli to divide from the "non-engaged" part of the membrane,

b) the signal is reinforced by a non-dialyzable serum factor,

c) the signal traverses the cytoplasm and acts upon the points at which chromatin is attached to the nuclear membrane,

d) when all parts of the cell periphery have been "engaged" no signal is emitted and cells remain arrested in the G1 phase of the cell cycle,

e) unrestrained growth transformation is primarily caused by impairment of mechanisms responsible for contact inhibition of membrane movement. This may either be due to failure of cell membranes to "engage" each other properly or a failure of the cell surface to become inactivated at the interface with the fluid medium or the contact area with its solid support. Increased sensitivity to a growth-promoting serum factor(s) is associated with such a failure. A close linkage seems to exist between irregular and unrestrained growth transformation (cf. CARTER, 1968). Apparent dissociations between the two phenomena may either depend on faulty techniques to observe cell membranes or possibly the development of autonomy at the proposed nuclear membrane sites controlling DNA synthesis.

Studies on loss of cell cycle inhibition *in vitro* have been critisized on the basis that they have failed to reveal qualitative differences in growth control between neoplastic and non-neoplastic cells. Even neoplastic cells will eventually reach a density where proliferation is strongly depressed and different kinds of non-neoplastic cells have different terminal cell densities. If all data on cell cycle inhibition in the literature are analysed with regard to the value for the terminal density obtained under comparable standard conditions a continuous series of figures is obtained. Neoplastic cells tend to give higher values but considerable overlapping with non-neoplastic cells is found. This has suggested that the difference between non-neoplastic and neoplastic cells may be an artifact or at most only a trivial quantitative difference.

Mass cultures of normal cells do not always fullfill the criterium of virtually complete suppression of mitosis at a defined local cell density. Glia-like cells from normal adult human brain are a prime example of nearly perfect cell cycle inhibition (PONTÉN, WESTERMARK, and HUGOSSON, 1969). At a density of about 75,000 cells per cm^2 less than 1% of the cells enter DNA synthesis in standard medium with 10% serum. Fibroblasts, particularly of embryonic lung origin, seem on the other hand to be less strict. They may reach 200,000 cells per cm^2 or more and still show $5-10\%$ non-inhibited cells. It is not known why fibroblasts are less strict but morphologic analysis of the local distribution of mitosis suggests that synthesis of extracellular collagen with retraction of cells and creation of empty clefts plays a decisive role. The holes stimulate the cell cycle in the surrounding cells and a correlation exists between the topographic microscopic density and inhibition of mitosis (PONTÉN and WESTERMARK, unpublished).

WESTERMARK (unpublished) presented strong evidence for the possibility that the relative absence of cell cycle inhibition among fibroblasts is an artifact. He seeded radioactively labeled fibroblasts onto stationary layer of human glia cells and found strong inhibition of proliferation among the fibroblasts under these conditions. These experiments involved various syngeneic, allogeneic and xenogeneic combinations. It seems probable that conditions can be defined where all normal cells which can adhere to solid substrates show virtually perfect cell cycle inhibition and that this inhibition operates across species and tissue borders.

Cells from neoplastic sources or cells exposed to transforming viruses have been brought in contact with normal cells under principally the same conditions as those used by WESTERMARK (unpublished). In most instances proliferation has then been observed (BOREK and SACHS, 1966a; TODARO and GREEN, 1966a; EAGLE and LEVINE, 1967; MACINTYRE and PONTÉN, 1967; EAGLE, LEVINE, and KOPROWSKI, 1968; POLLACK, GREEN, and TODARO, 1968; R. WEISS, 1969c; WESTERMARK, unpublished) although sometimes at a reduced rate (WESTERMARK, unpublished; PONTÉN and WESTERMARK, 1970). The only exception concerns certain polyoma virus transformed cells (page 64), but these have not been sufficiently well penetrated to exclude that they were caused by cell surface incompatibilities of an irrelevant nature. R. WEISS (1969c) also observed that PV transformed populations could change over to become insensitive to inhibition by normal cells. It is thus possible that transformed cells may show a quantitative reduction in the sensitivity against other transformed cells even reaching nearly complete inhibition of the cell cycle, but still an inability to respond properly to the presence of surrounding normal cells.

This hypothesis is compatible with tumor behavior *in vivo*. Many neoplastic cells are in contact with normal cells either in the periphery of the tumor or in the stroma. Proliferation is often most pronounced in these areas — an expected finding if tumor cells are imperfectly inhibited by normal cells.

Transformation should be regarded the tissue culture equivalent to carcinogenesis *in vivo*. This is well illustrated with viruses where only oncogenic viruses transform cells in culture. Tumor cells successfully grown *in vitro* also show the same growth control disturbances as cells transformed *in vitro*.

The precise relationship between transformation and neoplasia cannot be delineated, mainly because the latter term is not strictly defined. A neoplasm is composed of cells with more or less pronounced phenotypic and genotypic variation and large differences exist between individual tumors. No criterion has been found by which a single neoplastic cell can be defined. Neoplastic growth is always influenced by host factors. These exert their selective pressure on a heterogenous population and this combination of genetic variability and selection will lead to the emergence of variants with increasing malignancy — the well known phenomenon of tumor progression. The most important restraint may be an immunologic defense but such factors as hormones and chalones also play a role. It is probable that the proliferation of most neoplastic cells is completely suppressed and that many of them are even eliminated. Those few cells which manage to multiply to give rise to clinical cancers composed of progressively increasing populations with invasive properties are therefore highly select variants.

Restraining influences are either absent or weak *in vitro*. Immunologic defense is of course entirely lacking. Transformation therefore probably corresponds to the initial neoplastic event *in vivo* and primary populations of transformed cells to those tumors which never grew *in vivo* because the host managed to control them. It is therefore not surprising that many of the early transformed cells are only weakly or not at all tumorigenic.

Transformation *in vitro* thus reflects one of the primary events in carcinogenesis. It can be studied as an independent phenomenon intimately related to neoplasia. The characteristic growth control disturbance can be assessed without knowledge about the cells tumorigenicity after animal implantation.

B. Mechanisms for Viral Transformation

A distinction should be made between on one hand, the intimately linked irregular and unrestrained growth transformations, and on the other infinite growth transformation. Irregular and unrestrained growth transformation reflect the loss of contact (or density?) dependent locomotion and proliferation control. They can, in principle, be induced in any type of target cells. Established lines can react in the same manner as primary or secondary passage cells from stable species. Infinite growth transformation is, in contrast, strongly dependent on the character of cells exposed to a certain virus. It occurs readily if inherently unstable cells are infected but only rarely if cells from a stable source are employed.

1. Irregular and Unrestrained Growth Transformation

These interrelated alterations can be induced by DNA as well as RNA tumor viruses.

The most rewarding studies on *DNA viruses* have concerned PV and SV40. The smallness of their genomes and the availability of standard established lines of target cells have been of great advantage. A major difficulty has been the relatively low percentage of exposed cells which respond with permanent transformation. This has to a certain extent precluded any deeper insight into the early events during transformation.

It is definitely established that the SV40 or PV genomes cause the transformation. The alternative that it depends on selection of variant cells is excluded, because transformation can be induced in stable species even with purified SV40 DNA (AARONSON and TODARO, 1969) and at least part of the viral genome is regularly found in the transformed cells.

When PV or SV40 DNA has entered a cell many alternative results are possible dependent on the origin and physiologic state of the target cell but also on other unknown factors. Certain cells apparently ingest the virus, but its nucleic acid never becomes expressed or replicated. It is either excreted or degraded. This ill defined event has been referred to as "abortive infection". It is believed to occur in those susceptible target cells which do not respond with lysis or permanent transformation, but has not been critically distinguished from abortive transformation.

In the second type of interaction (abortive transformation), explored by STOKER (1968) in established hamster lines infected by PV, the viral genome is expressed. About one third of BHK-21 cells exposed to PV acquired a capacity to divide when suspended in a semisolid medium. The same cells also displayed an emigration pattern suggestive of irregular growth transformation when plated onto a solid substrate. Most of these colonies lost their capacity to grow in suspension after a few divisions. Simultaneously their emigration patterns returned to normal. STOKER suggested that the viral DNA caused the loss of surface dependent growth controls. The change was only temporary because viral DNA failed to become integrated with cellular DNA. It was instead lost either by degradation, excretion or dilution during cell division. When the cell had got rid of the foreign genome it returned to the same state as before the infection proving that PV DNA, even if it is expressed, does not need to incur any permanent alteration in the infected cell.

A third type of interaction may be illustrated by "pre-crisis" SV40 transformed cells originating from stable human fibroblasts. These show all the signs of unrestrained and irregular growth transformation. In spite of this most of them do not acquire the capacity to divide indefinitely but succumb in a delayed "phase III" without having returned to normal behavior. Although never experimentally demonstrated it is usually assumed that these cells contain the viral genome in an integrated form. They are probably not tumorigenic.

A fourth variety is the classical fully transformed cell which exhibits irregular and unrestrained growth transformation transmitted to an endless number of successive cell generations. These can regularly be obtained in unstable species, but only irregularly and with great difficulty from stable species. Only such lines have been used as prototypes in the experiments interpreted as demonstrations of integration of viral DNA into the cell's own genome. With a few possible exceptions these lines do not produce infectious virus spontaneously in spite of containing many viral genome equivalents which at least in the case of SV40 seem to be complete. The genomes are transmitted and distributed to the daughter cells presumably in association with the mitotic chromosomes. These cells are tumorigenic at least in immuno-depressed hosts.

A fifth type of interaction leads to rapid lysis of the target cells. This outcome is by far the most common response if cells from the natural virus host are employed. The lytic cycle is rather slow and the cell releases a large number of infectious virus particles when it succumbs to the infection. It is not known if any viral DNA becomes integrated with cellular DNA during this process but a major part is clearly dissociated from cellular DNA. In foreign cells lysis is rare as exemplified by infection of hamster cells by PV or SV40.

The factors which determine if any of five possible interactions will occur are largely unknown. It is for instance not known why SV40 gives a productive infection in GMK cells but a non-productive one in mouse cells. It is also not understood why only a minority of murine cells undergo detectable unrestrained growth transformation in contrast to the majority which react with lysis after exposure to PV. It is possible that all cells which contain viral DNA in an active form are in fact "transformed" in the sense that their locomotion and proliferation have escaped normal control, but that this is obscured by the rapid lysis which most cells show. The difference between detectable transformation and lysis among, for instance, PV exposed mouse cells may only reflect whether the cell survives the infection or not. Survival may be a chance phenomenon made possible by suppression of viral synthesis. The latter is dependent on the expression of "late" viral functions and it is possible that these can be repressed only if viral DNA is incorporated into cellular DNA in such a manner that the physiologic control of gene expression can operate. Survival may only be possible if the cell's mechanism for controlled transcription of DNA includes viral DNA.

How can the viral DNA genome cause the cell membrane alteration which is essential for irregular and/or unrestrained growth transformation? Theoretically this could be achieved specifically or non-specifically. A very direct variant of the former case would be if one of the proteins synthesized under the direction of viral DNA were transported to the cell membrane which is then altered to become unresponsive to the restraint normally exerted by surrounding normal cells. The best candidate for such a function is the S- or transplantation antigen which is located at the cell periphery. It would be important to know if this antigen is correlated with irregular growth and colony formation in agar in abortively as well as permanently transformed cells.

An indirect but specific effect by viral gene products is, however, also possible for instance by derepression of cellular genes influencing membrane motility. The uncovering of binding sites for the protein concanavalin A which occurs after transformation of rodent cells by PV or SV40 may be an example of such an indirect effect. These sites are present in a cryptic form on normal cells (BURGER, 1969; INBAR and SACHS, 1969) as is one S-antigen (HÄYRY and DEFENDI, 1970). Covering of the former sites on transformed cells with monovalent concanavalin

A normalized their growth control (BURGER and NOONAN, 1970). This strongly suggests that the concanavalin binding site may be the cell membrane structure responsible for the emittance of the "positive signal" discussed on page 171.

The reasoning above assumes that SV40 and PV contain a specific "transformation gene", which disrupts normal cellular growth control directly or indirectly by making the cell insensitive to the restraint on locomotion and division which should follow contact with other cells. As long as this gene is present and active the cells show unrestrained and irregular growth transformation. If the gene becomes associated with the cell's own genome it can only be lost by mutation or unequal distribution to daughter cells by a defunct division machinery. The latter events are so rare that transformation can be regarded a permanent change on the population level once the viral DNA has become associated with the cell's genome. A few cells which had lost their "transformation gene" would escape detection. Return to normal and survival as in abortive transformation is explained by the loss of the viral genome before it is incorporated into the cellular genome and/or replicated to the extent that the cell is irreversibly and lethally damaged.

To prove this idea it is probably necessary to positively identify a viral "transformation gene" or its end product. One copy of this gene should suffice to induce irregular and unrestrained growth transformation. Suppression or removal of the gene should bring the cell back to normal.

Clarification of whether a "transformation gene" exists may not be too far away, since SV40 and polyoma DNA only contain about 8 genes. Of these 3 or 4 "early" genes expressed in transformed cells have to be taken into account. Most (or all?) of their products have been descriptively identified but the function of, for instance, the T-antigen or the transplantation antigen is not understood. The gene for the T-antigen appears a somewhat unlikely candidate for a "transformation gene" because it is absent in 3T3 cells transformed by the temperature sensitive Ts-a mutant of PV, when these are kept at high temperature (VOGT cited by DULBECCO, 1969). The gene for the transplantation antigen may, as outlined above, be the "transformation gene", but no positive evidence exists.

The gene(s) responsible for stimulation of DNA synthesis in stationary cells may be identical to that causing synthesis of T-antigen or that which controls the transplantation antigen, but it can also be a separate entity. Stimulation of DNA synthesis disrupts a normal control of nucleic acid synthesis. It is logical to assume that it also destroys cell growth control but no proof of this assumption has been produced. The failure to dissociate the genetic function required for stimulation of DNA synthesis and growth in soft agar is compatible with the assumption that the DNA stimulation gene could be the "transformation gene".

Since no significant hybridization has been observed between PV and SV40 DNA, the hypothesis implies that the hypothetical "transformation genes" of both viruses are functionally similar but structurally different.

A second hypothesis denies the existence of a specific viral "transformation gene". Loss of growth control may only be an indirect effect caused by integration *per se*. This process must involve opening up of DNA chains, insertion of foreign DNA and repair. Breaks in DNA molecules can be induced by many other carcinogens such as X-rays, UV-light and mitogenic hydrocarbons. SV40 and PV could principally be of the same nature and owe the transforming and tumorigenic effect to random damage to DNA molecules. Any kind of such damage will lead to transformation because cells may only have a restricted number of responses. One of these is a loss of growth control which in itself may be just as non-specific as cell death. If an agent grossly severes the harmonic function of a cell, the cell may either die or lose its capacity to respond properly to the environment. The last response can be regarded a special case of the former where an odd cell manages to survive because the functional damage is not severe enough to cause lysis. If a virus or any other agent has caused breaks in DNA molecules and the cell still is capable of division it remains "transformed" regardless of the continued presence of viral genes. This is to be looked upon as a sign of past infection but is in itself irrelevant from the point of view of maintaining altered growth properties. The virus initiates irregular and unrestrained growth transformation non-specifically but is not necessary for its maintenance. The fact that transformed cell lines usually contain viral DNA is explained by the large number of integrated viral DNA equivalents. Neither mutation nor misdivision can remove all these "redundant" genes.

It is not definitely established that the viral DNA in transformed cells is present in an integrated state. The evidence rests entirely on DNA-DNA hybridization tests. The reliability of homology tests where one of the nucleic acid molecules is derived from the host cell has been questioned (McCARTHY, 1966) and there is no compelling evidence that SV40 and PV DNA has indeed been inserted into cellular DNA chains. This criticism does not affect the establishment of the presence of virus specific RNA in cells transformed by SV40 or PV by hybridization (BENJAMIN, 1966a) and viral DNA may be present in an episome-like non integrated state rather than being truly integrated.

Other indirect means by which oncogenic DNA viruses could cause unrestrained growth transformation have recently been suggested. CARTER (1968) in an original approach proposed that DNA viruses and other carcinogens owe their transforming effect to a capacity to induce heterogeneity in the cell membrane structure, because the cell division machinery is damaged and progeny cells will always display new membrane configurations. Any assembly of cells tends to arrange itself to give a minimum interfacial free energy for the system as a whole. Any population of cells has to reach this equilibrium for mitosis to stop. A prerequisite for the attainment of such a stationary state is a stability of the membranes of the individual cells. This can never be fulfilled in a population where cells with new surface properties are constantly created. Proliferation therefore never stops and the population of cells behaves as a neoplasm. The hypothesis predicts that both components of a mixture of transformed and normal cells should proliferate without restraint. This prediction has not yet been critically tested and CARTER's interesting speculation lacks experimental proof.

The finding that virtually all mice carry latent RNA tumor viruses belonging to the murine leukemia-sarcoma complex has inspired a third "indirect" hypothesis of transformation by DNA viruses (HUEBNER and TODARO, 1969). The authors pointed out that infective RNA tumor virus complexes have been definitely found in mice, chickens and cats. These physico-chemically related viruses have a broad host range and contain tumor genomes ("oncogenes") which are speculated to be present in practically every cell of all vertebrates. This is supported by the finding of C-type particles in tumor cells from hamsters, rats, cattle, swine, guinea-pigs, snakes, monkeys and humans. The oncogenes can be transmitted vertically through the zygote as well as horizontally as part of infective virus particles. The detection of the infectious particles or complement-fixing antigen may be exceedingly difficult and can require several months of serial tissue culture (AARONSON, HARTLEY, and TODARO, 1969). The long delay could not be explained by the presence of a few virus producing cells in the original explant but was probably due to a derepression of "virogenes" which made synthesis of infectious particles possible. Because of the common presence of oncogenic RNA viruses immunological tolerance is common and antibodies against their group specific complement fixing antigens can only exceptionally be detected in the species of origin. The omnipresent oncogenes are usually repressed and transmitted as part of the normal cell heritage. Radiation and carcinogenic hydrocarbones owe their oncogenicity to their capacity to derepress synthesis of these oncogenes. It has clearly been shown that lymphomas induced by ionizing irradiation in mice is accompanied by the appearance of increased amounts of MuLV (page 159). HUEBNER and TODARO (1969) propose that even the tumorigenic DNA viruses owe their transforming effect to a derepression of highly prevalent specific viral oncogenes. These would then in the manner outlined below direct synthesis of virus particles and induce the membrane alterations necessary for unrestrained and irregular growth transformation. Support for this thesis is derived from the common presence of C-particles in rodent cells transformed by adenovirus or SV40.

It is not possible to decide if DNA tumor viruses induce unrestrained and irregular growth by a specific "transformation gene" or indirectly. A pure indirect mechanism of the non-specific chromosome breaking type cannot, however, explain the typical dose-response curve with a ceiling at a transformation frequency of only a few per cent. The abortive infection observed by STOKER (1968) is also difficult to reconcile with a non-specific breaking effect of viral DNA on the cellular genome. It has been observed (WEINSTEIN and MOORHEAD, 1967) that loss of contact inhibition of mitosis occurs before visible chromosome damage in human cells exposed to SV40. If breaks in DNA chains cause unrestrained growth transformation, the reverse would have been expected.

RNA viruses from different species are remarkably similar morphologically and with respect to gross physico-chemical properties. Their capacity to induce irregular and/or unrestrained growth transformation is intimately correlated with their sarcomagenic potency *in vivo*. Under favorable laboratory conditions all fibroblasts exposed to virus respond with unrestrained and probably also irregular growth transformation within 24 hours. This property can be transmitted to the daughter cells and a cell lineage with transformed properties is then obtained. The induced loss of contact-dependent growth control will give the cell a selective advantage over normal fibroblasts. This is, however, often only temporary because cells showing pure unrestrained and irregular growth transformation, after a spurt of rapid proliferation, may decline in vitality and die after fewer cell cycles than their normal counterparts. A wide variation seems to exist with respect to capacity, to induce temporary loss of growth control and premature lysis between different viruses. *MuSV-M* transformed mouse cells seem to be entirely incapable of further division (BATHER, LEONARD, and YANG, 1968) in contrast to *MuSV-H* transformed mouse cells, and rat cells transformed by *MuSV-M* where there is suggestive evidence of proliferation (SIMONS, PEPPER, and BAKER, 1969; PARKMAN, LEVY, and TING, 1970). RSV transformed chicken and mammalian cells are also clearly capable of many successive generation cycles.

RNA viruses can probably infect and transform cells either from without or from within. The former type has been subject to much experimentation, whereas the existence of the latter mechanism is still speculative.

The cell membrane seems to be the principal barrier to transformation by RNA viruses from without. If this is penetrated, the viral genome is expressed and induces viral synthesis and in many instances also irregular and unrestrained growth transformation.

Viral synthesis may be complete and result in production of numerous infective virions or incomplete with production only of the internal group specific antigen, production of virus specific RNA or particles with highly reduced or absent infectious properties. Complete or incomplete synthesis seems to be largely dependent on the species employed. In the natural host complete synthesis is common, whereas distantly related hosts rarely produce infectious virions. Infection and complete or incomplete virus synthesis does not have to lead to unrestrained growth transformation. The latter event is instead largely dependent on the differentiated state of the target cell. Sarcoma viruses may be synthesized in practically all cells but induce unrestrained growth in a limited number of cell types, mainly fibroblasts. Leukemia viruses likewise multiply in a number of different cell types but are only oncogenic for hematopoietic cells.

Penetration of the cell membrane is determined by complex factors. An interaction between specific genetically controlled receptors and the corresponding specific antigens on the viral surface greatly facilitates infection. Of equal importance is the presence of leukemogenic RNA viruses. Preinfection with proper serotypes makes the cell resistant to infection and transformation by a related sarcomagenic RNA virus. These two factors together explain susceptibility and resistance in the natural hosts of avian and murine RNA sarcoma viruses. It is not known if the same factors also influence susceptibility in foreign hosts.

Synthesis of cellular DNA (*per se* or as an indication of progression through the cell cycle?) has also been shown to be a requirement for viral synthesis and transformation. Presumably this requirement involves an event after the virus has penetrated into the cell, but it has not been excluded that DNA synthesis influences viral penetration and also acts on the barrier function of cell membrane.

Induction of unrestrained and irregular growth transformation by RNA tumor viruses from within is still only a speculative hypothesis. As proposed by HUEBNER and TODARO (1969) it requires that cells carry silent viral genetic information in the form of "virogenes" and "oncogenes". These may be activated by various means such as carcinogenic hydrocarbons, other viruses, irradiation and even ageing as such. The factors which determine whether "virogenes" and "oncogenes" will be activated by a particular treatment are completely unknown. It can be speculated that these are different in different species. Mouse cells seem to be particularly prone to activation of these genes and murine leukemia/ sarcoma viruses are therefore often found in experimental tumors induced in various ways. Human cells may be much less susceptible to activation and C particles are not readily demonstrated. AARONSON, HARTLEY, and TODARO (1969) noted that the "spontaneous" production of MuLV in long term cultures was correlated with unrestrained growth transformation, which would support that the activation of "virogenes" is indeed responsible for this type of trans- formation. Chicken cells are, on the other hand, often loaded with leukemia virus but do not ever seem to transform spontaneously (PONTÉN, 1970).

The presence of the viral genetic message in a derepressed form is essential for the maintenance of a disturbed contact- dependent growth control. Cells can cease to express the viral genome even long after transformation, either because some progeny cells do not obtain any viral RNA at cell division or because viral genes are repressed. There is no definite evidence that the RNA tumor viruses can induce a permanent loss of surface-dependent growth control and then leave the cell in that state. Continuous expression of at least part of the viral genome seems to be required.

The nature and mode of transmission of the oncogenes and virogenes are not definitely known. TEMIN's (1964a) hypothesis of a "provirus" DNA complement- ary to the viral RNA and integrated with cellular DNA remains unproved, but is not opposed to the recent finding by DUESBERG and VOGT (1969). They observed that one of a pair of avian tumor RNA viruses could not substitute for the other with respect to the requirement for DNA synthesis early after infection. This may be so because the early DNA has a function which is temporally very limited, but it is also possible that each RNA genome is template for a specific DNA provirus and that therefore no substitution is possible.

One major obstacle — the apparent absence of the appropriate enzyme for TEMIN's provirus hypothesis has recently been removed. BALTIMORE (1970) and TEMIN and MIZUTANI (1970) established RNA-dependent DNA-polymerase activity associated with purified RLV, RSV and AMV virions. This indicates that information contained in the RNA of tumor viruses can be transcribed to specific DNA sequences. It remains to be shown that these are incorporated into cellular DNA, but if this is the case the process has to be „mild." No aneuploidy and mitotic aberrations are induced in cells from a stable species in contrast to the oncogenic DNA viruses. The dose-response curve for transformation is straight and has no ceiling. The newly created DNA genes may well occupy specific sites and could in fact be closely related to the normal genes for growth control. This implies that means for their specific repression should exist.

The mechanism by which RNA viruses induce and maintain irregular and unrestrained growth transformation is largely unknown. The cell surface contains an increased amount of acid mucopolysaccharides and at least the "basophilic round cells" have a decreased adhesion to other cells and solid surfaces. Moving "non-engaged" parts of the cell membrane may be the sites from which the signal is generated which tells the cell to continue to replicate. The decreased adhesive- ness may prevent the cell surface from making the intimate contacts necessary for "engagement" and thus also prevent regulation of the cell cycle. The problem can probably not be solved until normal cell cycle control is better understood.

2. Infinite Growth Transformation

The genuine capacity of a particular virus to induce this alteration can only be assessed with any confidence in stable species where there is no "background noise" in the form of spontaneous infinite growth transformation.

DNA viruses have clearly been shown to be capable of inducing a potential for endless cell doublings. Bovine cells exposed to SV40 form established lines with rather high regularity and an occasional line may also be obtained after infection with PV. SV40 exposed human fibroblasts form established lines with moderate frequency and in this species no authenticated instance of spontaneous infinite growth transformation has been recorded in spite of a very large number of attempts.

RNA tumor viruses readily induce unrestrained and irregular growth transformation in stable chicken, bovine and human cells as already described. In spite of this not a single case of an established line has been described. The transformed cells have all succumbed to a premature phase III.

Any explanation of the mechanism for infinite growth transformation has to take into account the difference between the effect of DNA and RNA viruses. Synthesis of the DNA viruses occurs in the nucleus whereas the RNA viruses, except possibly for a very early stage involving the nucleolus (SUSKIND, PRY, and RABOTTI, 1969), are synthesized and assembled in the cytoplasm.

RNA viruses have never been shown to have a chromosome breaking effect in stable species when critical experiments were performed where contamination by PPLO was excluded (SCHENDLER and HARRIS, 1967). SV40, in contrast, causes a severe chromosome damage in human cells and established PV or SV40 lines of bovine origin are grossly heteroploid. Spontaneous infinite growth transformation is, with the exception of lymphoblastoid cells, always accompanied by heteroploidy. The most likely hypothesis for the creation of established lines is that DNA viruses cause chromosome damage resulting in constant genetic recombination. This may lead to the emergence of populations capable of infinite survival because they can respond to different environmental conditions and stresses by selection of fit variants.

3. Model of Virus Transformation

If the facts discussed in this review are brought together a crude model of viral transformation may be formulated.

RNA and DNA tumor virus genomes induce — after they have been expressed — a change in the signal system leading from the cell membrane to the intracellular site for initiation of the cell cycle. Contacts with normal cells fail to inhibit cellular division properly. This alteration may be brought about by disruption of the cell's genetic control of the constituents of its periphery. Another alternative is that viral RNA interacts with a specific DNA sequence found in all animal cells. This DNA chain is associated with the nuclear membrane and could play an important role in cell proliferation control (YOSHIKAWA-FUKADA and EBERT, 1969). Signals initiated from the cell periphery would then not be properly perceived at an intranuclear control site for cell multiplication. The further fate of the infected cell may be widely different. With virulent DNA viruses acting upon cells from their natural hosts rapid disintegration follows in the vast majority of the infected cells. Cells which do not support complete virus synthesis and resist lysis usually inactivate the viral DNA and regain their normal cell cycle control. In very few cells viral DNA becomes associated with cellular DNA and cannot be eliminated. Most of these cells from a species capable of supporting viral synthesis may still succumb to the infection but a few survive because complete virus synthesis is curtailed to levels compatible with survival (usually no virus production). The survivors are still not ensured eternal life as transformed elements. If derived from unstable species they profit from the inherent tendency to form established lines and readily grow as "fully" transformed populations characterized by an infinite potential of cell doublings and a deficient response to the restraint on locomotion and proliferation exerted by normal cells. In stable species most of the transformed cells show a protracted mortality. Only a few cells manage to undergo the "right" genetic recombinations necessary to escape ageing and "phase III". Only then will they be "fully transformed" and entirely analogous to transformed cells of unstable origin. Viral genes may be specifically responsible for initiation and maintenance of cell surface-dependent irregular and unrestrained growth transformation; infinite growth transformation may on the other hand be a non-specific side effect of chromosome damage — easy to accomplish in unstable species but difficult to attain in stable species. The RNA virus genome resembles the DNA genome by causing unrestrained and irregular growth transformation presumably also by altering the cell membrane. The transformed state is maintained as long as the genetic message is expressed. Inherently unstable species form established lines, but no infinite growth transformation is obtained from stable species strongly suggesting that RNA viruses have no capacity to induce the genetic recombinations assumed to be required for infinite multiplication.

Unrestrained growth transformation seems to be directly correlated with tumorigenicity *in vivo*. Infinite growth transformation is probably not important *per se* since established mouse lines with a high degree of density-dependent cell cycle inhibition only have a low degree of tumorigenicity. The tumorigenic potential of most established lines is probably an indirect effect due to selection of variants lacking cell cycle inhibition which are automatically favored unless special precautions are taken such as the continued sparse seeding and repeated recloning used with 3T3 cells.

Irregular and unrestrained growth transformation can only be regarded a preliminary step towards high tumorigenicity. Further adaptations are probably necessary and a progressive tumor will also have to grow in spite of an immunological resistance. The cells of autonomous, malignant neoplasms have undergone a "full" transformation *i.e.* an infinite growth potential has been superimposed on irregular and unrestrained growth transformation.

A normal cell infected by a DNA tumor virus *in vivo* will thus have to go through a long series of difficult and uncommon stages before it can grow as a neoplastic cell. This probably explains why DNA viruses do not seem to act as natural carcinogens. Only virulent high titer laboratory strains inoculated into unnatural inherently unstable hosts are capable of inducing a malignant tumor. PV, SV40 and adenoviruses seem to be examples of well adapted agents which do little harm to their natural hosts and therefore are able to persist in nature. Induction of a tumor should be regarded an accident, which does not benefit the virus because the host is killed and the tumor cells usually do not yield any infectious virus.

RNA tumor viruses are natural carcinogens presumably because they can induce irregular and unrestrained growth transformation in a large proportion of exposed cells without any immediate lethal cell damage. The avian leukemia/sarcoma viruses probably manage to persist in nature because the tumors often do not appear until after the reproductive age and furthermore are relatively benign because of high antigenicity and incapacity for endless proliferation.

Acknowledgements

The difficult task of reviewing the rapidly enlarging literature on transformation *in vitro* would not have been possible without the many stimulating discussions which I have had with several colleagues and friends. I am particularly thankful to Heinz Bauer, Leonard Hayflick, Karl Lonberg-Holm, Paul Moorhead, Kenneth Nilsson, Lennart Philipson, Eero Saksela, Guy de Thé, Bengt Westermark and Anders Zetterberg for their help and constructive criticism. Excellent secretarial aid has been provided by Kerstin Lindberg. Experiments from our group quoted in this review have predominantly been supported by the Swedish Cancer Society.

References

AARONSON, S. A., J. W. HARTLEY, and G. J. TODARO: Mouse leukemia virus: "Spontaneous" release by mouse embryo cells after long-term *in vitro* cultivation. Proc. nat. Acad. Sci. (Wash.) 64, 87—94 (1969).

AARONSON, S. A., and G. J. TODARO: Development of 3T3-like lines from Balb/c mouse embryo cultures: Transformation susceptibility to SV40. J. cell. comp. Physiol. 72, 141—148 (1968a).

AARONSON, S. A., and G. J. TODARO: Basis for the acquisition of malignant potential by mouse cells cultivated *in vitro*. Science 162, 1024—1026 (1968b).

AARONSON, S. A., and G. J. TODARO: SV40 T antigen induction and transformation in human fibroblast cell strains. Virology 36, 254—261 (1968c).

AARONSON, S. A., and G. J. TODARO: Human diploid cell transformation by DNA extracted from the tumor virus SV40. Science 166, 390—391 (1969).

ABEL, P., and L. V. CRAWFORD: Physical characteristics of polyoma virus. III. Correlation with biological activities. Virology 19, 470—474 (1963).

ABERCROMBIE, M.: Behaviour of normal and malignant connective tissue cells *in vitro*. Canad. Cancer Conf. 4, 101—117 (1961).

ABERCROMBIE, M., and E. J. AMBROSE: Interference microscope studies of cell contacts in tissue culture. Exp. Cell Res. 15, 332—345 (1958).

ABERCROMBIE, M., and E. J. AMBROSE: The surface properties of cancer cells: a review. Cancer Res. 22, 525—548 (1962).

ABERCROMBIE, M., and G. GITLIN: The locomotory behaviour of small groups of fibroblasts. Proc. roy. Soc. B. 162, 289—302 (1965).

ABERCROMBIE, M., and J. E. H. HEAYSMAN: Observations on the social behaviour of cells in tissue culture. I. Speed of movement of chick heart fibroblasts in relation to their mutual contacts. Exp. Cell Res. 5, 111—113 (1953).

ABERCROMBIE, M., and J. E. H. HEAYSMAN: Observations on the social behaviour of cells in tissue culture. II. "Monolayering" of fibroblasts. Exp. Cell Res. 6, 293—306 (1954).

ABERCROMBIE, M., and J. E. H. HEAYSMAN: The directional movement of fibroblasts emigrating from cultured explants. Ann. Med. exp. Fenn. 44, 161—165 (1966).

ABERCROMBIE, M., J. E. H. HEAYSMAN, and H. M. KARTHAUSER: Social behaviour of cells in tissue culture. III. Mutual influence of sarcoma cells and fibroblasts. Exp. Cell Res. 13, 276—291 (1957).

AGEENKO, A. I.: Heterotransplantation of Rous chicken sarcoma in the brain. Probl. Oncol. (N.Y.) **3**, 170—175 (1957).

AHLSTRÖM, C. G., S. BERGMAN, and B. EHRENBERG: Neoplasms in Guinea pigs induced by an agent in Rous chicken sarcoma. Acta path. microbiol. scand. **58**, 177—178 (1963).

AHLSTRÖM, C. G., S. BERGMAN, N. FORSBY, and N. JONSSON: Rous sarcoma in mammals. Acta Un. int. Cancr. **19**, 294—298 (1963).

AHLSTRÖM, C. G., and N. FORSBY: Sarcomas in hamsters after injection with Rous chicken tumor material. J. exp. Med. **115**, 839—852 (1962).

AHLSTRÖM, C. G., and N. JONSSON: Induction of sarcoma in rats by a variant of Rous virus. Acta path. microbiol. scand. **54**, 145—172 (1962).

AHLSTRÖM, C. G., N. JONSSON, and N. FORSBY: Rous sarcoma in mammals. Acta path. microbiol. scand. **154**, 127—129 (1962).

ALBERT, D. M., A. S. RABSON, P. A. GRIMES, and L. VON SALLMAN: Neoplastic transformation *in vitro* of hamster lens epithelium by simian virus 40. Science **164**, 1077—1078 (1969).

ALFERT, M., and N. K. DAS: Evidence for control of the rate of nuclear DNA synthesis by the nuclear membrane in eukaryotic cells. Proc. nat. Acad. Sci. (Wash.) **63**, 123—128 (1969).

ALLEN, D. W.: Characterization of avian leukosis group-specific antigen from avian myeloblastosis virus. Biochim. biophys. Acta (Amst.) **154**, 388—396 (1968).

ALLEN, D. W.: The N terminal amino acid of an avian leukosis group-specific antigen from avian myeloblastosis virus. Virology **38**, 32—41 (1969).

ALLISON, A. C.: Interference with, and interferon production by, polyoma virus. Virology **15**, 47—51 (1961).

ALONI, Y., E. WINOCOUR, and L. SACHS: Characterization of the simian virus 40-specific RNA in virus-yielding and transformed cells. J. molec. Biol. **31**, 415—429 (1968).

ALTANER, C., and F. SVEC: Virus production in rat tumours induced by chicken sarcoma virus. J. nat. Cancer Inst. **37**, 745—752 (1966).

ALTSTEIN, A. D., and N. N. DODONOVA: Interaction between human and simian adenoviruses in simian cells. Complementation, phenotypic mixing and formation of monkey cell "adapted" virions. Virology **35**, 248—254 (1968).

ALTSTEIN, A. D., O. F. SÁRYCHEVA, and N. N. DODONOVA: Transforming activity of green monkey SA 7 (C 8) adenovirus in tissue culture. Science **158**, 1455—1456 (1967).

ALTSTEIN, A. D., E. N. TSETLIN, N. N. DODONOVA, N. N. VASILIEVA, I. S. LEVENBOOK, and A. E. CHIGIRINSKY: Oncogenicity of adenovirus isolated from green monkeys in hamsters. Neoplasma **15**, 113—116 (1968).

AMBS, E., und B. THORELL: Die Cytogenese bei der Virusleukämie des Huhnes. Acta haemat. (Basel) **21**, 284—294 (1959).

AMBS, E., and B. THORELL: On the type of hemoglobin in the virus-induced fowl erythroleukemia. J. nat. Cancer Inst. **25**, 685—695 (1960).

AMIES, C. R., and J. G. CARR: Immunological experiments with highly concentrated suspensions of the Rous I tumor-producing agent. J. Path. Bact. **49**, 497—513 (1939).

ANDERER, F. A., M. A. KOCH, and H. D. SCHLUMBERGER: Structure of simian virus. III. Alkaline degradation of the virus particle. Virology **34**, 452—458 (1968).

ANDERER, F. A., H. D. SCHLUMBERGER, M. A. KOCH, H. FRANK, and H. J. EGGERS: Structure of simian virus 40. II. Symmetry and components of the virus particle. Virology **32**, 511—523 (1967).

ANDRÉ, J. A., R. S. SCHWARTZ, W. J. MITUS, and W. DAMESHEK: The morphologic responses of the lymphoid system to homografts. I. First and second set responses in normal rabbits. Blood **19**, 313—333 (1962).

ANDRESEN, W. F., V. J. EVANS, F. M. PRICE, and TH. B. DUNN: Neoplastic transformations in cells explanted from the kidney of 3-day-old C3H mice. J. nat. Cancer Inst. **36**, 953—963 (1966).

ANDREWES, C. H.: The transmission of fowl tumours to pheasants. J. Path. Bact. 35, 407—413 (1932).

ANDREWES, C. H., and D. M. HORSTMANN: The susceptibility of viruses to ethyl ether. J. gen. Microbiol. 3, 290—297 (1949).

Anonymous: Suggestions for the classification of oncogenic RNA viruses. J. nat. Cancer Inst. 37, 395—397 (1966).

ARMSTRONG, D.: Multiple group-specific antigen components of avian tumor viruses detected with chicken and hamster sera. J. Virol. 3, 133—139 (1969).

ARMSTRONG, D., G. HENLE, and W. HENLE: Complement fixation tests with cell lines derived from Burkitt's lymphoma and acute leukemia. J. Bact. 91, 1257—1262 (1966).

ASHKENAZI, A., and J. L. MELNICK: Tumorigenicity of simian papovavirus SV 40 and of virus-transformed cells. J. nat. Cancer Inst. 30, 1227—1264 (1963).

AUER, C., A. ZETTERBERG, and D. KILLANDER: Changes in the binding between DNA and arginine residues in histone induced by cell crowding. Exp. Cell Res., 62, 32—38 (1970).

AXELROD, D., K. HABEL, and E. T. BOLTON: Polyoma virus genetic material in a virus-free polyoma-induced tumor. Science 146, 1466—1469 (1964).

AXLER, D. A., and R. L. CROWELL: Effect of anticellular serum on the attachment of enteroviruses to HeLa cells. J. Virol. 2, 813—821 (1968).

BADER, J. P.: Nucleic acids of Rous sarcoma virus and infected cells. Nat. Cancer Inst. Monogr. 17, 781—790 (1964a).

BADER, J. P.: The role of deoxyribonucleic acid in the synthesis of Rous sarcoma virus. Virology 22, 462—468 (1964b).

BADER, J. P.: Transformation by Rous sarcoma virus: A Requirement for DNA synthesis. Science 149, 757—758 (1965a).

BADER, J. P.: The requirement for DNA synthesis in the growth of Rous sarcoma and Rous-associated viruses. Virology 26, 253—261 (1965b).

BADER, J. P.: Metabolic requirements for infection by Rous sarcoma virus. In: Subviral Carcinogenesis, YOHEI ITO, ed., 1st Int. Symp. Tumor Viruses, Nagoya, "Nisshia" Print. Comp. Mibu, Nakagyo-ku, Kyoto, pp. 144—155 (1966a).

BADER, J. P.: Metabolic requirements for infection by Rous sarcoma virus. 1. The transient requirement for DNA synthesis. Virology 29, 444—451 (1966b).

BADER, J. P.: Metabolic requirements for infection by Rous sarcoma virus. II. The participation of cellular DNA. Virology 29, 452—461 (1966c).

BADER, J. P.: A change in growth potential of cells after conversion by Rous sarcoma virus. J. cell. comp. Physiol. 70, 301—308 (1968).

BAGDONAS, V., and C. OLSON: Observations on the epizootiology of cutaneous papillomatosis (Warts) of cattle. J. Amer. vet. med. Ass. 140, 50—52 (1953).

BALTIMORE, D.: Viral RNA-dependent DNA polymerase. Nature (Lond.) 226, 1209—1211 (1970).

BALUDA, M. A.: Properties of cells infected with avian myeloblastosis virus. Cold Spr. Harb. Symp. quant. Biol. 27, 415—425 (1962).

BALUDA, M. A.: Characterization of some properties of RNA from leukemia viruses. In: Subviral Carcinogenesis, YOHEI ITO, ed., 1st Int. Symp. Tumor Viruses, Nagoya, "Nisshia" Print. Comp., Mibu, Nakagyo-ku, Kyoto, pp. 19—35 (1966).

BALUDA, M. A., and I. E. GOETZ: Morphological conversion of cell cultures by avian myeloblastosis virus. Virology 15, 185—199 (1961).

BALUDA, M. A., I. E. GOETZ, and S. OHNO: Induction of differentiation in certain target cells by avian myeloblastosis virus: An in vitro study. In: Viruses, Nucleic Acids and Cancer, Williams and Wilkins Comp., Baltimore, pp. 387—400 (1963).

BALUDA, M. A., and P. P. JAMIESON: In vivo infectivity studies with avian myeloblastosis virus. Virology 14, 33—45 (1961).

BALUDA, M. A., C. MOSCOVICI, and I. E. GOETZ: Specificity of the in vitro inductive effect of avian myeloblastosis virus. Nat. Cancer Inst. Monogr. No. 17, 449—458 (1964).

BARBIERI, D., et G. BARSKI: Baisse de malignité d'une souche cellulaire de souris a la suite de l'infection chronique *in vitro* par le virus de la leucémie de Rauscher. Rev. franç. Étud. clin. biol. **13**, 703—706 (1968).

BARRY, R. D., D. R. IVES, and J. G. CRUICKSHANK: Participation of deoxyribonucleic acid in the multiplication of influenza virus. Nature (Lond.) **194**, 1139—1140 (1962).

BARSKI, G., C. BILLARDON, P. M. JULIEN et E. CARSWELL: Evolution *in vitro* et cancérisation des cellules pulmonaires de souris. Int. J. Cancer **1**, 541—556 (1966).

BARSKI, G., and R. CASSIGENA: Malignant transformation *in vitro* of cells from C5BL mouse normal pulmonary tissue. J. nat. Cancer Inst. **30**, 865—883 (1963).

BARSKI, G., S. MANDAL, J. BELEHRADEK, and D. BARBIERI: Search for antigenic modification in mouse cell lines spontaneously converted to malignancy *in vitro*. In: Specific Tumour Antigens, Munksgaard, Copenhagen **2**, 232—236 (1967).

BARSKI, G., and E. WOLFF: Malignancy evaluation of *in vitro* transformation of mouse cell lines in chick mesonephros organ cultures. J. Nat. Cancer Inst. **34**, 495—510 (1965).

BARSKI, G., and J. K. YOUN: Protective effect of specific immunization in Rauscher leukemia. Nat. Cancer Inst. Monogr. **22**, 659—669 (1966).

BARTH, P., J. RIMAN, and F. SORM: The problem of the phage-like structure of the avian leukosis virus. Experientia (Basel) **19**, 635—637 (1963).

BASES, R.: Antigenic differences between normal and polyoma virus-transformed cells. II. *In vitro* evidence for a virus-induced antigen. Cancer Res. **24**, 1216—1221 (1964).

BASES, R., and A. S. KING: Inhibition of Rauscher murine leukemia virus growth *in vitro* by actinomycin D. Virology **32**, 175—183 (1967).

BASILICO, C., and G. DI MAYORCA: Radiation target size of the lytic and the transforming ability of polyoma virus. Proc. nat. Acad. Sci. (Wash.) **54**, 125—127 (1965).

BASILICO, C., and G. MARIN: Susceptibility of cells in different stages of the mitotic cycle to transformation by polyoma virus. Virology **28**, 429—437 (1966).

BASILICO, C., Y. MATSUYA, and H. GREEN: Origin of the thymidine kinase induced by polyoma virus in productively infected cells. J. Virol. **3**, 140—145 (1969).

BASRUR, P. K., and J. P. W. GILMAN: Behavior of two cell strains derived from rat rhabdomyosarcomas. J. nat. Cancer Inst. **30**, 163—201 (1963).

BASSIN, R. H., P. J. SIMONS, F. C. CHESTERMAN, and J. J. HARVEY: Murine sarcoma virus (Harvey): Characteristics of focus formation in mouse embryo cell cultures, and virus production by hamster tumor cells. Int. J. Cancer **3**, 265—272 (1968).

BATHER, R.: The nucleic acid of partially purified Rous No. 1 sarcoma virus. Brit. J. Cancer **11**, 611—619 (1957).

BATHER, R., A. LEONARD, and J. YANG: Characteristics of the *in vitro* assay of murine sarcoma virus (Moloney) and virus-infected cells. J. nat. Cancer Inst. **40**, 551—560 (1968).

BAUER, H.: Untersuchungen über das Myeloblastose-Virus des Huhnes (BAI-Stamm A). II. Isolierung und Charakterisierung der im Virus enthaltenen Nucleinsäure. Z. Naturforsch. **21b**, 453—460 (1966).

BAUER, H., H. BAHNEMANN und W. SCHÄFER: Untersuchungen über das Myeloblastose-Virus des Huhnes (BAI-Stamm A). I. Nachweis und Vermehrungskinetik des Virus in Hühnerfibroblasten-Kulturen. Z. Naturforsch. **20b**, 959—965 (1965).

BAUER, H., and T. GRAF: Evidence for the possible existence of two envelope antigenic determinants and corresponding cell receptors for avian tumor viruses. Virology **37**, 157—161 (1969).

BAUER, H., and H.-G. JANDA: Group-specific antigen of avian leukosis viruses. Virus specificity and relation to an antigen contained in Rous mammalian tumor cells. Virology **33**, 483—490 (1967).

BAUER, H., und W. SCHÄFER: Isolierung eines gruppenspezifischen Antigens aus dem Hühnermyeloblastose-Virus (BAI-Stamm A). Z. Naturforsch. **20b**. 815—816 (1965).

BAUER, H., and W. SCHÄFER: Origin of group-specific antigen of chicken leukosis viruses. Virology **29**, 494—497 (1966).

BAUER, H., H. TOZAWA, D. P. BOLOGNESI, T. GRAF, and H. GELDERBLOM: Structure and specific antigens of avian leukosis viruses. Proc. 2nd Inst. Meeting on Comparative Leukemia Research, Cherry Hill, 1969.

BAUM, S. G., P. R. REICH, C. J. HYBNER, W. P. ROWE, and S. M. WEISSMAN: Biophysical evidence for linkage of adenovirus and SV40 DNA's in adenovirus 7-SV40 hybrid particles. Proc. nat. Acad. Sci. (Wash.) 56, 1509—1515 (1966).

BAUM, S. G., W. H. WIESE, and P. R. REICH: Studies on the mechanism of enhancement of adenovirus 7 infection in African green monkey cells by simian virus 40. Formation of adenovirus specific RNA. Virology 34, 373—376 (1968).

BEARD, J. W.: Avian virus growths and their etiologic agents. Advanc. Cancer Res. 7, 1—127 (1963).

BEARD, D., and J. W. BEARD: Virus of avian erythroblastosis. VI. Further studies on neutralization by antiserum to normal chicken protein. J. nat. Cancer Inst. 19, 923—939 (1957).

BEARD, D., G. S. BEAUDREAU, R. A. BONAR, D. G. SHARP, and J. W. BEARD: Virus of avian erythroblastosis. III. Antigenic constitution and relation to the agent of myeloblastosis. J. nat. Cancer Inst. 18, 231—259 (1957).

BEARDMORE, W. B., M. J. HAVLICK, A. SERAFINI, and I. W. MCLEAN, JR.: Interrelationship of adenovirus (type 4) and papovavirus (SV40) in monkey kidney cell cultures. J. Immunol. 95, 422—435 (1965).

BEAUDREAU, G. S., C. BECKER, R. A. BONAR, A. M. WALLBANK, D. BEARD, and J. W. BEARD: Virus of avian myeloblastosis. XIV. Neoplastic response of normal chicken bone marrow treated with the virus in tissue culture. J. nat. Cancer Inst. 24, 395—415 (1960).

BEAUDREAU, G. S., C. BECKER, D. G. SHARP, J. C. PAINTER, and J. W. BEARD: Virus of avian myeloblastosis. XI. Release of the virus by myeloblasts in tissue culture. J. nat. Cancer Inst. 20, 351—381 (1958).

BECKER, C., G. S. BEAUDREAU, W. CASTLE, B. W. GIBSON, D. BEARD, and J. W. BEARD: Virus of avian myeloblastosis. XXII. Influence of folic acid and other factors on growth of myeloblasts and virus synthesis in tissue culture. J. nat. Cancer Inst. 29, 455—481 (1962).

BECKER, Y.: RNA synthesis in cultured chick embryo fibroblasts. Exp. Cell Res. 47, 554—563 (1967).

BELL, T. M., A. MASSIE, M. G. R. ROSS, D. I. H. SIMPSON, and E. F. GRIFFIN: Further isolations of reovirus type 3 from cases of Burkitt's lymphoma. Brit. med. J. i, 1514—1517 (1966).

BENDICH, A., A. D. VIZOSO, and R. G. HARRIS: Intercellular bridges between mammalian cells in culture. Proc. nat. Acad. Sci. (Wash.) 57, 1029—1035 (1967).

BENEDETTI, E. L., et W. BERNHARD: Recherches ultrastructurales sur le virus de la leucémie érythroblastique du poulet. J. Ultrastruct. Res. 1, 309—336 (1958).

BENJAMIN, T. L.: Relative target sizes for the inactivation of the transforming and reproducing abilities of polyoma virus. Proc. nat. Acad. Sci. (Wash.) 54, 121—124 (1965).

BENJAMIN, T. L.: Virus-specific RNA in cells productively infected or transformed by polyoma virus. J. molec. Biol. 16, 359—373 (1966a).

BENJAMIN, T. L.: Radiobiological and biochemical investigations of polyoma virus-cell interactions. In: Subviral Carcinogenesis, YOHEI ITO, ed., 1st Int. Symp. Tumor Viruses, Nagoya, "Nisshia" Print. Comp. Mibu, Nakagyo-ku, Kyoto, pp. 62—81 (1966b).

BENJAMIN, T. L.: Absence of homology between polyoma or SV40 viral DNA and mitochondrial DNA from virus-induced tumors. Virology 36, 685—687 (1968).

BEN-PORAT, T., C. CATO, and A. S. KAPLAN: Unstable DNA synthesized by polyoma virus-infected cells. Virology 30, 74—81 (1966).

BEN-PORAT, T., and A. KAPLAN: The synthesis and fate of pseudorabies virus DNA in infected mammalian cells in the stationary phase of growth. Virology 20, 310—317 (1963).

BEN-PORAT, T., and A. KAPLAN: Correlation between replication and degradation of cellular DNA in polyoma virus-infected cells. Virology 32, 457—464 (1967).

BENYESH-MELNICK, M., D. J. FERNBACH, and R. T. LEWIS: Studies on human leukemia. I. Spontaneous lymphoblastoid transformation of fibroblastic bone marrow cultures derived from leukemic and nonleukemic children. J. nat. Cancer Inst. 31, 1311—1325 (1963).

BENYESH-MELNICK, M., O. K. SMITH, and D. J. FERNBACH: Studies on human leukemia. III. Electron microscopic findings in children with acute leukemia and in children with infectious mononucleosis. J. nat. Cancer Inst. 33, 571—579 (1964).

BERECZKY, E., and L. DMOCHOWSKI: Continuous phase contrast cinemicrophotography of polyoma infected mouse embryo cells. Virology 18, 320—324 (1962).

BERECZKY, E., L. DMOCHOWSKI, and C. E. GREY: Study of host-virus relationship. I. Light, phase and fluorescence microscopy of mouseembryo cells infected with polyoma virus. J. nat. Cancer Inst. 27, 99—113 (1961).

BERG, R., and U. STENRAM: Growth in rat of a sarcoma derived from rat kidney cells transformed in vitro by SV40. Acta path. microbiol. scand. 73, 305—315 (1968).

BERGMAN, S., and N. JONSSON: In vitro studies of Rous sarcoma. Acta path. microbiol. scand. 154, 130—133 (1962).

BERGS, V. V., and V. GROUPÉ: Role of humoral antibody and infecting dose in the recovery of Rous sarcoma virus from turkey tumor cells in tissue culture. J. nat. Cancer Inst. 29, 739—748 (1962).

BERGS, V. V., and V. GROUPÉ: Low malignancy of Rous sarcoma cells as evidenced by poor transplantability in turkeys. Science 139, 922—923 (1963).

BERMAN, L.: Comparative morphologic study of the virus-induced solid tumor of syrian hamsters. J. nat. Cancer Inst. 39, 847—901 (1967).

BERMAN, L., and C. S. STULBERG: Eight culture strains (Detroit) of human epitheliallike cells. Proc. Soc. exp. Biol. (N.Y.) 92, 730—735 (1956).

BERMAN, L., C. S. STULBERG, and F. H. RUDDLE: Human cell culture. Morphology of the Detroit strains. Cancer Res. 17, 668—676 (1957).

BERMAN, L. D., and A. C. ALLISON: Studies on murine sarcoma virus: A morphological comparison of tumorigenesis by the Harvey and Moloney strains in mice, and the establishment of tumor cell lines. Int. J. Cancer 4, 820—836 (1969).

BERNARD, C., M. BOIRON et J. LASNERET: Transformation et infection chronique de cellules embryonaires de rat par le virus du sarcoma de Moloney. C. R. Acad. Sci. (Paris) 264, 2170—2173 (1967).

BERNARD, C., B. GUILLEMAIN, J. PÉRIÈS et M. BOIRON: Conversion cellulaire provoquée in vitro par le virus du sarcoma murin (Moloney). Analyse de la courbe dose-réponse. Int. J. Cancer 3, 558—565 (1968).

BERNARD, C., D. SILVESTRE, J. TANZER, J. PÉRIÈS et M. BOIRON: Analyse d'une lignée cellulaire produisant de façon continue le virus leucémique de Rauscher in vitro. Exp. Cell Res. 40, 513—526 (1965).

BERNHARD, W.: Electron microscopy of tumor cells and tumor viruses: A review. Cancer Res. 18, 491—509 (1958).

BERNHARD, W.: The detection and study of tumor viruses with the electron microscope. Cancer Res. 20, 712—727 (1960).

BERNHARD, W., R. A. BONAR, D. BEARD, and J. W. BEARD: Ultrastructure of viruses of myeloblastosis and erythroblastosis isolated from plasma of leukemic chickens. Proc. Soc. exp. Biol. (N.Y.) 97, 48—52 (1958).

BERNHARD, W., H. L. FEBVRE et R. CRAMER: Mise en évidence au microscope électronique d'un virus dans les cellules infectées in vitro par l'agent du polyome. C. R. Acad. Sci. (Paris) 249, 483—485 (1959).

BERNHARD, W., CH. OBERLING et P. VIGIER: L'ultrastructure de virus dans le sarcome de Rous, leur rapport avec le cytoplasme des cellules tumorales. Bull. Cancer 43, 407—422 (1956).

BERNHARD, W., and P. TOURNIER: Ultrastructural cytochemistry applied to the study of virus infection. Cold Spr. Harb. Symp. quant. Biol. 27, 67—82 (1962).

BIBERFELD, P., and N. RINGERTZ: The application of immune electron microscopy to the demonstration of polyoma virus antigen in cultured mouse embryo cells. J. nat. Cancer Inst. 37, 451—465 (1966).

BICHEL, J.: Cultivation of leukemic cells in tissue culture. Acta path. microbiol. scand. **31**, 410—419 (1952).

BIGGS, P. M.: A discussion on the classification of the avian leukosis complex and fowl paralysis. Brit. vet. J. **117**, 326—334 (1961).

BIGGS, P. M., and L. N. PAYNE: Studies on Marek's disease. I. Experimental transmission. J. nat. Cancer Inst. **39**, 267—280 (1967a).

BIGGS, P. M., and L. N. PAYNE: The avian leukosis complex. Vet. Rec. (Suppl.) **80**, V-VII (1967b).

BILLINGHAM, R. E., L. W. FERRIGAN, and W. K. SILVERS: Cheek pouch of the Syrian hamster and tissue transplantation immunity. Science **132**, 1488 (only) (1960).

BIRNIE, G. D., and S. M. FOX: Deoxyribonucleic acid synthesis in polyoma-virus-infected mouse embryo cells. Biochem. J. **95**, 41P—42P (1965).

BISHUN, N. P., and R. N. P. SUTTON: Cytogenetic and other studies on the EB4 line of Burkitt tumour cells. Brit. J. Cancer **21**, 675—678 (1967).

BISSETT, M. L., and F. E. PAYNE: Development of antigens in human cells infected with simian virus 40. J. Bact. **91**, 743—749 (1966).

BLACK, P.: Studies on the genetic susceptibility of cells to polyoma virus transformation. Virology **24**, 179—185 (1964).

BLACK, P. H.: Transformation of mouse cell line 3T3 by SV40: Dose response relationship and correlation with SV40 tumor antigen production. Virology **28**, 760—763 (1966a).

BLACK, P. H.: A summary of studies on transformation by the adenovirus-SV40 hybrid viruses. In: Subviral Carcinogenesis, YOHEI ITO, ed., 1st. Int. Symp. Tumor Viruses, Nagoya, "Nisshia" Print. Comp. Mibu, Nakagyo-ku, Kyoto, pp. 220—234 (1966b).

BLACK, P. H.: The SV40 T-antigen. *In vitro*. Publication of the Tissue Culture Assoc., vol. II. Phenotypic Expression, Immunological, Biochemical, Morphological Annual Symposium May 31—June 3, 1966, c. Williams and Wilkins Comp.

BLACK, P. H.: An analysis of SV40-induced transformation of hamster kidney tissue *in vitro*. II. Persistence of SV40 viral genome in clones of transformed hamster cells. J. nat. Cancer Inst. **37**, 487—493 (1966d).

BLACK, P. H.: The oncogenic DNA viruses: A review of *in vitro* transformation studies. Ann. Rev. Microbiol. **22**, 391—426 (1968).

BLACK, P. H., L. D. BERMAN, and C. DIXON: *In vitro* transformation by adenovirus-simian virus 40 hybrid viruses. IV. Properties of clones isolated from cell lines transformed by adenovirus2-simian virus 40 and adenovirus12-simian virus 40 transcapsidant hybrid viruses. J. Virology **4**, 694—703 (1969).

BLACK, P. H., E. M. CRAWFORD, and L. V. CRAWFORD: The purification of simian virus 40. Virology **24**, 381—387 (1964).

BLACK, P. H., J. W. HARTLEY, W. P. ROWE, and R. J. HUEBNER: Transformation of bovine tissue culture cells by bovine papilloma virus. Nature (Lond.) **199**, 1016—1018 (1963).

BLACK, P. H., and H. IGEL: Studies on transformation by the adenovirus-SV40 hybrid viruses. In: Malignant Transformation by Viruses. W. H. KIRSTEN, ed., Springer-Verlag, Berlin—Heidelberg—New York, pp. 95—103 (1966).

BLACK, P., and W. P. ROWE: Transformation in hamster kidney monolayers by vacuolating virus SV40. Virology **19**, 107—109 (1963a).

BLACK, P. H., and W. P. ROWE: An analysis of SV40 induced transformation of hamster kidney tissue *in vitro*. I. General characteristics. Proc. nat. Acad. Sci. (Wash.) **50**, 606—613 (1963b).

BLACK, P. H., and W. P. ROWE: SV40-induced proliferation of tissue culture cells of rabbit, mouse and porcine origin. Proc. nat. Acad. Sci. (Wash.) **114**, 721—727 (1963c).

BLACK, P. H., and W. P. ROWE: Viral studies of SV40 tumorigenesis in hamsters. J. nat. Cancer Inst. **32**, 253—265 (1964).

BLACK, P. H., and W. P. ROWE: Induction of SV40 T-antigen with SV40 DNA. Virology **27**, 436—439 (1965).

BLACK, P., W. P. ROWE, and H. L. COOPER: An analysis of SV40 induced transformation of hamster kidney tissue *in vitro*. II. Studies of 3 clones derived from a
   continuous line of transformed cells. Proc. nat. Acad. Sci. (Wash.) **50**, 847—854
   (1963).
BLACK, P. H., W. P. ROWE, H. C. TURNER, and R. J. HUEBNER: A specific complement-fixing antigen present in SV40 tumor and transformed cells. Proc. nat. Acad.
   Sci. (Wash.) **50**, 1148—1156 (1963).
BLACK, P. H., and G. J. TODARO: *In vitro* transformation of hamster and human
   cells with the adeno 7-SV40 hybrid virus. Proc. nat. Acad. Sci. (Wash.) **54**, 374—
   381 (1965).
BLACK, P. H., and B. J. WHITE: *In vitro* transformation by the adenovirus-SV40
   hybrid viruses. II. Characteristics of the transformation of hamster cells by the
   adeno 2-, adeno 3-, and adeno 12-SV40 viruses. J. exp. Med. **125**, 629—646 (1967).
BLACKLOW, N. R., M. D. HOGGAN, and W. P. ROWE: Immunofluorescent studies of
   the potentiation of an adenovirus-associated virus by adenovirus 7. J. exp. Med.
   125, 755—765 (1967).
BLAIR, C. D., and P. H. DUESBERG: Structure of Rauscher mouse leukaemia virus
   RNA. Nature (Lond.) **220**, 396—399 (1968).
BLOOM, S., G. J. TODARO, and H. GREEN: RNA synthesis during preparation for
   growth in a resting population of mammalian cells. Biochem. biophys. Res.,
   Commun. **24**, 412—417 (1966).
BOEYÉ, A., J. L. MELNICK, and F. RAPP: SV40-adenovirus "hybrids": Presence of
   two genotypes and the requirement of their complementation for viral repli-
   cation. Virology **28**, 56—70 (1966).
BOIRON, M., J. P. LÉVY, J. LASNERET, S. OPPENHEIM, and J. BERNARD: Pathogenesis
   of Rauscher leukemia. J. nat. Cancer Inst. **35**, 865—884 (1965).
BOIRON, M., J. P. LÉVY, and J. PÉRIES: *In vitro* investigations on murine leukemia
   viruses. Progr. med. Virol. **9**, 341—391 (1967).
BOIRON, M., J. P. LÉVY, M. THOMAS, J. C. FRIEDMANN, and J. BERNARD: Some
   properties of bovine papilloma virus. Nature (Lond.) **201**, 423—424 (1964).
BOLUND, L., N. R. RINGERTZ, and H. HARRIS: Changes in the cytochemical properties
   of erythrocyte nuclei reactivated by cell fusion. J. Cell Sci. **4**, 71—87 (1969).
BONAR, R. A., and J. W. BEARD: Virus of avian myeloblastosis. XII. Chemical con-
   stitution. J. nat. Cancer Inst. **23**, 183—197 (1959).
BONAR, R. A., U. HEINE, and J. W. BEARD: Structure of BAI strain A (myeloblastosis)
   avian tumor virus. Nat. Cancer Inst. Monogr. No. **17**, 589—602 (1964).
BONAR, R. A., U. HEINE, D. BEARD, and J. W. BEARD: Virus of avian myeloblastosis
   (BAI strain A). XXIII. Morphology of virus and comparison with strain R (erythroblastosis). J. nat. Cancer Inst. **30**, 949—997 (1963).
BONAR, R. A., D. F. PARSONS, G. S. BEAUDREAU, C. BECKER, and J. W. BEARD:
   Ultrastructure of avian myeloblasts in tissue culture. J. nat. Cancer Inst. **23**,
   199—225 (1959).
BONAR, R. A., R. H. PURCELL, D. BEARD, and J. W. BEARD: Virus of avian myelo-
   blastosis (BAI strain A). XXIV. Nucleotide composition of the pentosenucleic
   acid and comparison with strain R (erythroblastosis). J. nat. Cancer Inst. **31**,
   705—716 (1963).
BONAR, R. A., L. SVERAK, D. P. BOLOGNESI, A. J. LANGLOIS, D. BEARD, and J. W.
   BEARD: Ribonucleic acid components of BAI strain A (myeloblastosis) avian
   tumor virus. Cancer Res. **27**, 1138—1157 (1967).
BONAR, R. A., D. WEINSTEIN, J. R. SOMMER, D. BEARD, and J. W. BEARD: Virus of
   avian myeloblastosis. XVII. Morphology of progressive virus myeloblast interactions *in vitro*. Nat. Cancer Inst. Monogr. No. **4**, 251—290 (1960).
BOREK, C., and L. SACHS: The difference in contact inhibition of cell replication
   between normal cells and cells transformed by different carcinogens. Proc. nat.
   Acad. Sci. (Wash.) **56**, 1705—1711 (1966a).
BOREK, C., and L. SACHS: *In vitro* cell transformation by X-irradiation. Nature (Lond.)
   210, 276—278 (1966b).

BOREK, C., and L. SACHS: Cell susceptibility to transformation by X-irradiation and fixation of the transformed state. Proc. nat. Acad. Sci. (Wash.) **57**, 1522—1527 (1967).

BORGES, P. R. F., and F. DURAN-REYNALS: On the induction of malignant tumors in pigeons by a chicken sarcoma virus after previous adaptation of the virus to ducks. Cancer Res. **12**, 55—58 (1952).

BORSOS, T.: Absence of neutralization of Rous sarcoma virus by antinormal-chicken embryo serum and complement. J. nat. Cancer Inst. **20**, 1215—1221 (1958).

BOURGAUX, P.: The fate of polyoma virus in hamster, mouse and human cells. Virology **23**, 46—55 (1964).

BOURGAUX, P., D. BOURGAUX-RAMOISY, and M. STOKER: Further studies on transformation by DNA from polyoma virus. Virology **25**, 364—371 (1965).

BOWER, R. K.: A quantitation of the influence of the chick embryo genotype on tumor production by Rous sarcoma virus on the chorioallantoic membrane. Virology 18, 372—377 (1962).

BOWER, R. K., N. R. GYLES, and C. J. BROWN: Tumor induction by Rous sarcoma virus on the chorioallantoic membranes of reciprocal crosses between resistant and susceptible strains of chickens. Virology **24**, 47—50 (1964).

BRAILOVSKY, C., R. WICKER, H. SUAREZ et R. CASSIGENA: Transformation *in vitro* de cellules de hamster chinois par l'adénovirus 12: Étude cytogénétique. Int. J. Cancer **2**, 133—142 (1967).

BRAND, K. G., and J. T. SYVERTON: Immunology of cultivated mammalian cells. I. Species specificity determined by hemagglutination. J. nat. Cancer Inst. **24**, 1007—1019 (1960).

BRAND, K. G., and J. T. SYVERTON: Results of species specific hemagglutination tests on "transformed" non-transformed, and primary cell cultures. J. nat. Cancer Inst. **28**, 147—157 (1962).

BRANTON, P. E., and R. SHEININ: Control of DNA synthesis in cells infected with polyoma virus. Virology **36**, 690—694 (1968).

BUBENÍK, J., L. DONNER, P. CHÝLE, P. KOLDOVSKÝ, V. MALASKOVÁ, J. LIBÁNSKÝ, L. VEPREK, and J. RÍMAN: Fibroblastoid transformation of malignant myeloblasts *in vitro*. Folia biol. (Praha) **14**, 359—371 (1968).

BUCCIARELLI, E., G. F. RABOTTI, and A. J. DALTON: Ultrastructure of meningeal tumors induced in dogs with Rous sarcoma virus. J. nat. Cancer Inst. **38**, 359—381 (1967).

BULLOUGH, W. S.: The chalones. Science J. **5**, 71—76 (1969).

BURDETTE, W. J., and J. S. YOON: Mutations, chromosomal aberrations and tumors in insects treated with oncogenic virus. Science **155**, 340—341 (1967).

BURGER M. M.: A difference in the architecture of the surface membrane of normal and virally transformed cells. Proc. nat. Acad. Sci. (Wash.), **62**, 994—1001 (1969).

BURGER, M. M. and K. D. NOONAN: Restoration of normal growth by covering of agglutinin sites on tumor cell surface. Nature (Lond.) **228**, 512 — 515 (1970).

BURKITT, D.: A sarcoma involving the jaws in African children. Brit. J. Surg. **46**, 218—223 (1958).

BURKITT, D.: A lymphoma syndrome in African children. Ann. roy. Coll. Surg. Engl. **30**, 211—219 (1962a).

BURKITT, D.: A tumor syndrome affecting children in tropical Africa. Postgrad. med. J. **38**, 71—79 (1962b).

BURKITT, D.: A children's cancer dependent on environment. In: Viruses, Nucleic Acids and Cancer, Williams and Wilkins Comp., pp. 615—629 (1963a).

BURKITT, D.: A lymphoma syndrome in tropical Africa. Int. Rev. exp. Path. **2**, 67—96 (1963b).

BURMESTER, B. R.: Transmission of tumor inducing avian viruses under natural conditions. Tex. Symp. Cancer Res. **15**, 92—110 (1957).

BURMESTER, B. R.: Current status of the avian leukosis complex. Agricult. Res. Serv. U.S. ARS 44—195 pp. 1—16 (1967).

BURMESTER, B. R., A. K. FONTES, and W. G. WALTER: Contact transmission of Rous sarcoma. J. nat. Cancer Inst. **25**, 307—313 (1960).

BURMESTER, B. R., A. K. FONTES, N. F. WATERS, W. R. BRYAN, and V. GROUPÉ: The response of several inbred lines of white leghorns to inoculation with the viruses of strain RPL 12 visceral lymphomatosis-erythroblastosis and of Rous sarcoma. Poultry Sci. **39**, 199—215 (1960).

BURMESTER, B. R., and T. N. FREDRICKSON: Some factors influencing the rate of contact transmission of Rous sarcoma virus. Avian Dis. **10**, 259—267 (1966).

BURMESTER, B. R., and R. L. WITTER: An outline of the diseases of the avian leukosis complex. Prod. Res. Rep. No. 94, Agricult. Res. Serv., 1—8 (1966).

BURNS, W. H., and P. H. BLACK: Analysis of simian virus 40-induced transformation of hamster kidney tissue *in vitro*. V. Variability of virus recovery from cell clones inducible with mitomycin C and cell fusion. J. Virol. **2**, 606—609 (1968).

BURNS, W. H., and P. H. BLACK: Induction experiments with adenovirus and polyoma virus transformed cell lines. Int. J. Cancer **4**, 204—211 (1969).

BUTEL, J. S., and F. RAPP: The effect of arabinofuranosylcytasine on the growth cycle of simian virus 40. Virology **27**, 490—495 (1965).

BUTEL, J. S., and F. RAPP: Complementation between a defective monkey cell-adapting component and human adenoviruses in simian cells. Virology **31**, 573—584 (1967).

BUTEL, J. S., F. RAPP, J. L. MELNICK, and B. A. RUBIN: Replication of adenovirus type 7 in monkey cells: A new determinant and its transfer to adenovirus type 2. Science **154**, 671—673 (1966).

BÜRK, R. R.: Growth inhibitor of hamster fibroblast cells. Nature (Lond.) **212**, 1261—1262 (1966).

BÜRK, R. R.: The detection and extraction of anomin, a growth inhibitor, from non-tumor cells. In: Growth Regulating Substances for Animal Cells in Culture. (DEFENDI and STOKER eds.), The Wistar Inst. Press. The Wistar Inst. Symp. Monogr. No. **7**, 39—50 (1967).

CALNEK, B. W.: Morphological alteration of RIF-infected chick embryo fibroblasts. Nat. Cancer Inst. Monogr. No. **17**, 425—447 (1964).

CAMPBELL, J. G.: A proposed classification of the leucosis complex and fowl paralysis. Brit. vet. J. **117**, 316—325 (1961).

CARP, R. I.: Thymidine kinase from normal, simian virus 40-transformed and simian virus 40-lytically infected cells. J. Virol. **1**, 912—919 (1967).

CARP, R. I., V. DEFENDI, R. V. GILDEN, A. J. GIRARDI, F. C. JENSEN, and R. KO-PROWSKI: SV 40 and Rous antigens in human and animal tissues. UICC Monograph Series **2**, 239—250 (1966).

CARP, R. I., and R. V. GILDEN: The inactivation of simian virus 40 infectivity and antigen-inducing capacity by ultraviolet light. Virology **27**, 639—641 (1965).

CARP, R. I., and R. V. GILDEN: A comparison of the replication cycles of simian virus 40 in human diploid and African green monkey kidney cells. Virology **28**, 150—162 (1966).

CARP, R. I., G. SAUER, and F. SOKOL: The effect of actinomycin D on the transcription and replication of simian virus 40 deoxyribonucleic acid. Virology **37**, 214—226 (1969).

CARP, R. I., and F. SOKOL: Further studies on the differences in the interaction of simian virus 40 with African green monkey kidney and human diploid cells. J. gen. Virol. **5**, 433—436 (1969).

CARREL, A.: On the permanent life of tissues outside of the organism. J. exp. Med. **15**, 516—528 (1912).

CARREL, A.: Present condition of a strain of connective tissue twenty-eight months old. J. exp. Med. **20**, 1—2 (1914).

CARREL, A.: Cultures pures de fibroblastes provenant de sarcomes fusocellulaires. C. R. Soc. Biol. (Paris) **90**, 1380—1382 (1924a).

CARREL, A.: La malignité des cultures pures de monocytes du sarcome du Rous. C. R. Soc. Biol. (Paris) **91**, 1067—1069 (1924b).

CARREL, A.: Some conditions of the reproduction *in vitro* of the Rous virus. J. exp. Med. **43**, 647—668 (1926).

CARREL, A., and A. H. EBELING: The transformation of monocytes into fibroblasts through the action of the Rous virus. J. exp. Med. **43**, 461—468 (1926).

CARTER, S. B.: Tissue homeostasis and the biological basis of cancer. Nature (Lond.) **220**, 970—974 (1968).

CASPAR, D. L. D., and A. KLUG: Physical principles in the construction of regular viruses. Cold Spr. Harb. Symp. quant. Biol. **27**, 1—24 (1962).

CASPERSSON, T., G. E. FOLEY, D. KILLANDER, and G. LOMAKKA: Cytochemical differences between mammalian cell lines of normal and neoplastic origins. Correlation with heterotransplantability in Syrian hamsters. Exp. Cell Res. **32**, 553—565 (1963).

CASPERSSON, T., and L. SANTESSON: Studies on protein metabolism in the cells of epithelial tumours. Acta radiol. Suppl. **46**, 1—105 (1942).

CASPERSSON, T. O.: Cell growth and cell function. A cytochemical study. New York, W. W. Norton & Co. Inc., p. 185 (1950).

CASTO, B. C.: Adenovirus transformation of hamster embryo cells. Virology **2**, 376—383 (1968a).

CASTO, B. C.: Effects of ultraviolet irradiation on the transforming and plaque-forming capacities of simian adenovirus SA7. J. Virol. **2**, 641—642 (1968b).

CASTO, B. C.: Transformation of hamster embryo cells and tumor induction in newborn hamsters by simian adenovirus SV11. J. Virol. **3**, 513—519 (1969).

CASTOR, L. N.: Contact regulation of cell division in an epithelial-like cell line. J. cell. comp. Physiol. **72**, 161—172 (1968).

CASTOR, L. N.: Flattening, movement and control of division of epithelial-like cells. J. cell. comp. Physiol. **75**, 57—64 (1970).

CHAMORRO, M. A.: Séparation par centrifugation différentielle de deux agents distincts de la leucémogénèse de la souris. C. R. Acad. Sci. (Paris) **265**, 649—651 (1967).

CHAMORRO, M. A., R. LATARJET, P. VIGIER, and F. ZAJDELA: New investigations on the Friend disease. In: Ciba Found. Symp. Tumor Viruses of Murine Origin, pp. 176—192. Churchill Ltd., London, 1962.

CHANG, Y. T.: Long-term cultivation of mouse peritoneal macrophages. J. nat. Cancer Inst. **32**, 19—35 (1964).

CHANG, R. S.: Continuous subcultivation of epithelial-like cells from normal human tissues. Proc. Soc. exp. Biol. (N.Y.) **87**, 440—443 (1954).

CHANG, R. S., and T. J. SINSKEY: Observations on the transformation of human amnion cell cultures by simian virus 40. J. nat. Cancer Inst. **40**, 505—512 (1968).

CHAPIN, M., and G. R. DUBES: Characteristics of five Rhesus monkey kidney cell lines. Proc. Soc. exp. Biol. (N.Y.) **115**, 965—970 (1964).

CHESTERMAN, F. C., J. J. HARVEY, R. R. DOURMASHKIN, and M. H. SALAMAN: The pathology of tumors and other lesions induced in rodents by virus derived from a rat with Moloney leukemia. Cancer Res. **26**, 1759—1768 (1966).

CHEVILLE, N. F.: Studies on connective tissue tumors in the hamster produced by bovine papilloma virus. Cancer Res. **26**, 2334—2339 (1966).

CHEVILLE, N. F., and C. OLSSON: Epithelial and fibroblastic proliferation in bovine cutaneous papillomatosis. Path. Vet. **1**, 248—257 (1964).

CHIRIGOS, M. A., D. SCOTT, W. TURNER, and K. PERK: Biological, pathological and physical characterization of a possible variant of a murine sarcoma virus (Moloney). Int. J. Cancer **3**, 223—237 (1968).

CHRISTOFINIS, G. J., and A. J. BEALE: Some biological characters of cell lines derived from normal rabbit kidney. J. Path. Bact. **95**, 377—381 (1968).

CHUBB, R. C., and P. M. BIGGS: The neutralization of Rous sarcoma virus. J. gen. Virol. **3**, 87—96 (1968).

CHURCHILL, A. E.: Herpes-type virus isolated in cell culture from tumors of chickens with Marek's disease. I. Studies in cell culture. J. nat. Cancer Inst. **41**, 939—950 (1968).

CHURCHILL, A. E., and P. M. BIGGS: Agent of Marek's disease in tissue culture. Nature (Lond.) **215**, 528—530 (1967).

CHÝLE, P.: Analysis of the virus-producing interaction of rat Rous sarcomas (XC and MR₅) with the fowl cell. Folia biol. (Praha) 10, 359—365 (1964).

CHÝLE, P., V. KLEMENT, and J. SVOBODA: Attempts to induce formation of Rous sarcoma virus in cells of tumour XC. Folia biol. (Praha) 9, 92—98 (1963).

CLARKSON, B., A. STRIFE, and E. DE HARVEN: Continuous culture of seven new cell lines (SK-L1 to 7) from patients with acute leukemia. Cancer 20, 926—947 (1967).

COGGIN, J. H., V. M. LARSON, and M. R. HILLEMAN: Prevention of SV40 virus tumorigenesis by irradiated disrupted and iododeoxyuridine treated tumor cell antigens. Proc. Soc. exp. Biol. (N.Y.) 124, 774—784 (1967).

COLBY, C., and H. RUBIN: Growth and nucleic acid synthesis in normal cells and cells infected with Rous sarcoma virus. J. nat. Cancer Inst. 43, 437—444 (1969).

COMINGS, D. E., and T. KAKEFUDA: Initiation of deoxyribonucleic acid replication at the nuclear membrane in human cells. J. molec. Biol. 33, 225—229 (1968).

CONE, C. D. JR.: Observations of self-induced mitosis and autosynchrony in sarcoma cell networks. Contribution L-6066, NASA Langley Res. Ctr., Langley Stn., Hampton, Virginia, U.S.A. (1968).

COOPER, E. H., D. T. HUGHES, and N. E. TOPPING: Kinetics and chromosome analyses of tissue culture lines derived from Burkitt lymphomata. Brit. J. Cancer 20, 102—113 (1966).

COOPER, H. L., and P. H. BLACK: Cytogenetic studies of hamster kidney cell cultures transformed by the simian vacuolating virus (SV40). J. nat. Cancer Inst. 30, 1015—1043 (1963).

COURINGTON, D., and P. K. VOGT: Electron microscopy of chick fibroblasts infected by defective Rous sarcoma virus and its helper. J. Virol. 1, 400—414 (1967).

COWLES, J., J. SAIKKONEN, and B. THORELL: On the presence of hemoglobin in erythroleukemia cells. Blood 13, 1176—1184 (1958).

CRAMER, R., et L. E. FEINENDEGEN: Incorporation de thymidine dans l'adn de cellules de souris et de hamsters infectés par le virus polyoma. Int. J. Cancer 1, 149—160 (1966).

CRAWFORD, L. V.: The adsorption of polyoma virus. Virology 18, 177—181 (1962).

CRAWFORD, L. V.: The physical characteristics of polyoma virus. II. The nucleic acid. Virology 19, 279—282 (1963).

CRAWFORD, L. V.: The physical characteristics of polyoma virus. IV. The size of the DNA. Virology 22, 149—152 (1964).

CRAWFORD, L. V., and P. H. BLACK: The nucleic acid of simian virus 40. Virology 24, 388—392 (1964).

CRAWFORD, L. V., and E. M. CRAWFORD: The properties of Rous sarcoma virus purified by density gradient centrifugation. Virology 13, 227—232 (1961).

CRAWFORD, L. V., and E. M. CRAWFORD: A comparative study of polyoma and papilloma viruses. Virology 21, 258—263 (1963).

CRAWFORD, L. V., E. M. CRAWFORD, and D. H. WATSON: The physical characteristics of polyoma virus. I. Two types of particle. Virology 18, 170—176 (1962).

CRAWFORD, L. V., R. DULBECCO, M. FRIED, L. MONTAGNIER, and M. STOKER: Cell transformation by different forms of polyoma virus DNA. Proc. nat. Acad. Sci. (Wash.) 52, 148—152 (1964).

CREECH, G. T.: Experimental studies of the etiology of common warts in cattle. J. agricult. Res. 29, 723—737 (1929).

CRITTENDEN, L. B.: Observations on the nature of a genetic cellular resistance to avian tumor viruses. J. nat. Cancer Inst. 41, 145—153 (1968).

CRITTENDEN, L. B., and W. OKAZAKI: Genetic influence of the Rs locus on susceptibility to avian tumor viruses. I. Neoplasms induced by RPL12 and three strains of Rous sarcoma virus. J. nat. Cancer Inst. 35, 857—863 (1965).

CRITTENDEN, L. B., and W. OKAZAKI: Genetic influence of the Rs locus on susceptibility to avian tumor viruses. II. Rous sarcoma virus antibody production after strain RPL12 virus inoculation. J. nat. Cancer Inst. 36, 299—303 (1966).

CRITTENDEN, L. B., W. OKAZAKI, and R. REAMER: Genetic resistance to Rous sarcoma virus in embryo cell cultures and embryos. Virology 20, 541—544 (1963).

CRITTENDEN, L. B., W. OKAZAKI, and R. REAMER: Genetic control of responses to Rous sarcoma and strain RPL12 viruses in the cells, embryos and chickens of two inbred lines. Nat. Cancer Inst., Monogr. 17, 161—177 (1964).

CRITTENDEN, L. B., H. A. STONE, R. REAMER, and W. OKAZAKI: Two loci controlling genetic cellular resistance to avian leukosis-sarcoma viruses. J. Virol. 1, 898—904 (1967).

CURTIS, A. S. G.: Control of some cell-contact reactions in tissue culture. J. nat. Cancer Inst. 26, 253—268 (1961).

DALLDORF, G., F. BERGAMINI, and P. FROST: Further observations of the lymphomas of African children. Proc. nat. Acad. Sci. (Wash.) 55, 297—302 (1966).

DALLDORF, G., C. LINSELL, F. BARNHART, and R. MARTYN: An epidemiologic approach to the lymphomas of African children and Burkitt's sarcoma of the jaws. Perspect. Biol. Med. 7, 435—449 (1964).

DALTON, A. J.: The Moloney agent. In: Tumors Induced by Viruses: Ultrastructural Studies (DALTON and HAGUENAU, eds.). New York. Academic Press Inc., pp. 207—217 (1962).

DALTON, A. J.: An electron microscopic study of a virus-induced murine sarcoma (Moloney). Nat. Cancer Inst. Monogr. 22, 143—168 (1966).

DALTON, A. J., F. HAGUENAU, and J. B. MOLONEY: Further electron microscope studies on the morphology of the Moloney agent. J. nat. Cancer Inst. 33, 255—275 (1964).

DALTON, A. J., L. W. LAW, J. B. MOLONEY, and R. A. MANAKER: An electron microscopic study of a series of murine lymphoid neoplasms. J. nat. Cancer Inst. 27, 747—791 (1961).

DARBYSHIRE, J. H.: Oncogenicity of bovine adenovirus type 3 in hamsters. Nature (Lond.) 211, 102 (1966).

DARBYSHIRE, J. H., L. D. BERMAN, F. C. CHESTERMAN, and H. G. PEREIRA: Studies on the oncogenicity of bovine adenovirus type 3. Int. J. Cancer 3, 546—557 (1968).

DAWE, C. J., L. W. LAW, W. D. MORGAN, and M. G. SHAW: Morphologic responses to tumor viruses. Fed. Proc. 21, 5—14 (1962).

DAWE, C. J., W. D. MORGAN, and M. S. SLATICK: Influence of epithelio-mesenchymal interactions on tumor induction by polyoma virus. Int. J. Cancer 1, 419—450 (1966).

DAWE, C. J., M. POTTER, and J. LEIGHTON: Progressions of a reticulum cell sarcoma of the mouse in vivo and in vitro. J. nat. Cancer Inst. 21, 753—781 (1958).

DEFENDI, V.: Effect of SV40 virus immunization on growth of transplantable SV40 and polyoma virus tumors in hamsters. Proc. Soc. exp. Biol. (N.Y.) 113, 12—16 (1963).

DEFENDI, V.: Transformation in vitro of mammalian cells by polyoma and simian 40 viruses. Progr. exp. Tumor Res. 8, 125—188 (1966).

DEFENDI, V., R. E. BILLINGHAM, W. K. SILVERS, and P. MOORHEAD: Immunological and karyological criteria for identification of cell lines. J. nat. Cancer Inst. 25, 359—386 (1960).

DEFENDI, V., B. EPHRUSSI, and H. KOPROWSKI: Expression of polyoma-induced cellular antigen(s) in hybrid cells. Nature (Lond.) 203, 495—496 (1964).

DEFENDI, V., and G. GASIC: Surface mucopolysaccharides of polyoma virus transformed cells. J. cell. comp. Physiol. 62, 23—31 (1963).

DEFENDI, V., F. JENSEN, and G. SAUER: Analysis of some viral functions related to neoplastic transformation. In: Molecular Biology of Viruses. (COLTER and PARANCHYCH, eds.). Academic Press, New York, pp. 645—663 (1967).

DEFENDI, V., and J. M. LEHMAN: Transformation of hamster embryo cells in vitro by polyoma virus: Morphological, karyological, immunological and transplantation characteristics. J. cell. comp. Physiol. 66, 351—410 (1965).

DEFENDI, V., and J. M. LEHMAN: Biological characteristics of primary tumors induced by polyoma virus in hamsters. Int. J. Cancer 1, 525—540 (1966).

DEFENDI, V., J. LEHMAN, and P. KRAEMER: "Morphological normal" hamster cells with malignant properties. Virology 19, 592—598 (1963).

DEFENDI, V., and F. TAGUCHI: Studies with the complement-fixing antigen induced by polyoma virus. Ann. Med. exp. Fenn. **44**, 232—241 (1966).

DEFTOS, L. J., A. S. RABSON, R. M. BUCKLE, G. D. AURBACH, and J. T. POTTS, JR.: Parathyroid hormone production *in vitro* by human parathyroid cells transformed by simian virus 40. Science **159**, 435—436 (1968).

DE HARVEN, E., and C. FRIEND: Electron microscope study of a cell-free induced leukemia of the mouse. A preliminary report. J. biophys. biochem. Cytol. **4**, 151—156 (1958).

DE HARVEN, E., and C. FRIEND: Further electron microscope studies of a mouse leukemia induced by cell-free filtrates. J. biophys. biochem. Cytol. **7**, 747—752 (1960).

DEICHMAN, G. I., and T. E. KLUCHAREVA: Loss of transplantation antigen in primary simian virus 40-induced tumors and their metastases. J. nat. Cancer Inst. **36**, 647—655 (1966).

DES LIGNERIS, M. J. A.: On the transplantation of Rous fowl sarcoma No. 1 into guinea-fowls and turkeys. Amer. J. Cancer **16**, p. 307 (1932).

DE THÉ, G.: Localization and origin of the adenosine triphosphatase activity of avian myeloblastosis virus. Nat. Cancer Inst. Monogr. No. 17, 651—671 (1964).

DE THÉ, G.: Association of enzymes: Adenosine triphosphatase and alkaline phosphatase with the virions of murine leukemias. Nat. Cancer Inst. Monogr. No. **22**, 169—189 (1966a).

DE THÉ, G.: Ultrastructural cytochemistry of enzymes associated with murine leukemia viruses. I. Adenosine triphosphatase. Int. J. Cancer **1**, 119—138 (1966b).

DE THÉ, G.: Action de la digitonine sur les virions leucémogènes murins. C. R. Acad. Sci. (Paris) **264**, 2347—2349 (1967).

DE THÉ, G., C. BECKER, and J. W. BEARD: Virus of avian myeloblastosis (BAI strain A). XXV. Ultracytochemical study of virus and myeloblast phosphatase activity. J. nat. Cancer Inst. **32**, 201—235 (1964).

DE THÉ, G., U. HEINE, J. R. SOMMER, L. ARVY, D. BEARD, and J. W. BEARD: Multiplicity of cell response to the BAI strain A (myeloblastosis) avian tumor virus. IV. Ultrastructural characters of the thymus in myeloblastosis and of the adenosine triphosphatase activity of thymic cells and associated virus. J. nat. Cancer Inst. **30**, 415—455 (1963).

DE THÉ, G., H. ISHIGURO, U. HEINE, D. BEARD, and J. W. BEARD: Multiplicity of cell response to the BAI strain A (myeloblastosis) avian tumor virus. VI. Ultrastructural aspects of adenosine triphosphatase activity of nephroblastoma cells and virus. J. nat. Cancer Inst. **30**, 1267—1301 (1963).

DE THÉ, G., and T. E. O'CONNOR: Structure of a murine leukemia virus after disruption with tween-ether and comparison with two myxoviruses. Virology **28**, vity. 713—728 (1966).

DHALIWAL, S. S.: Resistance of the chorioallantoic membrane of chick embryos to Rous sarcoma and MH_2 reticuloendothelioma viruses. J. nat. Cancer Inst. **30**, 323—336 (1963).

DIAMANDOPOULOS, G. T.: Histopathology of sarcomas induced in hamsters by clones of *in vitro* SV40-transformed homologous heart cells. Amer. J. Path. **53**, 753—767 (1968).

DIAMANDOPOULOS, G. T., M. F. DALTON-TUCKER, and V. VAN DER NOORDAA: Early *in vitro* SV40-mediated morphologic transformation of primary hamster cells. Amer. J. Path. **57**, 199—213 (1969).

DIAMANDOPOULOS, G. T., and J. F. ENDERS: Studies on transformation of syrian hamster cells by simian virus 40 (SV40): Acquisition of oncogenicity by virus-exposed cells apparently unassociated with the viral genome. Proc. nat. Acad. Sci. (Wash.) **54**, 1092—1099 (1965).

DIAMANDOPOULOS, G. T., and J. F. ENDERS: Comparison of the cytomorphologic characteristics of *in vitro* SV40 transformed hamster embryo cells with the histologic features of the neoplasms which they induce in the homologous host. Amer. J. Path. **49**, 397—417 (1966).

DIAMANDOPOULOS, G. T., S. S. TEVETHIA, F. RAPP, and J. F. ENDERS: Development of S and T antigens and oncogenicity in hamster embryonic cell lines exposed to SV 40. Virology **34**, 331—336 (1968).

DIAMOND, L.: Cell transformation *in vitro* and tumor induction *in vivo* by large- and small-plaque polyoma virus. Virology **23**, 73—80 (1964).

DIAMOND, L.: Two spontaneously transformed cell lines derived from the same hamster embryo culture. Int. J. Cancer **2**, 143—152 (1967a).

DIAMOND, L.: Transformation of Simian virus 40-resistant hamster cells with an adenovirus 7-Simian virus 40 hybrid. J. Virol. **1**, 1109—1116 (1967b).

DIAMOND, L., and L. V. CRAWFORD: Some characteristics of large-plaque and small-plaque lines of polyoma virus. Virology **22**, 235—244 (1964).

DIDERHOLM, H.: A fluorescent antibody study on the formation of simian virus 40 in monkey kidney cells. Acta path. microbiol. scand. **57**, 348—352 (1963).

DIDERHOLM, H.: Transformation of bovine cells *in vitro* by polyoma virus. and the properties of the transformed cells. Proc. Soc. exp. Biol. (N.Y.) **124**, 1197—1201 (1967).

DIDERHOLM, H., R. BERG, and T. WESSLÉN: Transformation of rat and guinea-pig cells *in vitro* by SV 40 and the transplantability of the transformed cells. Int. J. Cancer **1**, 139—148 (1966).

DIDERHOLM, H., and S. HERMODSSON: Two types of morphological transformation of bovine kidney cells infected *in vitro* with SV 40. Arch. ges. Virusforsch. **21**, 45—52 (1967).

DIDERHOLM, H., B. STENKVIST, J. PONTÉN, and T. WESSLÉN: Transformation of bovine cells *in vitro* after inoculation of simian virus 40 or its nucleic acid. Exp. Cell Res. **37**, 452—459 (1965).

DIDERHOLM, H., and T. WESSLÉN: Studies on the transformation of bovine cells *in vitro* by SV 40 and the properties of the transformed cells. Arch. ges. Virusforsch. **17**, 339—346 (1965).

DIEHL, V., G. HENLE, W. HENLE, and G. KOHN: Demonstration of a herpes group virus in cultures of peripheral leukocytes from patients with infectious mononucleosis. J. Virol. **2**, 663—669 (1968).

DIMAYORCA, G., J. CALLENDER, G. MARIN, and R. GIORDANO: Temperature-sensitive mutants of polyoma virus. Virology **38**, 126—133 (1969).

DIMAYORCA, G. A., B. E. EDDY, S. E. STEWART, W. S. HUNTER, C. FRIEND, and A. BENDICH: Isolation of infectious deoxyribonucleic acid from SE-polyoma infected tissue cultures. Proc. nat. Acad. Sci. (Wash.) **45**, 1805—1808 (1959).

DI STEFANO, H., and R. M. DOUGHERTY: Cytological observations of "nonproducer" Rous sarcoma cells. Virology **27**, 360—377 (1965).

DMOCHOWSKI, L.: The viral etiology of leukemia. Progr. med. Virol. **3**, 363—494 (1960).

DMOCHOWSKI, L., C. E. GREY, F. PADGETT, P. L. LANGFORD, and B. R. BURMESTER: Submicroscopic morphology of avian neoplasms. VI. Comparative studies on Rous sarcoma, visceral lymphomatosis, erythroblastosis, myeloblastosis and nephroblastoma. Tex. Rep. Biol. Med. **22**, 20—60 (1964).

DMOCHOWSKI, L., T. YUMOTO, C. E. GREY, R. L. HALES, L. LANGFORD, H. G. TAYLOR, E. J. FREIREICH, C. C. SHULLENBERGER, J. A. SHIVELY, and C. D. HOWE: Electron microscopic studies of human leukemia and lymphoma. Cancer **20**, 760—777 (1967).

DOERFLER, W.: The fate of the DNA of adenovirus type 12 in baby hamster kidney cells. Proc. nat. Acad. Sci. (Wash.) **60**, 636—643 (1968).

DOERFLER, W.: Non-productive infection of baby hamster kidney cells (BHK 21) with adenovirus type 12. Virology **38**, 587—606 (1969).

DOLJANSKI, L., and M. PIKOVSKI: Agent of fowl leukosis in tissue cultures. Cancer Res. **2**, 626—631 (1942).

DOLJANSKI, L., and E. TENENBAUM: Studies on Rous sarcoma cells cultivated *in vitro*. I. Cellular composition of pure cultures of Rous sarcoma cells. Cancer Res. **2**, 776—785 (1942).

DOUGHERTY, R. M., and H. S. DI STEFANO: Virus particles associated with "nonproducer" Rous sarcoma cells. Virology **27**, 351—359 (1965).

DOUGHERTY, R. M., and H. S. DI STEFANO: Lack of relationship between infection with avian leukosis virus and the presence of COFAL antigen in chick embryos. Virology **29,** 585—595 (1966).

DOUGHERTY, R. M., H. S. DI STEFANO, and F. K. ROTH: Virus particles and viral antigens in chicken tissues free of infectious avian leukosis virus. Proc. nat. Acad. Sci. (Wash.) **58,** 808—817 (1967).

DOUGHERTY, R. M., J. A. STEWART, and H. R. MORGAN: Quantitative studies of the relationships between infecting dose of Rous sarcoma virus, antiviral immune response and tumor growth in chickens. Virology **11,** 349—370 (1960).

DOURMASHKIN, R. R., and P. J. SIMONS: The ultrastructure of Rous sarcoma virus. J. Ultrastruct. Res. **5,** 505—522 (1961).

DRAYTON, H. A.: The inactivation by organic solvents and detergents of partially purified Rous I virus preparations. Brit. J. Cancer **15,** 348—353 (1961).

DREW, R. M.: Isolation and propagation of rabbit kidney epithelial like cells. Science **126,** 747—748 (1957).

DUBBS, D. R., S. KIT, R. A. DE TORRES, and M. ANKEN: Virogenic properties of bromodeoxyuridine-sensitive and bromodeoxyuridine-resistant simian virus 40-transformed mouse kidney cells. J. Virol. **1,** 968—979 (1967).

DUC-NGUYEN, H., E. N. ROSENBLUM, and R. F. ZEIGEL: Persistent infection of a rat kidney cell line with Rauscher murine leukemia virus. J. Bact. **92,** 1133—1140 (1966).

DUESBERG, P. H.: Physical properties of Rous sarcoma virus RNA. Proc. nat. Acad. Sci. (Wash.) **60,** 1511—1518 (1968).

DUESBERG, P. H., and W. S. ROBINSON: Nucleic acid and proteins isolated from the Rauscher mouse leukemia virus. Proc. nat. Acad. Sci. (Wash.) **55,** 219—227 (1966).

DUESBERG, P. H., H. L. ROBINSON, W. S. ROBINSON, R. J. HUEBNER, and H. C. TURNER: Proteins of Rous sarcoma virus. Virology **36,** 73—86 (1968).

DUESBERG, P. H., and P. K. VOGT: On the role of DNA synthesis in avian tumor virus infection. Proc. nat. Acad. Sci. (Wash.) **64,** 939—946 (1969).

DUFF, R. G.: Thesis, University of Colorado, Denver, 1968.

DUFF, R. G., and P. VOGT: Characteristics of two new avian tumor virus subgroups. Virology **39,** 18—30 (1969).

DUFFEL, D., R. HINZ, and E. NELSON: Neoplasms in hamsters induced by simian virus 40. Amer. J. Path. **45,** 59—73 (1964).

DULANEY, A. D.: Parotid gland tumor in AKR mice inoculated when newborn with cell-free AK leukemic extracts. Cancer Res. **16,** 877—879 (1956).

DULANEY, A. D., M. MAXEY, M. G. SCHILLING, and M. F. GOSS: Neoplasms in C3H mice which received AK-leukemic extracts. Cancer Res. **17,** 809—814 (1957).

DULBECCO, R.: Configurational and biological properties of polyoma virus DNA. Proc. roy. Soc. B 160, 423—431 (1964).

DULBECCO, R.: Mechanism of cell transformation by polyoma virus. Perspect. Biol. Med. **9,** 298—305 (1966).

DULBECCO, R.: Cell transformation by viruses. Science **166,** 962—968 (1969).

DULBECCO, R., and G. FREEMAN: Plaque production by the polyoma virus. Virology 8, 396—397 (1959).

DULBECCO, R., L. H. HARTWELL, and M. VOGT: Induction of cellular DNA synthesis by polyoma virus. Proc. nat. Acad. Sci. (Wash.) **53,** 403—410 (1965).

DULBECCO, R., and M. VOGT: Significance of continued virus production in tissue cultures rendered neoplastic by polyoma virus. Proc. nat. Acad. Sci. (Wash.) **46,** 1617—1623 (1960).

DULBECCO, R., and M. VOGT: Evidence for a ring structure of polyoma virus DNA. Proc. nat. Acad. Sci. (Wash.) **50,** 236—243 (1963).

DUNN, T. B.: The conservative pathologist and the morphology of tumors produced by three cancer viruses in mice. Tex. Rep. Biol. Med. **19,** 217—232 (1961).

DUNN, T. B., and M. K. DERINGER: Reticulum cell neoplasm, type B, or the "Hodgkin's-like lesion" of the mouse. J. nat. Cancer Inst. **40,** 771—821 (1968).

DUNN, T. B., J. B. MOLONEY, A. GREEN, and B. ARNOLD: Pathogenesis of a virus induced leukemia in mice. J. nat. Cancer Inst. **26,** 189—221 (1961).

DURAN-REYNALS, F.: The reciprocal infection of ducks and chickens with tumor-inducing viruses. Cancer Res. **2**, 343—369 (1942).

DURAN-REYNALS, F.: The infection of turkeys and guinea fowls by the Rous sarcoma virus and the accompanying variations of the virus. Cancer Res. **3**, 569—577 (1943).

DURAN-REYNALS, F.: Virus-induced tumors and the virus theory of cancer. In: The Physiopathology of Cancer (HAMBURGER and FISHMAN, eds.), New York, Paul B. Hoeber Inc., p. 298—337 (1953).

DUTCHER, R. M.: Bovine leukemia. Growth **29**, 1—5 (1965).

EAGLE, H.: Nutrition needs of mammalian cells in tissue culture. Science **122**, 501—504 (1955).

EAGLE, H.: Amino acid metabolism in mammalian cell cultures. Science **130**, 432—437 (1959).

EAGLE, H., and E. M. LEVINE: Growth regulatory effects of cellular interaction. Nature (Lond.) **213**, 1102—1106 (1967).

EAGLE, H., E. M. LEVINE, and H. KOPROWSKI: Species specificity in growth regulatory effects of cellular interaction. Nature (Lond.) **220**, 266—269 (1968).

EARLE, W. R.: Production of malignancy *in vitro*. IV. The mouse fibroblasts, cultures and changes seen in the living cells. J. nat. Cancer Inst. **4**, 165—212 (1943).

EARLE, W. R., and A. NETTLESHIP: Production of malignancy *in vitro*. V. Results of injections of cultures into mice. J. nat. Cancer Inst. **4**, 213—228 (1943).

EARLE, W. R., E. SHELTON, and E. L. SCHILLING: Production of malignancy *in vitro*. XI. Further results from reinjection of *in vitro* cell strains into strain C3H mice. J. nat. Cancer Inst. **10**, 1105—1113 (1950).

EAST, J., and J. J. HARVEY: The differential action of neonatal thymectomy in mice infected with murine sarcoma virus-Harvey (MSV-H). Int. J. Cancer **3**, 614—627 (1968).

EASTON, J. M.: Potentiation of SV40 replication in human cells by adenovirus 4. Cancer **19**, 1486—1488 (1966).

EASTON, J. M., and C. W. HIATT: Possible incorporation of SV40 genome within capsid proteins of adenovirus 4. Proc. nat. Acad. Sci. (Wash.) **54**, 1100—1104 (1965).

EBELING, A. H.: The permanent life of connective tissue outside of the organism. J. exp. Med. **17**, 237—285 (1913).

EBELING, A. H.: A ten year old strain of fibroblasts. J. exp. Med. **35**, 755—759 (1922).

ECKERT, E. A., D. G. SHARP, D. BEARD, I. GREEN, and J. W. BEARD: Virus of avian erythromyeloblastic leukosis. IX. Antigenic constitution and immunologic characterization. J. nat. Cancer Inst. **16**, 593—643 (1955).

ECKERT, E. A., R. ROTT, and W. SCHÄFER: Myxovirus-like structure of avian myeloblastosis virus. Z. Naturforsch. **21b**, 339—341 (1963).

ECKERT, E. A., R. ROTT, and W. SCHÄFER: Studies on the BAI strain A (Avian myeloblastosis) virus. I. Production and examination of potent virus-specific complement-fixing antisera. Virology **24**, 426—433 (1964a).

ECKERT, E. L., R. ROTT, and W. SCHÄFER: Studies on the BAI strain A (Avian myeloblastosis) virus. II. Some properties of viral split products. Virology **24**, 434—440 (1964b).

ECKHART, W.: Complementation and transformation by temperature-sensitive mutants of polyoma virus. Virology **38**, 120—125 (1969).

EDDY, B. E.: Simian virus 40 (SV40): An oncogenic virus. Progr. exp. Tumor Res. **4**, 1—26 (1964).

EDDY, B. E., G. S. BORMAN, W. H. BERKELEY, and R. D. YOUNG: Tumors induced in hamsters by injection of rhesus monkey kidney cell extracts. Proc. Soc. exp. Biol. (N.Y.) **107**, 191—197 (1961).

EDDY, B. E., G. S. BORMAN, G. E. GRUBBS, and R. D. YOUNG: Identification of the oncogenic substance in rhesus monkey kidney cell culture as simian virus 40. Virology **17**, 65—75 (1962).

EDDY, B. E., G. E. GRUBBS, and R. D. YOUNG: Persistent infection of human carcinoma and primary chick embryo cell cultures with simian virus 40. Proc. Soc. exp. Biol. (N.Y.) **111**, 718—722 (1962).

EDDY, B. E., W. P. ROWE, J. W. HARTLEY, S. E. STEWART, and R. J. HUEBNER: Hemagglutination with the SE polyoma virus. Virus **6**, 290—291 (1958).

EDDY, B. E., S. E. STEWART, and W. BERKELEY: Cytopathogenicity in tissue cultures by a tumor virus from mice. Proc. Soc. exp. Biol. (N.Y.) **98**, 848—851 (1958).

EDDY, B. E., S. E. STEWART, R. YOUNG, and G. B. MIDER: Neoplasms in hamsters induced by mouse tumor agent passed in tissue culture. J. nat. Cancer Inst. **20**, 747—761 (1958).

ELSDALE, T. : Parallel orientation of fibroblasts *in vitro*. Exp. Cell Res. **51**, 439—450 (1968).

ELSDALE, T., and R. FOLEY: Morphogenetic aspects of multilayering in Petri dish cultures of human fetal lung fibroblasts. J. Cell Biol. **41**, 298—311 (1969).

ENDERS, J. F., and G. TH. DIAMANDOPOULOS: A study of variation and progression in oncogenicity in an SV 40-transformed hamster heart cell line and its clones. Proc. roy. Soc. B **171**, 431—443 (1969).

ENGELBRETH-HOLM, J.: Tumor-producing viruses in fowls. Acta path. microbiol. scand. (Suppl.) **38**, 26—36 (1938).

EPHRUSSI, B., and H. M. TEMIN: Infection of chick iris epithelium with the Rous sarcoma virus *in vitro*. Virology **11**, 547—552 (1960).

EPSTEIN, M. A.: Composition of the Rous virus nucleoid. Nature (Lond.) **181**, 1808 (only) (1958).

EPSTEIN, M. A., and B. G. ACHONG: Fine structural organization of human lymphoblasts of a tissue culture strain EB 1 from Burkitt's lymphoma. J. nat. Cancer Inst. **34**, 241—253 (1965).

EPSTEIN, M. A., B. G. ACHONG, and Y. M. BARR: Virus particles in cultured lymphoblasts from Burkitt's lymphoma. Lancet **1**, 702—703 (1964).

EPSTEIN, M. A., B. G. ACHONG, Y. M. BARR, B. ZAJAC, G. HENLE, and W. HENLE: Morphological and virological investigations on cultured Burkitt tumor lymphoblasts (strain Raji). J. nat. Cancer Inst. **37**, 547—559 (1966).

EPSTEIN, M. A., B. G. ACHONG, and J. H. POPE: Virus in cultured lymphoblasts from a New Guinea Burkitt lymphoma. Brit. med. J. **2**, 290—291 (1967).

EPSTEIN, M. A., and Y. M. BARR: Cultivation *in vitro* of human lymphoblasts from Burkitt's malignant lymphoma. Lancet **1**, 252—253 (1964).

EPSTEIN, M. A., and Y. M. BARR: Characteristics and mode of growth of a tissue culture strain (EB 1) of human lymphoblasts from Burkitt's lymphoma. J. nat. Cancer Inst. **34**, 232—238 (1965).

EPSTEIN, M. A., Y. M. BARR, and B. G. ACHONG: The behaviour and morphology of a second tissue culture strain (EB 2) of lymphoblasts from Burkitt's lymphoma. Brit. J. Cancer **19**, 108—115 (1965).

EPSTEIN, M. A., Y. M. BARR, and B. G. ACHONG: Preliminary observations on new lymphoblast strains (EB 5) from Burkitt tumours in a British and a Ugandan patient. Brit. J. Cancer **20**, 475—479 (1966).

EPSTEIN, M. A., G. HENLE, B. G. ACHONG, and Y. M. BARR: Morphological and biological studies on a virus in cultured lymphoblasts from Burkitt's lymphoma. J. exp. Med. **121**, 761—770 (1965).

ERICHSEN, S., J. ENG, and H. R. MORGAN: Comparative studies in Rous sarcoma with virus, tumor cells and chick embryo cells transformed *in vitro* by virus. I. Production of mucopolysaccharides. J. exp. Med. **114**, 435—440 (1961).

ERIKSSON, J. U. BRUNK, B. WESTERMARK, and R. HUGOSSON: Modulation of the cell surface of human glia-like cells *in vitro*. Acta path. microbiol. scand. In press.

ERIKSON, R. L.: Studies on the RNA from avian myeloblastosis virus. Virology **37**, 124—131 (1969).

ESPMARK, J. Å., and A. FAGRAEUS: Identification of the species of origin of cells by mixed hemadsorption: A mixed antiglobulin reaction applied to monolayer cell cultures. J. Immunol. **94**, 530—537 (1965).

EVANS, V. J., N. M. HAWKINS, B. B. WESTFALL, and W. R. EARLE: Studies on culture lines derived from mouse liver parenchymatous cells grown in long-term tissue culture. Cancer Res. **18**, 261—266 (1958).

Evans, V. J., J. L. Jackson, W. F. Andresen, and J. T. Mitchell: Chromosomal characteristics and neoplastic transformation of C3H mouse embryo cells *in vitro* in horse and fetal calf serum. J. nat. Cancer Inst. **38**, 761—769 (1967).

Evans, V. J., G. A. Parker, and Th. B. Dunn: Neoplastic transformation in C3H mouse embryonic tissue *in vitro* determined by intraocular growth. I. Cells from chemically defined medium with and without serum supplement. J. nat. Cancer Inst. **32**, 89—107 (1964).

Fahey, J. L., I. Finegold, A. S. Rabson, and R. A. Manaker: Immunoglobulin synthesis *in vitro* by established human cell lines. Science **152**, 1259—1261 (1966).

Febvre, H., L. Rothschild, J. Arnoult, and F. Haguenau: *In vitro* malignant conversion of rat embryonic cell lines with the Bryan strain of Rous sarcoma virus. Nat. Cancer Inst., Monogr. **17**, 459—477 (1964).

Fedoroff, S.: Proposed usage of animal tissue culture terms. J. nat. Cancer Inst. **38**, 607—611 (1967).

Feldman, L. A., J. S. Butel, and F. Rapp: Interaction of papovavirus SV40 and adenoviruses. I. Induction of adenovirus tumor antigen during abortive infection of simian cells. J. Bact. **91**, 813—818 (1966).

Ferguson, J., and G. A. Tomkins: Chromosome studies during long-term cultivation of epithelioid cercopithecus and cynomolgus monkey kidney cell lines. J. nat. Cancer Inst. **33**, 619—630 (1964).

Ferguson, J., and A. Wansbrogh: Isolation and long-term culture of diploid mammalian cell lines. Cancer Res. **22**, 556—562 (1962).

Fernandes, M., and P. Moorhead: Transformation of African green monkey kidney cultures infected with simian vacuolating virus (SV40). Tex. Rep. Biol. Med. **23**, 242—258 (1965).

Fieldsteel, A. H., P. J. Dawson, and W. L. Bostick: Quantitative aspects of Friend leukemia virus in various murine hosts. Proc. Soc. exp. Biol. (N.Y.) **108**, 826—829 (1961).

Fine, R., M. Mass, and W. T. Murakami: Protein composition of polyoma virus. J. molec. Biol. **36**, 167—177 (1968).

Finegold, I., J. L. Fahey, and H. Granger: Synthesis of immunoglobulins by human cell lines in tissue culture. J. Immunol. **99**, 839—848 (1967).

Finegold, I., Y. Hirshaut, and J. L. Fahey: Immunochemical and morphologic comparison of donor tissue with immunoglobulin-producing tissue culture lines from two patients with malignancies. Cancer Res. **28**, 1538—1549 (1968).

Fink, M. A., and R. A. Malmgren: Fluorescent antibody studies of the viral antigen in a murine leukemia (Rauscher). J. nat. Cancer Inst. **31**, 1111—1121 (1963).

Firor, W. M., and G. O. Gey: Observations on the conversions of normal into malignant cells. Ann. Surg. **121**, 700—703 (1945).

Fischer, A., and R. C. Parker: The occurrence of mitoses in normal and malignant tissues *in vitro*. Brit. J. exp. Path. **10**, 312—321 (1929).

Fischinger, P. J., C. O. Moore, and T. E. O'Connor: Isolation and identification of a helper virus found in the Moloney sarcoma-leukemia virus complex. J. nat. Cancer Inst. **42**, 605—622 (1969).

Fischinger, P. J., and T. E. O'Connor: Tissue culture assay of helper activity of murine leukemia virus for murine sarcoma virus. J. nat. Cancer Inst. **40**, 1199—1212 (1968).

Fischinger, P. J., and T. E. O'Connor: Viral infection across species barriers: Reversible alteration of murine sarcoma virus for growth in cat cells. Science **165**, 714—716 (1969a).

Fischinger, P. J., and T. E. O'Connor: Radiation leukemia virus: Quantitative tissue culture assay. Science **165**, 306—309 (1969b).

Fisher, H. W., and J. Yeh: Contact inhibition in colony formation. Science **155**, 581—582 (1967).

Fogel, M., and V. Defendi: Infection of muscle cultures from various species with oncogenic DNA viruses (SV40 and polyoma). Proc. nat. Acad. Sci. (Wash.) **58**, 967—973 (1967).

FOGEL, M., and V. DEFENDI: Internuclear transfer of SV40-T-antigen in absence of infectious virus. Virology **34**, 370–373 (1968).

FOGEL, M., R. GILDEN, and V. DEFENDI: Polyoma virus-induced "complement-fixing antigen" in tumors and infected cells as detected by immunofluorescence. Proc. Soc. exp. Biol. (N.Y.) **124**, 1047–1052 (1967).

FOGEL, M., and L. SACHS: The in vitro and in vivo analysis of mammalian tumour viruses. Brit. J. Cancer **13**, 266–281 (1959).

FOGEL, M., and L. SACHS: The activation of virus synthesis in polyoma-transformed cells. Virology **37**, 327–334 (1969).

FOGH, J., and H. FOGH: Chromosome changes in PPLO-infected FL human amnion cells. Proc. Soc. exp. Biol. (N.Y.) **119**, 233–238 (1965).

FOGH, J., and R. O. LUND: Continuous cultivation of epithelial cell strain (FL) from human amniotic membrane. Proc. Soc. exp. Biol. (N.Y.) **94**, 532–537 (1957).

FOLEY, G. E., and A. H. HANDLER: Differentiation of "normal" and neoplastic cells maintained in tissue culture by implantation into normal hamsters. Proc. Soc. exp. Biol. (N.Y.) **94**, 661–664 (1957).

FOLEY, G. E., and A. H. HANDLER: Tumorigenic activity of tissue cell cultures. Ann. N.Y. Acad. Sci. **76**, 506–512 (1958).

FORCE, E. E., and R. C. STEWART: Effect of 5-iodo-2-deoxyuridine on multiplication of Rous sarcoma virus. Proc. Soc. exp. Biol. (N.Y.) **116**, 803–806 (1964).

FORD, D. K.: Chromosomal changes occurring in Chinese hamster cells during prolonged culture in vitro. Canad. Cancer Conf. **3**, 171–188 (1959).

FORD, D. K., C. BOQUSZEWSKI, and N. AUERSPERG: Chinese hamster cell strains in vitro. Spontaneous chromosome changes and latent polyoma-virus infection. J. nat. Cancer Inst. **26**, 691–706 (1961).

FORD, D. K., and G. YERGANIAN: Observations on the chromosomes of Chinese hamster cells in tissue culture. J. nat. Cancer Inst. **21**, 393–425 (1958).

FORRESTER, J. A., E. J. AMBROSE, and I. A. MACPHERSON: Electrophoretic investigations of a clone of hamster fibroblasts and polyoma-transformed cells from the same population. Nature (Lond.) **196**, 1068–1070 (1962).

FOULDS, L.: The filterable tumours of fowls. A critical review. Suppl. to Eleventh Sci. Rep. Invest. Imp. Cancer Res. Found., London, England: Taylor and Francis, pp. 1–41 (1934).

FOULDS, L.: The experimental study of tumor progression: a review. Cancer Res. **14**, 327–339 (1954).

FRANKS, D.: Antigenic heterogeneity in cultures of mammalian cells. In vitro, Publ. Tissue Culture Assoc., Vol. II, Phenotypic Expression. Immunol. Biochem. Morphol., Williams and Wilkins Co. 74–81 (1966).

FRASER, K. B., and E. M. CRAWFORD: Immunofluorescent and electron-microscopic studies of polyoma virus in transformation reactions with BHK21 cells. Exp. molec. Path. **4**, 51–65 (1965).

FRASER, K. B., and M. GHARPURE: Immunofluorescent tracing of polyoma virus in transformation experiments with BHK21 cells. Virology **18**, 505–507 (1962).

FREARSON, P. M., S. KIT, and D. R. DUBBS: Deoxythymidylate synthetase and deoxythymidine kinase activities of virus-infected animal cells. Cancer Res. **25**, 737–744 (1965).

FREARSON, P. M., S. KIT, and D. R. DUBBS: Induction of dihydrofolate reductase activity by SV40 and polyoma virus. Cancer Res. **26**, 1653–1660 (1966).

FREEMAN, A. E., P. H. BLACK, E. A. VANDERPOOL, P. H. HENRY, J. B. AUSTIN, and R. J. HUEBNER: Transformation of primary rat embryo cells by adenovirus type 2. Proc. nat. Acad. Sci. (Wash.) **58**, 1205–1212 (1967).

FREEMAN, A. E., P. H. BLACK, R. WOLFORD, and R. J. HUEBNER: The adenovirus type 12 – rat embryo transformation system. J. Virol. **1**, 362–367 (1967).

FREEMAN, A. E., C. H. CALISHER, P. J. PRICE, H. C. TURNER, and R. J. HUEBNER: Calcium sensitivity of cell cultures derived from adenovirus-induced tumors. Proc. Soc. exp. Biol. (N.Y.) **122**, 835–840 (1966).

FREEMAN, A. E., S. HOLLINGER, P. J. PRICE, and C. H. CALISHER: The effect of calcium on cell lines derived from adenovirus type 12-induced hamster tumors. Exp. Cell Res. **39**, 259—264 (1965).

FREEMAN, A. E., E. A. VANDERPOOL, P. H. BLACK, H. C. TURNER, and R. J. HUEBNER: Transformation of primary rat embryo cells by a weakly oncogenic adenovirus type 3. Nature (Lond.) **216**, 171—173 (1967).

FREEMAN, A. E., P. J. PRICE, H. J. IGEL, J. C. YOUNG, J. M. MARYAK, and R. J. HUEBNER: Morphological transformation of rat embryo cells induced by diethyl-nitrosamine and murine leukemia viruses. J. nat. Cancer Inst. **44**, 65—78 (1970).

FREEMAN, G.: Focus formation by Japanese quail cells infected with Rous sarcoma virus. J. nat. Cancer Inst. **31**, 761—767 (1963).

FRIED, M.: Cell-transforming ability of a temperature sensitive mutant of polyoma virus. Proc. nat. Acad. Sci. (Wash.) **53**, 486—491 (1965).

FRIED, M., and J. D. PITTS: Replication of polyoma virus DNA. I. A resting cell system for biochemical studies on polyoma virus. Virology **34**, 761—770 (1968).

FRIEDMAN, R. M., and A. S. RABSON: Possible role of interferon in determining the oncogenic effect of polyoma virus variants. J. exp. Med. **119**, 71—81 (1964a).

FRIEDMAN, R. M., and A. S. RABSON: Polyoma virus strains of differing oncogenicity: Transplantation immunity in mice. Virology **23**, 273—274 (1964b).

FRIEDMAN, R. M., A. S. RABSON, and W. R. KIRKHAM: Variation in interferon production by polyoma virus strains of differing oncogenicity. Proc. Soc. exp. Biol. (N.Y.) **112**, 347—349 (1963).

FRIEDMANN, J. C., J. P. LEVY, J. LASNERET, M. THOMAS, M. BOIRON et J. BERNARD: Induction de fibromes sous-cutanés chez le hamster doré par inoculation d'extrait cellulaire de papillome bovin. C. R. Acad. Sci. (Paris) **257**, 2328—2331 (1963).

FRIEDMANN, TH., J. E. SEEGMILLER, and J. H. SUBAK-SHARPE: Metabolic cooperation between genetically marked human fibroblasts in tissue culture. Nature (Lond.) **220**, 272—274 (1968).

FRIEND, C.: Cell-free transmission in adult Swiss mice of a disease having the character of a leukemia. J. exp. Med. **105**, 307—318 (1957a).

FRIEND, C.: Leukemia of adult mice caused by a transmissible agent. Ann. N.Y. Acad. Sci. **68**, 522—532 (1957b).

FRIEND, C., M. C. PATULEIA, and E. DE HARVEN: Erythrocytic maturation *in vitro* of murine (Friend) virus-induced leukemic cells. Nat. Cancer Inst., Monogr. **22**, 505—514 (1966),

FRIESEN, B., and H. RUBIN: Some physicochemical and immunological properties of an avian leukosis virus (RIF). Virology **15**, 387—396 (1961).

FRITZ, R. B., A. J. LANGLOIS, D. BEARD, and J. W. BEARD: Strain MC29 avian leukosis virus: Immunologic relationships to other avian tumor viruses. J. Immunol. **101**, 1199—1206 (1968).

FUJINAGA, K., and M. GREEN: The mechanism of viral carcinogenesis by DNA mammalian viruses: I. Viral-specific RNA in polyribosomes of adenovirus tumor and transformed cells. Proc. nat. Acad. Sci. (Wash.) **55**, 1567—1574 (1966).

FUJINAGA, K., and M. GREEN: Mechanism of viral carcinogenesis by DNA mammalian viruses. II. Viral-specific RNA in tumor cells induced by "weakly" oncogenic human adenoviruses. Proc. nat. Acad. Sci. (Wash.) **57**, 806—812 (1967).

FUJINAGA, K., and M. GREEN: Mechanism of viral carcinogenesis by DNA mammalian viruses. V. Properties of purified viral-specific RNA from human adenovirus-induced tumor cells. J. molec. Biol. **31**, 63—73 (1968).

FUJINAMI, A.: A pathological study in chicken sarcoma. Trans. Jap. Path. Soc. **20**, p. 3 (1930).

FUJINAMI, A., and S. HATANO: Contribution of the pathology of hetero-transplantation of tumour. A duck sarcoma from chicken sarcoma. Gann **23**, 67—75 (1929).

FUJINAMI, A., and K. SUZUE: Contribution to the pathology of tumour growth. Experiments on the growth of chicken sarcoma in the case of hetero-transplantation. Trans. Jap. Path. Soc. **15**, p. 616 (1928).

FURSPHAN, E. J., and D. D. POTTER: Low-resistance junctions between cells in embryos and tissue culture. Curr. Top. develop. Biol. **3**, 95—127 (1968).

GALIBERT, F., C. BERNARD, P. CHENAILLE et M. BOIRON: Acide ribonucléique de haut poids moléculaire isolé du virus leucémogène de Rauscher. C. R. Acad. Sci. (Paris) **261**, 1771—1774 (1965).

GARRIGA, S., and W. H. CROSBY: The incidence of leukemia in families of patients with hypoplasia or the marrow. Blood **14**, 1008—1014 (1959).

GARTLER, S. M.: Apparent HeLa cell contamination of human heteroploid cell lines. Nature (Lond.) **217**, 750—751 (1968).

GAYLORD, W. H.: Virus-like particles associated with the Rous sarcoma as seen in sections of the tumor. Cancer Res. **15**, 80—83 (1955).

GAYLORD, W. H., and G. D. HSIUNG: The vacuolating virus of monkeys: II. Virus morphology and intranuclear distribution with some histochemical observations. J. exp. Med. **114**, 987—996 (1961).

GEERING, G., L. J. OLD, and A. BOYSE: Antigens of leukemias induced by naturally occurring murine leukemia virus: Their relation to the antigens of Gross virus and other murine leukemia viruses. J. exp. Med. **124**, 753—772 (1966).

GERBER, P.: Tumors induced in hamsters by simian virus 40: persistent subviral infection. Science **140**, 889—890 (1963).

GERBER, P.: Virogenic hamster tumor cells: Induction of virus synthesis. Science **145**, 833 (only) (1964).

GERBER, P.: Studies on the transfer of subviral infectivity from SV 40-induced hamster tumor cells to indicator cells. Virology **28**, 501—509 (1966).

GERBER, P., and S. M. BIRCH: Complement-fixing antibodies in sera of human and nonhuman primates to viral antigens derived from Burkitt's lymphoma cells. Proc. nat. Acad. Sci. (Wash.) **58**, 478—484 (1967).

GERBER, P., D. HAMRE, R. A. MOY, and E. N. ROSENBLUM: Infectious mononucleosis: Complement-fixing antibodies to herpes-like virus associated with Burkitt lymphoma. Science **161**, 173—175 (1968).

GERBER, P., and R. L. KIRSCHSTEIN: SV 40-induced ependymomas in newborn hamsters. I. Virus-tumor relationships. Virology **18**, 582—588 (1962).

GERBER, P., and J. H. MONROE: Studies on leukocytes growing in continuous culture derived from normal human donors. J. nat. Cancer Inst. **40**, 855—866 (1968).

GERSHON, D., P. HAUSEN, L. SACHS, and E. WINOCOUR: On the mechanism of polyoma virus-induced synthesis of cellular DNA. Proc. nat. Acad. Sci. (Wash.) **54**, 1584—1592 (1965).

GERSHON, D., and L. SACHS: The temporal relationships of protein and DNA synthesis in polyoma virus development. Virology **24**, 604—609 (1964).

GERSHON, D., and L. SACHS: The early synthesis of RNA in polyoma virus development. Virology **29**, 44—48 (1966).

GERSHON, D., L. SACHS, and E. WINOCOUR: The induction of cellular DNA synthesis by simian virus 40 in contact-inhibited and in X-irradiated cells. Proc. nat. Acad. Sci. (Wash.) **56**, 918—925 (1966).

GEY, G. O.: Some aspects of the constitution and behavior of normal and malignant cells maintained in continuous culture. Harvey Lect. **50**, 154—229 (1955).

GEY, G. O., M. K. GEY, W. M. FIROR, and W. O. SELF: Cultural and cytologic studies on autologous normal and malignant cells of specific *in vitro* origin. Acta Un. int. Cancr. **6**, 706—712 (1949).

GILDEN, R. V., and R. I. CARP: Effects of cycloheximide and puromycin on synthesis of simian virus 40 T antigen in green monkey kidney cells. J. Bact. **91**, 1295—1297 (1966).

GILDEN, R. V., R. I. CARP, F. TAGUCHI, and V. DEFENDI: The nature and localization of the SV 40-induced complement-fixing antigen. Proc. nat. Acad. Sci. (Wash.) **53**, 684—692 (1965).

GILEAD, Z., and H. S. GINSBERG: Characterization of the tumorlike (T) antigen induced by type 12 adenovirus. I. Purification of the antigen from infected KB cells and a hamster tumor cell line. J. Virol. **2**, 7—14 (1968a).

GILEAD, Z., and H. S. GINSBERG: Characterization of the tumorlike (T) antigen induced by type 12 adenovirus. II. Physical and chemical properties. J. Virol. **2**, 15—20 (1968b).

GINSBURGH, H., and L. SACHS: *In vitro* culture of a mammalian leukemia virus. Virology **13**, 380—382 (1961a).

GINSBURGH, H., and L. SACHS: Long-term cultivation in tissue culture of leukemic cells from mouse leukemia induced by Moloney virus or by X-rays. J. nat. Cancer Inst. **27**, 1153—1171 (1961b).

GINSBURGH, H., and L. SACHS: Leukemia induction in mice by Moloney virus from long-and short-term tissue cultures, and attempts to detect a leukemogenic virus in cultures from X-ray-induced leukemia. J. nat. Cancer Inst. **28**, 1391—1410 (1962).

GIRARDI, A. J.: Prevention of SV40 virus oncogenesis in hamsters. I. Tumor resistance induced by human cells transformed by SV40. Proc. nat. Acad. Sci. (Wash.) **54**, 445—451 (1965).

GIRARDI, A. J.: Tumor resistance and tumor enhancement with SV40 virus-induced tumors. Germinal Centers in Immune Responses, Proc. Symp. Univ. Bern, Switzerland, June 22—24, 1966 (Springer-Verlag, Berlin, Heidelberg, New York). pp. 422—427 (1966).

GIRARDI, A. J., M. R. HILLEMAN, and R. E. ZWICKEY: Tests in hamsters for oncogenic quality of ordinary viruses including adenovirus type 7. Proc. Soc. exp. Biol. (N.Y.) **115**, 1141—1150 (1964).

GIRARDI, A. J., F. C. JENSEN, and H. KOPROWSKI: SV40-induced transformation of human diploid cells: Crisis and recovery. J. cell. comp. Physiol. **65**, 69—84 (1965).

GIRARDI, A. J., V. M. LARSON, and M. R. HILLEMAN: Further tests in hamsters for oncogenic quality of ordinary viruses and Mycoplasma, with correlative review. Proc. Soc. exp. Biol. (N.Y.) **118**, 173—179 (1965).

GIRARDI, A. J., B. H. SWEET, V. B. SLOTNICK, and M. R. HILLEMAN: Development of tumors in hamsters inoculated in the neonatal period with vacuolating virus SV40. Proc. Soc. exp. Biol. (N.Y.) **109**, 649—660 (1962).

GIRARDI, A. J., D. WEINSTEIN, and P. S. MOORHEAD: SV40 transformation of human diploid cells. A parallel study of viral and karyologic parameters. Ann. Med. exp. Fenn. **44**, 242—254 (1966).

GLADE, P. R., Y. HIRSHAUT, D. P. STITES, and L. N. CHESSIN: Infectious mononucleosis. *In vitro* evidence for limited lymphoproliferation. Blood **33**, 292—299 (1969).

GLADE, P. R., J. A. KASEL, H. L. MOSES, J. WHANG-PENG, P. F. HOFFMAN, J. K. KAMMERMEYER, and L. N. CHESSIN: Infectious mononucleosis continuous suspension culture of peripheral blood leukocytes. Nature (Lond.) **217**, 564—565 (1968).

GLYNN, J. P., J. L. McCOY, and A. FEFER: Cross-resistance to the transplantation of syngeneic Friend, Moloney and Rauscher virus-induced tumors. Cancer Res. **28**, 434—439 (1968).

GOLDBLATT, H., and G. CAMERON: Induced malignancy in cells from rat myocardium subjected to intermittent anaerobiosis during long propagation *in vitro*. J. exp. Med. **97**, 525—552 (1953).

GOLDBLUM, N., Z. RAVID, and Y. BECKER: Effect of withdrawal of arginine and other amino acids on the synthesis of tumour and viral antigens of SV40 virus. J. gen. Virol. **3**, 143—146 (1968).

GOLDÉ, A.: Chemical changes in chick embryo cells infected with Rous sarcoma virus *in vitro*. Virology **16**, 9—20 (1962).

GOLDÉ, A., and P. VIGIER: Growth of Rous sarcoma virus and cells in non-confluent chick embryo monolayers. Virology **15**, 36—46 (1961).

GOLDNER, H., A. J. GIRARDI, V. M. LARSON, and M. R. HILLEMAN: Interruption of SV40 virus tumorigenesis using irradiated homologous tumor antigen. Proc. Soc. exp. Biol. (N.Y.) **117**, 851—857 (1964).

GOTLIEB-STEMATSKY, T., and S. LEVENTON: Studies on the biological properties of two plaque variants isolated from SE polyoma virus. Brit. J. exp. Path. **41**, 507—519 (1960).

GOTLIEB-STEMATSKY, T., Z. ROTEM, and S. KARBY: Production and susceptibility to interferon of polyoma virus variants of high and low oncogenic properties. J. nat. Cancer Inst. **37**, 99—103 (1966).

GOTLIEB-STEMATSKY, T., and R. SHILO: Studies on the tumorigenic properties of baby hamster kidney cell lines and a method of selection of high and low tumorigenic clones. Virology **22**, 314—320 (1964).

GOTLIEB-STEMATSKY, T., A. YANIV, and A. GAZITH: Spontaneous, malignant transformation of hamster embryo cell *in vitro*. J. nat. Cancer Inst. **36**, 477—482 (1966).

GRAFFI, A.: Chloroleukemia of mice. Ann. N.Y. Acad. Sci. **68**, 540—558 (1957).

GRANBOULAN, N., J. HUPPERT et F. LACOUR: Examen au microscope électronique du RNA du virus de la myéloblastose aviaire. J. molec. Biol. **16**, 571—575 (1966).

GRANBOULAN, N., et P. TOURNIER: Horaire et localisation de la synthèse des acides nucléiques pendant la phase d'éclipse du virus SV40. Ann. Inst. Pasteur **109**, 837—854 (1965).

GRANBOULAN, N., P. TOURNIER, R. WICKER, and W. BERNHARD: An electron microscope study of the development of SV40 virus. J. Cell Biol. **17**, 423—441 (1963).

GREEN, M., and K. FUJINAGA: Genetic relatedness of viral specific RNA's in tumor cells induced by highly and weakly oncogenic human adenovirus. In: Subviral Carcinogenesis (YOHEI ITO, ed.), 1st Int. Symp. Tumor Viruses, Nagoya, "Nisshia" Print. Comp. Mitu, Nakagyo-ku, Kyoto, pp. 89—95 (1966).

GREENE, H. S. N.: The transplantation of tumours to the brains of heterologous species. Cancer Res. **11**, 529—534 (1951).

GREENE, H. S. N.: Compatibility and non-compatibility in tissue transplantation. In: Biological Specificity and Growth (E. G. BUTLER, ed.). Princeton: Princeton University Press, pp. 177—194 (1955).

GRIFFIN, E. R., D. H. WRIGHT, T. M. BELL, and M. G. R. ROSS: Demonstration of virus particles in biopsy material from cases of Burkitt's tumour. Europ. J. Cancer **2**, 353—358 (1966).

GROSS, L.: "Spontaneous" leukemia developing in C3H mice following inoculation in infancy with AK leukemic extracts, or AK embryos. Proc. Soc. exp. Biol. (N.Y.) **76**, 27—32 (1951a).

GROSS, L.: Pathogenic properties, and "vertical" transmission of the mouse leukemia agent. Proc. Soc. exp. Biol. (N.Y.) **78**, 342—348 (1951b).

GROSS, L.: Mouse leukemia. Ann. N. Y. Acad. Sci. **54**, 1184—1196 (1952).

GROSS, L.: A filterable agent, recovered from AK leukemic extracts, causing salivary gland carcinomas in C3H mice. Proc. Soc. exp. Biol. (N.Y.) **83**, 414—421 (1953a).

GROSS, L.: Biological properties of the mouse leukemia agent. Cancer **6**, 153—158 (1953b).

GROSS, L.: Development and serial cell-free passage of a highly potent strain of mouse leukemia virus. Proc. Soc. exp. Biol. (N.Y.) **94**, 767—771 (1957).

GROSS, L.: Viral etiology of "spontaneous" mouse leukemia. A review. Cancer Res. **18**, 371—381 (1958).

GROSS, L.: Biological and pathogenic properties of a mouse leukemia virus. Acta haemat. (Basel) **23**, 259—275 (1960).

GROSS, L.: Oncogenic Viruses. Pergamon Press, New York, pp. 153—281 (1961a).

GROSS, L.: Viral etiology of mouse leukemia. Advanc. Cancer Res. **6**, 149—180 (1961b).

GROSS, L.: How many different viruses causing leukemia in mice? Acta haemat. (Basel) **32**, 44—62 (1964a).

GROSS, L.: Attempt at classification of mouse leukemia viruses. Mouse leukemia virus type A and the Friend virus. Acta haemat. (Basel) **32**, 81—88 (1964b).

GROSS, L.: The Rauscher virus: A mixture of the Friend virus and of the mouse leukemia virus (Gross)? Acta haemat. (Basel) **35**, 200—213 (1966).

GROSS, L., Y. DREYFUSS, and L. A. MOORE: Attempt to propagate "passage A" mouse leukemia virus on normal mouse embryo cells in tissue culture. Proc. Amer. Ass. Cancer Res. **3**, 231 (only) (1961).

GROSSFELD, H.: Production of hyaluronic acid in tissue culture of Rous sarcoma. Nature (Lond.) **196**, 782—783 (1962).

GURNEY, T.: Local stimulation of growth in primary cultures of chick embryo fibroblasts. Proc. nat. Acad. Sci. (Wash.) **62**, 906—911 (1969).

GYE, W. E., and W. J. PURDY: The infective agent in tumor filtrates: A further investigation by means of antisera to normal tissues. Brit. J. exp. Path. **14**, 250—259 (1933).

HABEL, K.: Resistance of polyoma virus immune animals to transplanted polyoma tumors. Proc. Soc. exp. Biol. (N.Y.) **106**, 722—725 (1961).

HABEL, K.: Immunological determinants of polyoma virus oncogenesis. J. exp. Med. **115**, 181—193 (1962).

HABEL, K.: Common antigen in polyoma tumors (Abstracts). Fed. Proc. **22**, 438 (only) (1963).

HABEL, K.: Specific complement-fixing antigens in polyoma tumors and transformed cells. Virology **25**, 55—61 (1965).

HABEL, K., and B. E. EDDY: Specificity of resistance to tumor challenge of polyoma and SV40 virus-immune hamsters. Proc. Soc. exp. Biol. (N.Y.) **113**, 1—4 (1963).

HABEL, K., J. JENSEN, J. PAGANO, and H. KOPROWSKI: Specific complement-fixing tumor antigen in SV40 transformed human cells. Proc. Soc. exp. Biol. (N.Y.) **118**, 4—9 (1965).

HABEL, K., and R. J. SILVERBERG: Relationship of polyoma virus and tumor *in vivo*. Virology **12**, 463—476 (1960).

HAFF, R. F., and H. E. SWIM: Serial propagation of three stains of rabbit fibroblasts: Their susceptibility to infection with vaccinia virus. Proc. Soc. exp. Biol. (N.Y.) **93**, 200—204 (1956).

HAGUENAU, F., and J. W. BEARD: The avian sarcoma-leukosis complex, its biology and ultrastructure. In: Ultrastructure in Biological Systems. Tumors induced by Viruses: Ultrastructural Studies (DALTON and HAGUENAU, eds.). New York: Academic Press, Inc. **1**, 1—59 (1962).

HAGUENAU, F., A. J. DALTON, and J. B. MOLONEY: A preliminary report of electron microscopy and bioassay studies on the Rous sarcoma I virus. J. nat. Cancer Inst. **20**, 633—649 (1958).

HAGUENAU, F., H. FEBVRE et J. ARNOULT: Mode de formation intracellulaire du virus du sarcome de Rous. Étude ultrastructurale. J. Microscopie **1**, 445—454 (1962).

HAGUENAU, F., and H. HANAFUSA: A quantitative electron microscopic study of the virus particles found in cells infected with Rous sarcoma virus. Virology **34**, 275—281 (1968).

HAHN, G. M., J. R. STEWART, S. J. YANG, and V. PARKER: Chinese hamster cell monolayer cultures. 1. Changes in cell dynamics and modifications of the cell cycle with the period of growth. Exp. Cell Res. **49**, 285—292 (1968).

HALBERSTAEDTER, L., L. DOLJANSKI, and E. TENENBAUM: Experiments on the cancerization of cells *in vitro* by means of Rous sarcoma agent. Brit. J. exp. Path. **22**, 179—187 (1941).

HALL, W. T., J. W. HARTLEY, and K. K. SANFORD: Characteristics of and relationship between C particles and intracisternal A particles in cloned cell strains. J. Virol. **2**, 238—247 (1968).

HAMPTON, E. G., and M. L. EIDINOFF: Transformation of chick embryo cells in culture by Rous sarcoma virus — cytochemical studies. Cancer Res. **22**, 1061—1066 (1962).

HANAFUSA, H.: Nature of the defectiveness of Rous sarcoma virus. Nat. Cancer Inst. Monogr. **17**, 543—556 (1964).

HANAFUSA, H.: Analysis of the defectiveness of Rous sarcoma virus. III. Determining influence of a new helper virus on the host range and susceptibility to interference of RSV. Virology 25, 248—255 (1965).

HANAFUSA H.: Rapid transformation of cells by Rous saracoma virus. Proc. nat. Acad. Sci. (Wash.), 63, 318—325 (1969).

HANAFUSA, H., and T. HANAFUSA: Analysis of defectiveness of Rous sarcoma virus. IV. Kinetics of RSV production. Virology 28, 369—378 (1966a).

HANAFUSA, H., and T. HANAFUSA: Determining factor in the capacity of Rous sarcoma virus to induce tumors in mammals. Proc. nat. Acad. Sci. (Wash.) 55, 532—538 (1966b).

HANAFUSA, H., and T. HANAFUSA: Further studies on RSV production from transformed cells. Virology 34, 630—636 (1968).

HANAFUSA, H., T. HANAFUSA, and H. RUBIN: The defectiveness of Rous sarcoma virus. Proc. nat. Acad. Sci. (Wash.) 49, 572—580 (1963).

HANAFUSA, H., T. HANAFUSA, and H. RUBIN: Analysis of the defectiveness of Rous sarcoma virus. I. Characterization of the helper virus. Virology 22, 591—601 (1964a).

HANAFUSA, H., T. HANAFUSA, and H. RUBIN: Differential responsiveness of Rous sarcoma virus stocks to specific cellular resistance induced by avian leukosis viruses. Virology 22, 643—645 (1964b).

HANAFUSA, H., T. HANAFUSA, and H. RUBIN: Analysis of the defectiveness of Rous sarcoma virus. II. Specification of RSV antigenicity by helper virus. Proc. nat. Acad. Sci. (Wash.) 51, 41—48 (1964c).

HARDING, C. V., W. L. WILSON, J. R. WILSON, J. R. REDDAN, and V. N. REDDY: Triggering of the cell cycle in an organized tissue in vitro. J. cell. comp. Physiol. 72, 213—220 (1968).

HARE, J. D.: Transplant immunity to polyoma virus induced tumors. I. Correlations with biological properties of virus strains. Proc. Soc. exp. Biol. (N.Y.) 115, 805—810 (1964a).

HARE, J. D.: Transplant immunity to polyoma virus-induced tumors. II. Evidence for host-dependent immunogenic variation of polyoma virus. Proc. Soc. exp. Biol. (N.Y.) 117, 598—603 (1964b).

HARE, J. D.: Transplant immunity to polyoma virus-induced tumor cells. IV. A. polyoma strain defective in transplant antigen induction. Virology 31, 625—632 (1967a).

HARE, J. D.: Analysis of hamster tumor cell lines isolated from single colonies soon after induction by polyoma strains differing in transplant antigen production. J. Virol. 1, 905—911 (1967b).

HARE, D. J., and H. R. MORGAN: A polyoma virus variant with new antigenic determinants. Virology 19, 105—107 (1963).

HARE, J. D., P. BALDUZZI, and H. R. MORGAN: Polyoma virus and L cell relationship. I. Some characteristics of a cell line persistently infected with polyoma virus. J. nat. Cancer Inst. 30, 45—56 (1963).

HARE, J. D., and T. GODAL: Transplant immunity to polyoma virus induced tumors. III. Evidence for heterogeneity among transplant antigens in vivo. Proc. Soc. exp. Biol. (N.Y.) 118, 632—636 (1965).

HAREL, J., L. HAREL, A. GOLDÉ et P. VIGIER: Homologie entre génome du virus du sarcome de Rous (RSV) et génome cellulaire. C. R. Acad. Sci. (Paris) 263, 745—748 (1966).

HAREL, L., A. GOLDÉ, J. HAREL, L. MONTAGNIER et P. VIGIER: Isolement d'un acide ribonucléique de haut poids moléculaire de deux souches différentes du virus de Rous. C. R. Acad. Sci. (Paris) 261, 4559—4562 (1965).

HAREL, L., J. HAREL, F. LACOUR et J. HUPPERT: Homologie entre génome du virus de la myéloblastose aviaire (AMV) et génome cellulaire. C. R. Acad. Sci. (Paris) 263, 616—619 (1966).

HARRIS, H., and J. F. WATKINS: Hybrid cells derived from mouse and man: Artificial heterokaryons of mammalian cells from different species. Nature (Lond.) **205,** 640—646 (1965).

HARRIS, M.: Quantitative growth studies with chick myoblasts in glass substrate cultures. Growth **21,** 149—166 (1957).

HARRIS, M.: Evaluation of drug resistance in cell cultures by differential toxicity tests. J. nat. Cancer Inst. **26,** 13—18 (1961).

HARRIS, M.: Cell Culture and Somatic Variation. Holt, Rinehart and Winston, New York (1964).

HARRIS, R. J. C.: A virus etiology for Bukitt's tumor? Int. J. Cancer **2,** 559—561 (1967).

HARRIS, R. J. C., H. MALMGREN, and B. SYLVÉN: The polysaccharides of Rous sarcoma No. 1. Brit. J. Cancer **8,** 141—146 (1954).

HARRISON, R. G.: The reactions of embryonic cells to solid substrates. J. exp. Zool. **17,** 521—544 (1914).

HARTLEY, J. W., and W. P. ROWE: Production of altered cell foci in tissue culture by defective Moloney sarcoma virus particles. Proc. nat. Acad. Sci. (Wash.) **55,** 780—786 (1966).

HARTLEY, J. W., W. P. ROWE, W. I. CAPPS, and R. J. HUEBNER: Complement fixation and tissue culture assays for mouse leukemia viruses. Proc. nat. Acad. Sci. (Wash.) **53,** 931—938 (1965).

HARTWELL, L. H., M. VOGT, and R. DULBECCO: Induction of cellular DNA synthesis by polyoma virus. II. Increase in the rate of enzyme synthesis after infection with polyoma virus in mouse kidney cells. Virology **27,** 262—272 (1965).

HARVEY, J. J.: An unidentified virus which causes the rapid production of tumours in mice. Nature (Lond.) **204,** 1104—1105 (1964).

HARVEY, J. J.: Susceptibility of *Praomys (Mastomys) natalensis* to the murine sarcoma virus-Harvey (MSV-H). Int. J. Cancer **3,** 634—643 (1968).

HARVEY, J. J., and J. EAST: Biological activity and separation of a leukaemogenic virus from murine sarcoma virus-Harvey (MSV-H). Int. J. Cancer **4,** 655—665 (1969).

HATANAKA, M., and R. DULBECCO: Induction of DNA synthesis by SV40. Proc. nat. Acad. Sci. (Wash.) **56,** 736—740 (1966).

HATANAKA, M., and R. DULBECCO: SV40-specific thymidine kinase. Proc. nat. Acad. Sci. (Wash.) **58,** 1888—1894 (1967).

HAUSCHKA, T. S., and A. LEVAN: Characterization of five ascites tumors with respect to chromosome ploidy. Anat. Rec. **111,** 467 (only) (1951).

HAUSCHKA, T. S., and A. LEVAN: Inverse relationship between chromosome ploidy and host-specificity of sixteen transplantable tumors. Exp. Cell Res. **4,** 457—467 (1953).

HAYFLICK, L.: The establishment of a line (WISH) of human amnion cells in continuous cultivation. Exp. Cell Res. **23,** 14—20 (1961).

HAYFLICK, L.: The limited *in vitro* lifetime of human diploid cell strains. Exp. Cell Res. **37,** 614—636 (1965).

HAYFLICK, L.: Senescence and cultured cells. Perspect. exp. Gerontol. **14,** 195—211 (1966).

HAYFLICK, L., and P. MOORHEAD: The serial cultivation of human diploid cell strains. Exp. Cell Res. **25,** 585—621 (1961).

HEINE, U., G. S. BEAUDREAU, C. BECKER, D. BEARD, and J. W. BEARD: Virus of avian erythroblastosis. VII. Ultrastructure of erythroblasts from the chicken and from tissue culture. J. nat. Cancer Inst. **26,** 359—388 (1961).

HEINE, U., R. A. BONAR, C. BECKER, and J. W. BEARD: Lysosome system of avian leukemic myeloblasts. Nat. Cancer Inst., Monogr. **17,** 677—689 (1964).

HEINE, U., A. GRAFFI, J. G. HELMCKE und A. RANDT: Virusartige Partikeln in zellfrei übertragbaren Mäuseleukämien. Naturwissenschaften **44,** 449—450 (1957).

HEINE, U., A. J. LANGLOIS, J. ŘÍMAN, and J. W. BEARD: Ultrastructure of chick embryo cells altered by strain MC 29 avian leukosis virus. Cancer Res. **29**, 442—458 (1969).

HEINE, U., G. DE THÉ, H. ISHIGURO, J. R. SOMMER, D. BEARD, and J. W. BEARD: Multiplicity of cell response to the BAI strain A (myeloblastosis) avian tumor virus. II. Nephroblastoma (Wilms' tumor): Ultrastructure. J. nat. Cancer Inst. **29**, 41—105 (1962).

HELLSTRÖM, I.: Resistance levels of polyoma tumors and tumors of other origin to the cytopathogenic effect of polyoma virus. J. nat. Cancer Inst. **31**, 1511—1523 (1963).

HELLSTRÖM, I.: Distinction between the effects of antiviral and anticellular polyoma antibodies on polyoma tumour cells. Nature (Lond.) **208**, 652—653 (1965).

HELLSTRÖM, I., K. E. HELLSTRÖM, and H. O. SJÖGREN: Further studies on superinfection of polyoma-induced mouse tumors with polyoma virus *in vitro*. Virology **16**, 282—300 (1962).

HELLSTRÖM, I., K. E. HELLSTRÖM, H. O. SJÖGREN, and G. KLEIN: Superinfection of polyoma-induced mouse tumors with polyoma virus *in vitro*. Exp. Cell Res. **21**, 255—259 (1960).

HELLSTRÖM, I., and H. O. SJÖGREN: Demonstration of H-2 isoantigens and polyoma specific tumor antigens by measuring colony formation *in vitro*. Exp. Cell Res. **40**, 212—215 (1965).

HELLSTRÖM, I., and H. O. SJÖGREN: Demonstration of common antigen(s) in mouse and hamster polyoma tumors. Int. J. Cancer **1**, 481—489 (1966).

HENLE, G., and F. DEINHARDT: The establishment of strains of human cells in tissue culture. J. Immunol. **79**, 54—59 (1957).

HENLE, G., F. DEINHARDT, and J. RODRIGUEZ: The development of polyoma virus in mouse embryo cells as revealed by fluorescent antibody staining. Virology **8**, 388—391 (1959).

HENLE, G., and W. HENLE: Evidence for a persistent viral infection in a cell line derived from Burkitt's lymphoma. J. Bact. **89**, 252—258 (1965).

HENLE, G., and W. HENLE: Studies on cell lines derived from Burkitt's lymphoma. Trans. N.Y. Acad. Sci. Ser. II, **29**, 71—79 (1966a).

HENLE, G., and W. HENLE: Immunofluorescence in cells derived from Burkitt's lymphoma. J. Bact. **91**, 1248—1256 (1966b).

HENLE, G., W. HENLE, and V. DIEHL: Relation of Burkitt's tumor-associated herpes-type virus to infectious mononucleosis. Proc. nat. Acad. Sci. (Wash.) **59**, 94—101 (1968).

HENLE, G., H. C. HINZE, and W. HENLE: Persistent infection of L cells with polyoma virus: periodic destruction and repopulation of the cultures. J. nat. Cancer Inst. **31**, 125—141 (1963).

HENLE, W., G. DIEHL, G. KOHN, H. ZUR HAUSEN, and G. HENLE: Herpes-type virus and chromosome marker in normal leukocytes after growth with irradiated Burkitt cells. Science **157**, 1064—1065 (1967).

HENLE, W., and G. HENLE: Effect of arginine-deficient media on the herpes-type virus associated with cultured Burkitt tumor cells. J. Virol. **2**, 182—191 (1968).

HENLE, W., K. HUMMELER, and G. HENLE: Antibody coating and agglutination of virus particles separated from the EB 3 line of Burkitt lymphoma cells. J. Bact. **92**, 269—271 (1966).

HENRY, P., P. H. BLACK, M. N. OXMAN, and S. M. WEISSMAN: Stimulation of DNA synthesis in mouse cell line 3T3 by simian virus 40. Proc. nat. Acad. Sci. (Wash.) **56**, 1170—1176 (1966).

HINUMA, Y., and J. T. GRACE: Cloning of immunoglobulin-producing human leukemic and lymphoma cells in long-term culture. Proc. Soc. exp. Biol. (N.Y.) **124**, 107—111 (1967).

HINUMA, Y., and J. T. GRACE: Cloning of Burkitt lymphoma cells cultured *in vitro*. Cancer **22**, 1089—1095 (1969).

HOGGAN, D. M., W. P. ROWE, P. H. BLACK, and R. J. HUEBNER: Production of "tumor-specific" antigens by oncogenic viruses during acute cytolytic infections. Proc. nat. Acad. Sci. (Wash.) **53**, 12—19 (1965).

HOLLEY, R. W., and J. A. KIERNAN: "Contact inhibition" of cell division in 3T3 cells. Proc. nat. Acad. Sci. (Wash.) **60**, 300—304 (1968).

HOLLINSHEAD, A. C., B. BUNNAG, and T. C. ALFORD: Relationship between the subunits and "T" antigens of adenovirus type 12. Nature (Lond.) **215**, 397—399 (1967).

HOLT, S. J., and R. M. HICKS: The localization of acid phosphatase in rat liver cells as revealed by combined cytochemical staining and electron microscopy. J. biophys. biochem. Cytol. **11**, 47—66 (1961).

HOPPS, H. E., E. L. BERNHEIM, A. NISALAK, J. H. TJIO, and J. E. SMADEL: Biological characteristics of a continuous kidney cell line derived from African green monkey. J. Immunol. **91**, 416—424 (1963).

HOUSE, W. M., and M. STOKER: Structure of normal and polyoma virus-transformed hamster cell cultures. J. Cell Sci. **1**, 169—173 (1966).

HOWATSON, A. F., and J. D. ALMEIDA: Observations on the fine structure of polyoma virus. J. biophys. biochem. Cytol. **8**, 828—834 (1960).

HOWATSON, A. F., and L. V. CRAWFORD: Direct counting of the capsomeres in polyoma and papilloma viruses. Virology **21**, 1—6 (1963).

HSIUNG, G., and W. H. GAYLORD: The vacuolating virus of monkeys. I. Isolation, growth characteristics and inclusion body formation. J. exp. Med. **114**, 975—985 (1961).

HSU, T. C.: Chromosomal evolution in cell populations. Int. Rev. Cytol. **12**, 69—161 (1961).

HSU, T. C., and O. KLATT: Mammalian chromosomes *in vitro*. IX. On genetic polymorphism in cell populations. J. nat. Cancer Inst. **21**, 437—473 (1958).

HUEBNER, R. J.: The murine leukemia-sarcoma virus complex. Proc. nat. Acad. Sci. (Wash.) **58**, 835—842 (1967).

HUEBNER, R. J., D. ARMSTRONG, M. OKUYAN, P. S. SARMA, and H. C. TURNER: Specific complement-fixing viral antigens in hamster and guinea pig tumors induced by the Schmidt-Ruppin strain of avian sarcoma. Proc. nat. Acad. Sci. (Wash.) **51**, 742—750 (1964).

HUEBNER, R. J., M. J. CASEY, R. M. CHANOCK, and K. SCHELL: Tumors induced in hamsters by a strain of adenovirus type 3: Sharing of tumor antigens and "neoantigens" with those produced by adenovirus type 7 tumors. Proc. nat. Acad. Sci. (Wash.) **54**, 381—388 (1965).

HUEBNER, R. J., R. M. CHANOCK, B. A. RUBIN, and M. J. CASEY: Induction by adenovirus type 7 of tumors in hamsters having the antigenic characteristics of SV40 virus. Proc. nat. Acad. Sci. (Wash.) **52**, 1333—1340 (1964).

HUEBNER, R. J., J. W. HARTLEY, W. P. ROWE, W. T. LANE, and W. I. CAPPS: Rescue of the defective genome of Moloney sarcoma virus from a non-infectious hamster tumor and the production of pseudotype sarcoma viruses with various murine leukemia viruses. Proc. nat. Acad. Sci. (Wash.) **56**, 1164—1169 (1966).

HUEBNER, R. J., H. G. PEREIRA, A. C. ALLISON, A. C. HOLLINSHEAD, and H. C. TURNER: Production of type-specific C antigen in virus-free hamster tumor cells induced by adenovirus type 12. Proc. nat. Acad. Sci. (Wash.) **51**, 432—439 (1964).

HUEBNER, R. J., W. P. ROWE, and W. T. LANE: Oncogenic effects in hamsters of human adenovirus type 12 and 18. Proc. nat. Acad. Sci. (Wash.) **48**, 2051—2058 (1962).

HUEBNER, R. J., W. P. ROWE, H. C. TURNER, and W. T. LANE: Specific adenovirus complement-fixing antigens in virus-free hamster and rat tumors. Proc. nat. Acad. Sci. (Wash.) **50**, 379—389 (1963).

HUEBNER, R. J., and G. TODARO: Oncogenes of RNA tumor viruses as determinants of cancer. Proc. nat. Acad. Sci. (Wash.) **64**, 1087—1094 (1969).

HULL, R. N., I. S. JOHNSON, C. G. CULBERTSON, C. B. REIMER, and H. WRIGHT: Oncogenicity of the simian adenoviruses. Science **150**, 1044—1046 (1965).

HUMMELER, K., G. HENLE, and W. HENLE: Fine structure of a virus in cultured lymphoblasts from Burkitt lymphoma. J. Bact. **91**, 1366—1368 (1966).

HUPPERT, J., F. LACOUR, J. HAREL, and L. HAREL: High molecular weight RNA from avian myeloblastosis virus. Cancer Res. 26, 1561—1568 (1966).

HÄYRY, P., and V. DEFENDI: Use of mixed hemagglutination technique in detection of virus-induced antigen(s) on SV 40-transformed cell surface. Virology 36, 317—321 (1968).

HÄYRY, P., and V. DEFENDI: Demonstration of specific antigen(s) on the surface of SV 40-transformed cells using the mixed hemagglutination technique. Transpl. Proc. 1, 119—121 (1969).

HÄYRY, P., and V. DEFENDI: Surface antigen(s) of SV 40-transformed tumor cells. Virology 41, 22—29 (1970).

IGEL, H. J., and P. H. BLACK: In vitro transformation by the adenovirus-SV 40 hybrid viruses. III. Morphology of tumors induced with transformed cells. J. exp. Med. 125, 647—656 (1967).

IGEL, H. J., R. J. HUEBNER, B. DEPPA, and S. BUMGARNER: Rescue of the defective murine sarcoma virus genome by radiation-induced leukemia virus from C57B1 mice. Proc. nat. Acad. Sci. (Wash.) 58, 1870—1877 (1967).

INBAR, M., and L. SACHS: Structural difference in sites on the surface membrane of normal and transformed cells. Nature (Lond.) 223, 710—712 (1969).

INGRAM, V. M.: A side view of moving fibroblasts. Nature (Lond.) 222, 641—644 (1969).

IOACHIM, H. L.: Continuous formation of plasma cells in long-term cultures of spleen. Exp. Cell Res. 38, 247—263 (1965).

IOACHIM, H. L., and L. BERWICK: Continuous viral replication and cellular neoplastic transformation in cultures of normal rat thymus infected with Gross leukemia virus. Int. J. Cancer 3, 61—73 (1968).

IOACHIM, H. L., L. BERWICK, and J. FURTH: Replication of Gross leukemia virus in long-term cultures of rat thymomas: Bioassays and electron microscopy. Cancer Res. 26, 803—811 (1966).

IOACHIM, H. L., and J. FURTH: Intrareticular cell multiplication of leukemic lymphoblasts in thymic tissue cultures. J. nat. Cancer Inst. 32, 339—359 (1964).

IRLIN, I. S.: Immunofluorescent demonstration of a specific surface antigen in cells infected or transformed by polyoma virus. Virology 32, 725—728 (1967).

ISHIGURO, H., D. BEARD, J. R. SOMMER, U. HEINE, G. DE THÉ, and J. W. BEARD: Multiplicity of cell response to the BAI strain A (myeloblastosis) avian tumor virus. I. Nephroblastoma (Wilms' tumor): Gross and microscopic pathology. J. nat. Cancer Inst. 29, 1—39 (1962).

ISHIMOTO, N., H. M. TEMIN, and J. L. STROMINGER: Studies of carcinogenesis by avian sarcoma viruses. II. Virus-induced increase in hyaluronic acid synthetase in chicken fibroblasts. J. biol. Chem. 241, 2052—2057 (1966).

ISHIZAKI, R., and P. K. VOGT: Immunological relationships among envelope antigens of avian tumor viruses. Virology 30, 375—387 (1966).

ISING-IVERSEN, U.: Heterologous growth of Rous sarcoma. Acta path. microbiol. scand. 50, 145—155 (1960).

IWAKATA, S., and J. T. GRACE, JR.: Cultivation in vitro of myeloblasts from human leukemia. N.Y. State J. Med. 64, 2279—2282 (1964).

JAINCHILL, J. L., and G. J. TODARO: Stimulation of cell growth in vitro by serum with and without growth factor. Relation to contact inhibition and viral transformation. Exp. Cell Res. 59, 137—146 (1970).

JARRETT, O., and I. MACPHERSON: The basis of the tumorigenicity of BHK 21 cells. Int. J. Cancer 3, 654—662 (1968).

JENSEN, E. M., W. KOROL, S. L. DITTMAR, and T. J. MEDREK: Virus containing lymphocyte cultures from cancer patients. J. nat. Cancer Inst. 39, 745—754 (1967).

JENSEN, F., A. GIRARDI, R. GILDEN, and H. KOPROWSKI: Infection of human and simian tissue cultures with Rous sarcoma virus. Proc. nat. Acad. Sci. (Wash.) 52, 53—59 (1964).

JENSEN, F., R. B. L. GWATKIN, and J. D. BIGGERS: A simple organ culture method which allows simultaneous isolation of specific types of cells. Exp. Cell Res. **34**, 440—447 (1964).

JENSEN, F., H. KOPROWSKI, J. PAGANO, J. PONTÉN, and G. RAVDIN: Autologous and homologous implantation of human cells transformed *in vitro* by simian virus 40. J. nat. Cancer Inst. **32**, 917—937 (1964).

JENSEN, F., H. KOPROWSKI, and J. PONTÉN: Rapid transformation of human fibroblast cultures by simian virus 40. Proc. nat. Acad. Sci. (Wash.) **50**, 343—348 (1963).

JENSEN, F. C., and H. KOPROWSKI: Absence of repressor in SV40-transformed cells. Virology **37**, 687—690 (1969).

JOHNSON, M., and P. MORA: Lipids of the Rauscher mouse leukemia virus. Virology **31**, 230—237 (1967).

KABATH, E. A., and J. FURTH: Neutralization of the agent causing leukosis and sarcoma of fowls by rabbit antisera. J. exp. Med. **74**, 257—261 (1941).

KAHLER, H., W. P. ROWE, B. J. LLOYD, and J. W. HARTLEY: Electron microscopy of mouse parotid tumour (polyoma) virus. J. nat. Cancer Inst. **22**, 647—653 (1959).

KAIGHN, M. E., J. D. EBERT, and P. M. STOTT: The susceptibility of differentiating muscle clones to Rous sarcoma virus. Proc. nat. Acad. Sci. (Wash.) **56**, 133—140 (1966).

KAMEI, H., and G. E. MOORE: Production of phage-neutralizing activity *in vitro* by cells derived from Burkitt's lymphoma. Nature (Lond.) **215**, 860—861 (1967).

KÁRA, J.: Induction of cellular DNA synthesis in chick embryo fibroblasts infected with Rous sarcoma virus in culture. Biochem. biophys. Res. Commun. **32**, 817—824 (1968a).

KÁRA, J.: Induction of deoxycytidylate deaminase and uridine kinase and activation of cellular DNA synthesis in the course of transformation of chicken embryo cells infected by Rous sarcoma virus *in vitro* Folia biol. (Praha) **14**, 249—265 (1968b).

KÁRA, J., and R. WEIL: Specific activation of the DNA-synthesizing apparatus in contact-inhibited mouse kidney cells by polyoma virus. Proc. nat. Acad. Sci. (Wash.) **57**, 63—70 (1967).

KASTEN, F. H., C. VENDRELY, P. TOURNIER, and R. WICKER: DNA lesions induced in nuclei and nucleoli of monkey kidney cells in tissue culture by simian vacuolating virus (SV40). J. cell. comp. Physiol. **66**, 33—48 (1965).

KATO, R.: Localization of "spontaneous" and Rous sarcoma virus-induced breakage in specific regions of the chromosomes of the Chinese hamster. Hereditas **58**, 221—247 (1967).

KELLOFF, G., and P. K. VOGT: Localization of avian tumor virus group-specific antigen in cell and virus. Virology **29**, 377—384 (1966).

KEOGH, E. V.: Ectodermal lesions produced by the virus of Rous sarcoma. Brit. J. exp. Path. **19**, 1—9 (1938).

KHARE, G. P., and R. A. CONSIGLI: Multiplication of polyoma virus. I. Use of selectively labeled (H3) virus to follow the course of infection. J. Bact. **90**, 819—821 (1965).

KHERA, K. S., A. ASHKENAZI, F. RAPP, and J. L. MELNICK: Immunity in hamsters to cells transformed *in vitro* and *in vivo* by SV40. J. Immunol. **91**, 604—613 (1963).

KHOURY, G., and J. VAN DER NOORDAA: Absence of infectious virus from a line of SV40-transformed human liver cells. Proc. Soc. exp. Biol. (N.Y.) **131**, 297—300 (1969).

KILLANDER, D.: Intercellular variations in generation time and amounts of DNA, RNA and mass in a mouse leukemia population *in vitro*. Exp. Cell Res. **40**, 21—31 (1965).

KILLANDER, D.: Intercellular variability in normal and neoplastic cell populations *in vitro*. Almqvist and Wiksell, 1—32 (1966).

KILLANDER, D.: Intercellular variation in the amounts of DNA, RNA and mass in a leukemia and a fibroblast cell population *in vitro*. Europ. J. Cancer **3**, 309—314 (1967).

KILLANDER, D., and R. RIGLER: Initial changes of deoxyribonucleoprotein and synthesis of nucleic acid in phytohemagglutinin stimulated human leucocytes *in vitro*. Exp. Cell Res. **39**, 701—704 (1965).

KILLANDER, D., and R. RIGLER: Activation of deoxyribonucleoprotein in human leucocytes stimulated by phytohemagglutinin. I. Kinetics of the binding of acridine orange to deoxyribonucleoprotein. Exp. Cell Res. **54**, 163—170 (1969).

KIRSCHSTEIN, R. L., A. S. RABSON, and G. T. O'CONOR: Ependymomas produced in Syrian hamsters by adenovirus 7, strain E 46 ("hybrid" of adenovirus 7 and SV 40). Proc. Soc. exp. Biol. (N.Y.) **120**, 484—487 (1965).

KIRSCHSTEIN, R. L., A. S. RABSON, F. J. PAUL, and E. A. PETERS: Double infection of newborn syrian hamsters with simian virus 40 and human adenovirus 12. Cancer Res. **26**, 1361—1364 (1966).

KIT, S.: Enzyme inductions in cell cultures during productive and abortive infections by papovavirus SV 40. In: The Molecular Biology of Viruses. Colter and Paranchych (eds.), Academic Press, New York-London. pp. 495—525 (1967).

KIT, S., D. R. DUBBS, R. A. DE TORRES, and J. L. MELNICK: Enhanced thymidine kinase activity following infection of green monkey kidney cells by simian adenoviruses, simian papovavirus SV 40 and an adenovirus-SV 40 "Hybrid". Virology **27**, 453—457 (1965).

KIT, S., D. R. DUBBS, and P. M. FREARSON: Enzymes of nucleic acid metabolism in cells infected with polyoma virus. Cancer Res. **26**, 638—646 (1966).

KIT, S., D. R. DUBBS, P. M. FREARSON, and J. L. MELNICK: Enzyme induction in SV 40-infected green monkey kidney cultures. Virology **29**, 69—83 (1966).

KIT, S., D. R. DUBBS, L. J. PIEKARSKI, R. A. DE TORRES, and J. L. MELNICK: Acquisition of enzyme function by mouse kidney cells abortively infected with papovavirus SV 40. Proc. nat. Acad. Sci. (Wash.) **56**, 463—470 (1966).

KIT, S., T. KURIMURA, and D. R. DUBBS: Transplantable mouse tumor line induced by injection of SV 40-transformed mouse kidney cells. Int. J. Cancer **4**, 384—392 (1969).

KIT, S., T. KURIMURA, M. L. SALVI, and D. R. DUBBS: Activation of infectious SV 40 DNA synthesis in transformed cells. Proc. nat. Acad. Sci. (Wash.) **60**, 1239—1246 (1968).

KITAHARA, T., and J. L. MELNICK: Thermal separation of the synthesis of papovavirus SV 40 tumor and virus antigens. Proc. Soc. exp. Biol. (N.Y.) **120**, 709—712 (1965).

KLEIN, E., P. CLIFFORD, G. KLEIN, and C. A. HAMBERGER: Further studies on the membrane immunofluorescence reaction of Burkitt lymphoma cells. Int. J. Cancer **2**, 27—36 (1967).

KLEIN, E., and G. KLEIN: Antigenic properties of lymphomas induced by the Moloney agent. J. nat. Cancer Inst. **32**, 547—568 (1964).

KLEIN, G.: Genetics of Somatic Cells. (V. J. BURDETTE, ed.). Holden Day, Inc. San Francisco, pp. 646 (407—468) (1963).

KLEIN, G., P. CLIFFORD, E. KLEIN, R. T. SMITH, J. MINOWADA, M. KOURILSKY, and J. H. BURCHENAL: Membrane immunofluorescence reactions of Burkitt lymphoma cells from biopsy specimens and tissue cultures. J. nat. Cancer Inst. **39**, 1027—1044 (1967).

KLEIN, G., P. CLIFFORD, E. KLEIN, and J. STJERNSWÄRD: Search for tumor specific immune reactions in Burkitt lymphoma patients by the membrane immunofluorescence reaction. Proc. nat. Acad. Sci. (Wash.) **55**, 1628—1635 (1966).

KLEIN, G., G. GEERING, L. J. OLD, G. HENLE, W. HENLE, and P. CLIFFORD: Comparison of the anti-EBV titer and the EBV-associated membrane reactive and precipitating antibody levels in the sera of Burkitt lymphoma and nasopharyngeal carcinoma patients and controls. Int. J. Cancer **5**, 185—194 (1970).

KLEIN, G., and E. KLEIN: Antigenic behavior of Moloney lymphomas: Independence of virus release and immunosensitivity. Science **145**, 1316—1317 (1964).

KLEIN, G., G. PEARSON, G. HENLE, W. HENLE, G. GOLDSTEIN, and P. CLIFFORD: Relation between Epstein-Barr viral and cell membrane immunofluorescence in Burkitt tumor cells. III. Comparison of blocking of direct membrane immunofluorescence and anti-EBV-reactivities of different sera. J. exp. Med. 129, 697—705 (1969).

KLEIN, G., G. PEARSON, J. S. NADKARNI, J. J. NADKARNI, E. KLEIN, G. HENLE, W. HENLE, and P. CLIFFORD: Relation between Epstein-Barr viral and cell membrane immunofluorescence of Burkitt tumor cells. I. Dependence of cell membrane immunofluorescence on presence of EB virus. J. exp. Med. 128, 1011—1020 (1968).

KLEIN, G., H. O. SJÖGREN, and E. KLEIN: Demonstration of host resistance against isotransplantation of lymphomas induced by the Gross agent. Cancer Res. 22, 955—961 (1962).

KLEIN, J. C.: Absence of malignant transformation after weekly irradiation of a mouse spleen cell culture. J. nat. Cancer Inst. 37, 655—661 (1966).

KLEMENT, V., P. CHÝLE, and J. SVOBODA: Comparison of the biological properties of tumours formed in rats after the administration of two variants of Rous sarcoma. Folia biol. (Praha) 9, 412—419 (1963).

KLEMENT, V., J. W. HARTLEY, W. ROWE, and R. J. HUEBNER: Recovery of a hamster-specific, focus-forming, and sarcomagenic virus from a "non-infectious" hamster tumor induced by the Kirsten mouse sarcoma virus. J. nat. Cancer Inst. 43, 925—934 (1969).

KLEMENT, V., W. P. ROWE, J. W. HARTLEY, and W. E. PUGH: Mixed culture cytopathogenicity: A new test for growth of murine leukemia viruses in tissue culture. Proc. nat. Acad. Sci. (Wash.) 63, 753—758 (1969).

KLEMENT, V., and J. SVOBODA: Induction of tumours in Syrian hamsters by two variants of Rous sarcoma virus. Folia biol. (Praha) 9, 181—188 (1963).

KNOWLES, B. B., F. JENSEN, Z. STEPLEWSKI, and H. KOPROWSKI: Rescue of infectious SV40 after fusion between different SV40-transformed cells. Proc. nat. Acad. Sci. (Wash.) 61, 42—45 (1968).

KOCH, M. A., and H. J. EGGERS: Mutants of simian virus 40 which differ in cell-transforming activity. Nature (Lond.) 214, 178 (only) (1967).

KOCH, M. A., H. J. EGGERS, F. A. ANDERER, H. D. SCHLUMBERGER, and H. FRANK: Structure of simian virus 40. I. Purification and physical characterization of the virus particle. Virology 32, 503—510 (1967).

KOCH, M. A., and A. B. SABIN: Specificity of virus-induced resistance to transplantation of polyoma and SV40 tumors in adult hamsters. Proc. Soc. exp. Biol. (N.Y.) 113, 4—12 (1964).

KOHN, A., and E. MOSES: Clonal response of chick fibroblasts to infection with Rous sarcoma virus (RSV). Israel J. med. Sci. 1, 194—200 (1965).

KOHN, G., W. J. MELLMAN, P. S. MOORHEAD, J. LOFTUS, and G. HENLE: Involvement of C group chromosomes in five Burkitt lymphoma cell lines. J. nat. Cancer Inst. 38, 209—222 (1967).

KOPROWSKI, H., F. JENSEN, A. GIRARDI, and I. KOPROWSKA: Neoplastic transformation. Cancer Res. 26, 1980—1993 (1966).

KOPROWSKI, H., F. JENSEN, and Z. STEPLEWSKI: Activation of production of infectious tumor virus SV40 in heterokaryon cultures. Proc. nat. Acad. Sci. (Wash.) 58, 127—133 (1967).

KOPROWSKI, H., J. PONTÉN, F. JENSEN, R. RAVDIN, P. MOORHEAD, and E. SAKSELA: Transformation of cultures of human tissue infected with simian virus SV40. J. cell. comp. Physiol. 59, 281—292 (1962).

KRIVIT, W., and R. A. GOOD: Simultaneous occurrence of mongolism and leukemia. J. Dis. Child. 94, 289—293 (1957).

KROOTH, R. S., M. W. SHAW, and B. K. CAMPBELL: A persistent strain of diploid fibroblasts. J. nat. Cancer Inst. 32, 1031—1044 (1964).

KRUSE, P. F., W. WHITTLE, and E. MIEDEMA: Mitotic and non-mitotic multiple-layered perfusion cultures. J. Cell Biol. 42, 113—121 (1969).

KRUSE, P. F., JR., and E. J. MIEDEMA: Production and characterization of multiple layered populations of animal cells. J. Cell Biol. 27, 273—279 (1965).

KUSANO, T., and I. YAMANE: Transformation *in vitro* of the embryonal hamster brain cells by human adenovirus type 12. Tohoku J. exp. Med. 92, 141—150 (1967a).

KUSANO, T., and I. YAMANE: General characteristics of the cells transformed *in vitro* by human adenovirus type 12. Tohoku J. exp. Med. 92, 151—160 (1967b).

KUWATA, T.: Studies on the growth of Rous sarcoma and its variant strains in corti-sone-treated hamsters. Cancer Res. 20, 170—171 (1960).

KUWATA, T., Y. YASUMURA, and M. KANISAWA: Effect of viral infection and cortisone upon tumour growth in homologous and heterologous hosts. Nature (Lond.) 182, 1678—1679 (1958).

LACY, S., and M. GREEN: Adenovirus multiplication: Genetic relatedness of tumori-genic human adenovirus types 7, 12 and 18. Science 150, 1296—1298 (1965).

LACY, S. S., and M. GREEN: Biochemical studies on adenovirus multiplication. VII. Homology between DNA's of tumorigenic and non-tumorigenic human adenoviruses. Proc. nat. Acad. Sci. (Wash.) 52, 1053—1059 (1964).

LAGERLÖF, B.: *In vitro* investigations of the virus-induced fowl erythro-leukemia. I. Long-term cultivation of normal and leukemic bone marrow cells. Acta path. microbiol. scand. 49, 344—360 (1960a).

LAGERLÖF, B.: *In vitro* investigations of the virus-induced fowl erythro-leukemia. II. The production of virus by cultured erythroleukemia cells and cultured normal cells exposed to the erythroleukemia virus *in vitro*. Acta path. microbiol. scand. 49, 361—372 (1960b).

LAGERLÖF, B.: Transplantation studies with chromosome-labelled erythroleukemia cells of fowl. Acta path. microbiol. scand. 56, 135—142 (1962).

LANGLOIS, A. J., and J. W. BEARD: Converted-cell focus formation in culture by strain MC29 avian leukosis virus. Proc. Soc. exp. Biol. (N.Y.) 126, 718—722 (1967).

LANGLOIS, A. J., D. P. BOLOGNESI, R. B. FRITZ, and J. W. BEARD: Strain MC29 avian leukosis virus release by chick embryo cells infected with the agent. Proc. Soc. exp. Biol. (N.Y.) 131, 138—143 (1969).

LANGLOIS, A. J., S. SANKARAN, P. H. L. HSIUNG, and J. W. BEARD: Massive direct conversion of chick embryo cells by strain MC29 avian leukosis virus. J. Virol. 1, 1082—1084 (1967).

LASFARGUES, E. Y., and D. H. MOORE: Cell transfusion by the mammary tumor virus *in vitro*. Recent results in cancer research. In: Malignant Transformation by Viruses. (W. H. KIRSTEN, ed.). Springer-Verlag, Berlin—Heidelberg—New York, pp. 44—55 (1966).

LATARJET, R., R. CRAMER, and L. MONTAGNIER: Inactivation, by UV-, X- and γ-radiations, of the infecting and transforming capacities of polyoma virus. Virology 33, 104—111 (1967).

LAUSCH, R. N., S. S. TEVETHIA, and F. RAPP: Evidence of SV40-specific transplantation and surface antigens in cells transformed by para-adenovirus 12. J. Immunol. 101, 645—649 (1968).

LAVER, W. G., H. G. PEREIRA, W. C. RUSSEL, and R. C. VALENTINE: Isolation of an internal component from adenovirus type 5. J. molec. Biol. 37, 379—386 (1968).

LAW, L. W., and J. B. MOLONEY: Studies of congenital transmission of a leukemia virus in mice. Proc. Soc. exp. Biol. (N.Y.) 108, 715—723 (1961).

LAW, L. W., A. S. RABSON, and C. J. DAWE: Variant of a parotid tumor (polyoma) virus showing a change in oncogenic properties. Nature (Lond.) 190, 97—98 (1961).

LAW, L. W., R. C. TING, and M. F. STANTON: Some biologic, immunogenic and morphologic effects in mice after infection with a murine sarcoma virus. I. Biologic and immunologic studies. J. nat. Cancer Inst. 40, 1101—1112 (1968).

LECLERC, J. C., D. SILVESTRE, J. P. LÉVY, S. OPPENHEIM et B. VARET: Étude ultra-structurale d'un virus murin sarcomatogène (MSV). Int. J. Cancer 2, 475—488 (1967).

Leduc, E. H., T. Wicker, S. Avrameas, and W. Bernhard: Ultrastructural localization of SV40T antigen with enzyme-labelled antibody. J. gen. Virol. 4, 609—614 (1969).

Lee, H. H., M. E. Kaighn, and J. D. Ebert: Viral antigens in differentiating muscle colonies after infection with Rous sarcoma virus in vitro. Proc. nat. Acad. Sci. (Wash.) 56, 521—525 (1966).

Lee, H. H., M. E. Kaighn, and J. D. Ebert: Induction of thymidine-3 H incorporation in multinucleated myotubes by Rous sarcoma virus. Int. J. Cancer 3, 126—136 (1968).

Lennert, K.: Bildung und Differenzierung der Blutzellen, insbesonders der Lymphocyten. Verh. dtsch. path. Ges. 50, 163—213 (1966).

Leuchtenberger, C., R. Leuchtenberger, C. Vendrely, and R. Vendrely: The quantitative estimation of deoxyribose nucleic acid (DNA) in isolated individual animal nuclei by the Casperson ultraviolet method. Exp. Cell Res. 3, 240—244 (1951).

Levan, A.: Chromosome studies on some human tumors and tissues of normal origin, grown in vivo and in vitro at the Sloan-Kettering institute. Cancer 9, 648—663 (1956).

Levan, A.: Cancerogenesis. A genetic adaptation on the cellular level. In: Achste Jaarbook van Kankeronderzoek en Kankerbestrijding in Nederland, Lund: Carl Bloms Boktryckeri, pp. 110—126 (1958).

Levan, A., and J. J. Biesele: Role of chromosomes in cancerogenesis, as studied in serial tissue culture of mammalian cells. Ann. N.Y. Acad. Sci. 71, 1022—1053 (1958).

Levan, A., and T. S. Hauschka: Chromosome numbers of three mouse ascites tumors. Hereditas 38, 251—255 (1952).

Levin, M. J., M. N. Oxman, G. Th. Diamandopoulos, A. S. Levine, P. H. Henry, and J. F. Enders: Virus-specific nucleic acids in SV40-exposed hamster embryo cell lines: Correlation with S and T antigens. Proc. nat. Acad. Sci. (Wash.) 62, 589—596 (1969).

Levine, E. M., Y. Becker, Ch. W. Boone, and H. Eagle: Contact inhibition, macromolecular synthesis and polyribosomes in cultured human diploid fibroblasts. Proc. nat. Acad. Sci. (Wash.) 53, 350—356 (1965).

Levine, M.: The cytology of the tumor cell in the Rous chicken sarcoma. Amer. J. Cancer 36, 276—302 (1939a).

Levine, M.: The cytology of the tumor cell in the Rous chicken sarcoma. Amer. J. Cancer 36, 386—430 (1939b).

Levine, M.: The cytology of the tumor cell in the Rous chicken sarcoma. Amer. J. Cancer 36, 581—602 (1939c).

Levine, M.: The cytology of the tumor cell in the Rous chicken sarcoma. Amer. J. Cancer 37, 69—107 (1939d).

Levinson, W.: Fragmentation of the nucleus in Rous sarcoma virus-infected chick embryo cells. Virology 32, 74—83 (1967).

Levinthal, J. D., M. Jakobovitz, and M. D. Eaton: Polyoma disease and tumors in mice: the distribution of viral antigen detected by immunofluorescence. Virology 16, 314—319 (1962).

Levinthal, J. D., and H. M. Shein: Morphology and distribution of SV40-infected cells in human simian and hamster renal cell cultures as detected by immunofluorescence. Proc. Soc. exp. Biol. (N.Y.) 112, 405—409 (1963).

Levy, J. A., and G. Henle: Indirect immunofluorescence tests with sera from African children and cultured Burkitt lymphoma cells. J. Bact. 92, 275—276 (1966).

Levy, J. A., M. Virolainen, and V. Defendi: Human lymphoblastoid lines from lymph node and spleen. Cancer 22, 517—524 (1968).

Lévy, J. P., et J. Périès: La culture in vitro des virus leucémogènes. Path. et Biol. 12, 1158—1162 (1964).

Lewis, A. M., Jr., S. G. Baum, K. O. Prigge, and W. P. Rowe: Occurrence of adenovirus-SV40 hybrids among monkey kidney cell adapted strains of adenovirus. Proc. Soc. exp. Biol. (N.Y.) 122, 214—218 (1966).

LEWIS, M., JR., M. J. LEVIN, W. H. WIESE, C. S. CRUMPACKER, and P. H. HENRY: A non-defective (competent) adenovirus-SV40 hybrid isolated from the AD 2-SV40 hybrid population. Proc. nat. Acad. Sci. **63**, 1128—1135 (1969).

LIEBERMAN, M., and H. S. KAPLAN: Leukemogenic activity of filtrates from radiation-induced lymphoid tumors of mice. Science **130**, 387—388 (1959).

LILLY, F., E. A. BOYSE, and L. J. OLD: Genetic basis of susceptibility to viral leukaemogenesis. Lancet **2**, 1207—1209 (1964).

LITHNER, F., and J. PONTÉN: Bovine fibroblasts in long-term tissue culture. Chromosome studies. Int. J. Cancer **1**, 579—588 (1966).

LITTLEFIELD, J. W., and C. BASILICO: Infection of thymidine kinase-deficient BHK cells with polyoma virus. Nature (Lond.) **211**, 250—252 (1966).

LO, W. H. Y., G. O. GEY, and P. SHAPRAS: The cytopathogenic effect of the Rous sarcoma virus on chicken fibroblasts in tissue culture. Bull. Johns Hopk. Hosp. **96**, 248—266 (1955).

LOEB, L., and M. S. FLEISHER: The growth of tissues in the test tube under experimentally varied conditions with special reference to mitotic cell proliferation. J. med. Res. **40**, 509—550 (1919).

LOEWENSTEIN, W. R.: Some reflections on growth and differentiation. Perspect. Biol. Med. **11**, 260—272 (1968a).

LOEWENSTEIN, W. R.: III. Emergence of order in tissues and organs. Communication through cell junctions. Implications in growth control and differentiation. Development. Biologia Suppl. (Bratislava) **2**, 151—183 (1968b).

LOVE, R., and M. V. FERNANDES: Cytological and cytochemical studies of green monkey kidney cells infected *in vitro* with simian virus 40. J. Cell Biol. **25**, 529—543 (1965).

LOVE, R., and A. RABSON: Cytochemistry of nucleic acids in murine lymphoma cells infected with polyoma virus *in vitro*. Nature (Lond.) **188**, 1039—1040 (1960).

LUCAS, A. M., and C. JAMROZ: Atlas of Avian Hematology. Agricult. Monogr. **25**, 1—271 (1961).

LUDFORD, R. J.: The production of tumors by cultures of normal cells treated with filtrates of filterable fowl tumors. Amer. J. Cancer **31**, 414—429 (1937).

MACIEIRA-COELHO, A.: Dissociation between inhibition of movement and inhibition of division in RSV transformed human fibroblasts. Exp. Cell Res. **47**, 193—200 (1967a).

MACIEIRA-COELHO, A.: Relationship between DNA synthesis and cell density in normal and virus transformed cells. Int. J. Cancer **2**, 297—303 (1967b).

MACIEIRA-COELHO, A.: The influence of cell density on growth inhibition of human fibroblasts *in vitro*. Proc. Soc. exp. Biol. (N.Y.) **125**, 548—552 (1967c).

MACIEIRA-COELHO, A., and J. PONTÉN: Analogy in growth between late passage human embryonic and early passage human adult fibroblasts. J. Cell Biol. **43**, 374—377 (1969).

MACIEIRA-COELHO, A., J. PONTÉN, and L. PHILIPSON: The division cycle and RNA-synthesis in diploid human cells at different passage levels *in vitro*. Exp. Cell Res. **42**, 673—684 (1966a).

MACIEIRA-COELHO, A., J. PONTÉN, and L. PHILIPSON: Inhibition of the division cycle in confluent cultures of human fibroblasts *in vitro*. Exp. Cell Res. **43**, 20—29 (1966b).

MACINTYRE, E. H., R. A. GRIMES, and A. E. VATTER: Cytology and growth characteristics of human tumour astrocytes transformed by Rous sarcoma virus. J. Cell Sci. **5**, 583—602 (1969).

MACINTYRE, E. H., and J. PONTÉN: Interaction between normal and transformed bovine fibroblasts in culture. I. Cells transformed by Rous sarcoma virus. J. Cell Sci. **2**, 309—322 (1967).

MACPHERSON, I.: Characteristics of a hamster cell clone transformed by polyoma virus. J. nat. Cancer Inst. **30**, 795—815 (1963).

MACPHERSON, I.: Reversion in hamster cells transformed by Rous sarcoma virus. Science **148**, 1731—1733 (1965).

MACPHERSON, I.: Malignant transformation and reversion in virus infected cells. In: Recent Results in Cancer Research, Malignant Transformation by Viruses. (W. H. KIRSTEN, ed.). Springer-Verlag, Berlin—Heidelberg—New York, pp. 1—8 (1966).

MACPHERSON, I., and L. MONTAGNIER: Agar suspension culture for the selective assay of cells transformed by polyoma virus. Virology **23**, 291—294 (1964).

MACPHERSON, I., and M. STOKER: Polyoma transformation of hamster cell clones — an investigation of genetic factors affecting cell competence. Virology **16**, 147—151 (1962).

MADIN, S. H., and N. J. DARBY, JR.: Established kidney cell lines of normal adult bovine and ovine origin. Proc. Soc. exp. Biol. (N.Y.) **98**, 574—576 (1958).

MAIZEL, J. V. JR., D. O. WHITE, and M. D. SCHARFF: The polypeptides of adenovirus. II. Soluble proteins, cores, top components and the structure of the virion. Virology **36**, 126—136 (1968).

MAKINO, S.: A cytological study of the Yoshida sarcoma, an ascites tumor of white rats. Chromosoma **4**, 649—674 (1952).

MALMGREN, R. A., M. A. FINK, and W. MILLS: Demonstration of the intracellular location of Rous sarcoma virus antigen by fluorescein-labeled antiserums. J. nat. Cancer Inst. **24**, 995—1001 (1960).

MALMGREN, R. A., G. RABOTTI, and A. S. RABSON: Intracellular localization of polyoma virus antigen demonstrated with fluorescein-labeled antiserums. J. nat. Cancer Inst. **24**, 581—587 (1960).

MALMGREN, R. A., A. S. RABSON, P. G. CARNEY, and F. J. PAUL: Immunofluorescence of green monkey kidney cells infected with adenovirus 12 and adenovirus 12 plus simian virus 40. J. Bact. **91**, 262—265 (1966).

MALMGREN, R. A., K. K. TAKEMOTO, and P. G. CARNEY: Immunofluorescent studies of mouse and hamster cell surface antigens induced by polyoma virus. J. nat. Cancer Inst. **40**, 263—268 (1968).

MANAKER, R. A., and V. GROUPÉ: Discrete foci of altered chicken embryo cells associated with Rous sarcoma virus in tissue culture. Virology **2**, 838—840 (1956).

MANAKER, R. A., E. M. JENSEN, and W. KOROL: Long-term propagation of a murine leukemia virus in an established cell line. J. nat. Cancer Inst. **33**, 363—371 (1964).

MANAKER, R. A., P. C. STROTHER, A. A. MILLER, and C. V. PICZAK: Behavior *in vitro* of a mouse lymphoid-leukemia virus. J. nat. Cancer Inst. **25**, 1411—1419 (1960).

MANNWEILER, K., et W. BERNHARD: L'ultrastructure du myxosarcome de Fujinami. Bull. Cancer **45**, 223—236 (1958).

MARIN, G., and C. BASILICO: Transformation of polyploid hamster cells by polyoma virus. Nature (Lond.) **216**, 62—63 (1967).

MARIN, G., and J. M. LITTLEFIELD: Selection of morphologically normal cell lines from polyoma-transformed BHK21/13 hamster fibroblasts. J. Virol. **2**, 69—77 (1968).

MARIN, G., and I. MACPHERSON: Reversion in polyoma-transformed cells: Retransformation, induced antigens and tumorigenicity. J. Virol. **3**, 146—149 (1969).

MARTINEZ-PALOMO, A., and C. BRAILOVSKY: Surface layer in tumor cells transformed by adeno-12 and SV40 viruses. Virology **34**, 379—382 (1968).

MATSUOKA, Y., M. TAKAHASHI, Y. YAGI, G. E. MOORE, and D. PRESSMAN: Synthesis and secretion of immunoglobulins by established cell lines of human hematopoietic origin. J. Immunol. **101**, 1111—1120 (1968).

MATSUYA, Y., and I. YAMANE: Serial culture of Syrian hamster fibroblasts in albumin fortified medium and their regular development into established lines. Exp. Cell Res. **50**, 652—654 (1968).

MATTERN, C. F. T.: Polyoma and papilloma viruses: Do they have 42 or 92 subunits? Science **137**, 612—613 (1962).

MATTERN, C. F. T., K. TAKEMOTO, and A. M. DE LEVA: Electron microscopic observations on multiple polyoma virus-related particles. Virology **32**, 378—392 (1967).

MAURER, B. A., K. WERTMAN, and I. YALL: Fate of labeled polyoma virus DNA in cell cultures. Proc. Soc. exp. Biol. (N.Y.) **131**, 93—97 (1969).

MAYOR, H. D., R. M. JAMISON, and L. E. JORDAN: Biophysical studies on the nature of the simian papova virus particle (vacuolating SV40 virus). Virology **19**, 350—366 (1963).

MAYOR, H. D., S. E. STINEBAUGH, R. M. JAMISON, L. E. JORDAN, and J. L. MELNICK: Immunofluorescent, cytochemical and microcytological studies on the growth of the simian vacuolating virus (SV40) in tissue culture. Exp. molec. Path. **1**, 397—416 (1962).

McALLISTER, R. M., and I. MACPHERSON: Transformation of a hamster cell line by adenovirus type 12. J. gen. Virol. **2**, 99—106 (1968).

McALLISTER, R. M., M. O. NICOLSON, A. M. LEWIS, JR., I. MACPHERSON, and R. J. HUEBNER: Transformation of rat embryo cells by adenovirus type 1. J. gen. Virol. **4**, 29—36 (1969).

McBRIDE, W. D., and A. WIENER: *In vitro* transformation of hamster kidney cells by human adenovirus type 12. Proc. Soc. exp. Biol. (N.Y.) **115**, 870—874 (1964).

McCARTHY, B. J.: General aspects of homology studies with nucleic acids. In: Subviral Carcinogenesis. (Yohei Ito, ed.). "Nissha" Printing Company, Mibu, Nakagyo-ku, Kyoto, pp. 43—61 (1966).

McCARTHY, R. E.: "Spontaneous" malignant transformation *in vitro* of adult mouse cell lines: Effect of site of implantation on expression of cell antigenicity and malignancy. Cancer Res. **28**, 615—621 (1968).

McCARTHY, R. E., V. JUNIUS, S. FARBER, H. LAZARUS, and G. E. FOLEY: Cytogenetic analysis of human lymphoblasts in continuous culture. Exp. Cell Res. **40**, 197—200 (1965).

McCORMICK, K. J., W. A. STENBACK, J. J. TRENTIN, G. KLEIN, J. S. NADKARNI, J. J. NADKARNI, and P. CLIFFORD: Complement-fixation test for detection of herpes-like viruses in cell cultures of Burkitt's lymphoma. J. Virol. **3**, 525—527 (1969).

McCULLOCH, E. A., and R. C. PARKER: Continuous cultivation of cells of hemic origin. Canad. Cancer Conf. **2**, 152—167 (1957).

MEDINA, D., and L. SACHS: The *in vitro* formation of a stable cell-virus association with polyoma virus. Virology **10**, 387—388 (1960).

MEDINA, D., and L. SACHS: Cell-virus interaction with the polyoma virus: the induction of cell transformation and malignancy *in vitro*. Brit. J. Cancer **15**, 885—904 (1961).

MEDINA, D., and L. SACHS: Studies on the lytic interaction and cell transformation with a large — and a small — plaque mutant of polyoma virus. Virology **19**, 127—139 (1963).

MEIER, H., D. D. MYERS, and R. J. HUEBNER: Genetic control by the hr-locus of susceptibility and resistance to leukemia. Proc. nat. Acad. Sci. (Wash.) **63**, 759—766 (1969).

MELLORS, R. C., and J. S. MUNROE: Cellular localization of Rous sarcoma virus as studied with fluorescent antibody. J. exp. Med. **112**, 963—974 (1960).

MELNICK, J.: Papova virus group. Science **135**, 1128—1130 (1962).

MELNICK, J., and F. RAPP: Possible relationship between primate papovaviruses, human wart and simian SV40. J. nat. Cancer Inst. **34**, 529—534 (1965).

MENDELSOHN, C. G., and A. M. KLIGMAN: Isolation of wart virus in tissue culture. Successful reinoculation into humans. Arch. Derm. **83**, 559—562 (1961).

METZGAR, R. S., and S. R. OLEINICK: The study of normal and malignant cell antigens by mixed agglutination. Cancer Res. **28**, 1366—1371 (1968).

MEYER, H. M., H. E. HOPPS, N. G. ROGERS, B. E. BROOKS, B. C. BERNHEIM, W. P. JONES, A. NISALAK, and R. D. DOUGLAS: Studies on simian virus 40. J. Immunol. **88**, 796—805 (1962).

MICHEL, M. R., B. HIRT, and R. WEIL: Mouse cellular DNA enclosed in polyoma viral capsids (pseudovirions). Proc. nat. Acad. Sci. (Wash.) **58**, 1381—1388 (1967).

MILES, C. P., and F. O'NEILL: Chromosome studies of 8 *in vitro* lines of Burkitt's lymphoma. Cancer Res. **27**, 392—402 (1967).

MILES, C. P., F. O'NEILL, D. ARMSTRONG, B. CLARKSON, and J. KEANE: Chromosome patterns of human leukocyte established cell lines. Cancer Res. **28**, 481—490 (1968).

MILES, C. P., and S. D. STOREY: Nuclear chromocenters of cultured chicken cells. Exp. Cell Res. **27**, 377—381 (1962).

MILLER, J. F.: Etiology and pathogenesis of mouse leukemia. Advanc. Cancer Res. **6**, 291—368 (1961).

MILONE, S.: Innesti a zig-zag sarcoma di Peyton Rous fra pollo e ratto. Arch. Sci. med. **52**, 362—368 (1928).

MINOWADA, J.: Effect of X-irradiation on DNA-synthesis in polyoma virus-infected cultures. Exp. Cell Res. **33**, 161—175 (1964).

MINOWADA, J., L. CHAI, and G. E. MOORE: Studies of Burkitt lymphoma cells. III. Equilibrium density gradient centrifugation of virus particles isolated from Burkitt lymphoma cell lines. Cancer **23**, 300—305 (1969).

MINOWADA, J., G. KLEIN, P. CLIFFORD, E. KLEIN, and G. E. MOORE: Studies of Burkitt lymphoma cells. I. Establishment of a cell line (B 35M) and its characteristics. Cancer **20**, 1430—1437 (1967).

MITCHELL, J. T., W. F. ANDRESEN, and V. J. EVANS: Comparative effects of horse, calf and fetal serums on chromosomal characteristics and neoplastic conversion of mouse embryo cells *in vitro*. J. nat. Cancer Inst. **42**, 709—721 (1969).

MITCHINER, M. B.: Ultrastructural observations on the Harvey mouse leukaemia-sarcoma virus. J. Path. Bact. **93**, 593—600 (1967).

MLADENOV, Z., U. HEINE, D. BEARD, and J. W. BEARD: Strain MC29 avian leukosis virus. Myelocytoma, endothelioma, and renal growths: Pathomorphological and ultrastructural aspects. J. nat. Cancer Inst. **38**, 251—285 (1967).

MOBERGER, G.: Malignant transformation of squamous epithelium. Acta radiol. Suppl. **112**, 1—75 (1954).

MÖLLER, G.: Demonstration of mouse isoantigens at the cellular level by the fluorescent antibody technique. J. exp. Med. **114**, 415—434 (1961).

MOLONEY, J. B.: Biological studies on a lymphoid leukemia virus extracted from sarcoma 37. I. Origin and introductory investigations. J. nat. Cancer Inst. **24**, 933—951 (1960).

MOLONEY, J. B.: The murine leukemias. Fed. Proc. **21**, 19—31 (1962).

MOLONEY, J. B.: The rodent leukemias: Virus-induced murine leukemias. Ann. Rev. Med. **15**, 383—392 (1964).

MOLONEY, J. B.: A virus-induced rhabdomyosarcoma of mice. Nat. Cancer Inst. Monogr. **22**, 139—142 (1966).

MOMMAERTS, E. B., E. A. ECKERT, D. BEARD, D. G. SHARP, and J. W. BEARD: Dephosphorylation of adenosine triphosphate by concentrates of the virus of avian erythromyeloblastic leukosis. Proc. Soc. exp. Biol. (N.Y.) **79**, 450—455 (1952).

MOMMAERTS, E. B., D. G. SHARP, E. A. ECKERT, D. BEARD, and J. W. BEARD: Virus of avian erythromyeloblastic leukosis. I. Relation of specific plasma particles to the dephosphorylation of adenosine triphosphate. J. nat. Cancer Inst. **14**, 1011—1025 (1954).

MONTAGNIER, L., et I. MACPHERSON: Croissance sélective en gélose de cellules de hamster transformées par le virus du polyome. C. R. Acad. Sci. (Paris) **258**, 4171—4173 (1964).

MONTAGNIER, L., I. MACPHERSON, and O. JARRETT: An epithelioid variant of the BHK21 hamster fibroblast line and its transformation by polyoma virus. J. nat. Cancer Inst. **36**, 503—512 (1966).

MOORE, A. E.: Growth and persistence of Friend leukemia virus in tissue culture. J. nat. Cancer Inst. **30**, 885—895 (1963).

MOORE, A. E., and C. FRIEND: Attempts at growing the mouse leukemia virus in tissue culture. Proc. Amer. Ass. Cancer Res. **2**, 328 (only) (1958).

MOORE, G. E., R. E. GERNER, and H. A. FRANKLIN: Culture of normal human leukocytes. J. Amer. med. Ass. **199**, 519—524 (1967).

MOORE, G. E., J. T. GRACE, P. CITRON, R. GERNER, and A. BURNS: Leukocyte cultures of patients with leukemia and lymphomas. N.Y. St. J. Med. **66**, 2757—2764 (1966).

MOORE, G. E., E. ITO, K. ULRICH, and A. A. SANDBERG: Culture of human leukemia cells. Cancer **19**, 713—723 (1966).

MOORE, G. E., H. KITAMURA, and S. TOSHIMA: Morphology of cultured hematopoietic cells. Cancer **22**, 245—267 (1968).

MOORE, G. E., and W. F. MCLIMANS: The life span of the cultured normal cell: Concepts derived from studies of human lymphoblasts. J. theor. Biol. **20**, 217—226 (1968).

MOORE, G. E., and J. W. PICKREN: Study of a virus containing hematopoietic cell line and a melanoma cell line derived from a patient with a leukemoid reaction. Lab. Invest. **16**, 882—891 (1967).

MOORE, R., and J. UREN: Permanent cell lines of the marsupial mouse, *Antechnius swainsonii*. Exp. Cell Res. **44**, 273—282 (1966).

MOORHEAD, P. S., and E. SAKSELA: Non-random chromosomal aberrations in SV40-transformed human cells. J. cell. comp. Physiol. **62**, 57—83 (1963).

MOORHEAD, P. S., and E. SAKSELA: The sequence of chromosome aberrations during SV40 transformation of a human diploid cell strain. Hereditas **52**, 271—284 (1965).

MORA, P. T., and V. W. MCFARLAND: The nucleic acid of a murine leukemia virus. Proc. nat. Acad. Sci. (Wash.) **54**, 756—763 (1965).

MORA, P. T., V. W. MCFARLAND, and S. W. LUBORSKY: Nucleic acid of the Rauscher mouse leukemia virus. Proc. nat. Acad. Sci. (Wash.) **55**, 438—445 (1966).

MORGAN, H. R.: Ultrastructure of the surfaces of cells infected with avian leukosis-sarcoma viruses. J. Virol. **2**, 1113—1146 (1968).

MORGAN, H. R., and A. P. ANDRESE: Malignancy *in vivo* of chick embryo cells infected with Rous sarcoma virus *in vitro*. Experientia (Basel) **17**, 309—310 (1961).

MORGAN, H. R., and A. P. ANDRESE: Comparative studies in Rous sarcoma with virus, tumor cells and chick embryo cells transformed *in vitro* by virus. J. exp. Med. **116**, 329—336 (1962).

MORGAN, H. R., and S. GANAPATHY: Comparative studies in Rous sarcoma. IV. Glucose metabolism of normal and Rous sarcoma virus-infected cells. Proc. Soc. exp. Biol. (N.Y.) **113**, 312—315 (1963).

MORGAN, H. R., and W. TRAUB: In: Avian Tumor Viruses. Nat. Cancer Inst. Monogr. **17**, 392 (1964).

MORRIS, J. A., M. J. CASEY, B. E. EDDY, W. T. LANE, and R. J. HUEBNER: Occurrence of SV40 neoplastic and antigenic information in vaccine strains of adenovirus type 3. Proc. Soc. exp. Biol. (N.Y.) **122**, 679—684 (1966).

MORRIS, J. A., K. M. JOHNSON, C. G. AULISIO, R. M. CHANOCK, and V. KNIGHT: Clinical and serologic responses in volunteers given vacuolating virus (SV40) by respiratory route. Proc. Soc. exp. Biol. (N.Y.) **108**, 56—59 (1961).

MORTON, D. L., W. T. HALL, and R. A. MALMGREN: Human liposarcomas. Tissue cultures containing foci of transformed cells with viral particles. Science **165**, 813—816 (1969).

MOSCONA, A., and H. MOSCONA: The dissociation and aggregation of cells from organ rudiments of the early chick embryo. J. Anat. **86**, 287—301 (1952).

MOSCOVICI, C.: A quantitative assay for avian myeloblastosis virus. Proc. Soc. exp. Biol. (N.Y.) **125**, 1213—1215 (1967).

MOSCOVICI, C., M. G. MOSCOVICI, and M. ZANETTI: Transformation of chick fibroblast cultures with avian myeloblastosis virus. J. cell. comp. Physiol. **73**, 105—108 (1969).

MOSCOVICI, C., and P. K. VOGT: Effects of genetic cellular resistance on cell transformation and virus replication in chicken hematopoietic cell cultures infected with avian myeloblastosis virus (RAI-A). Virology **35**, 487—497 (1968).

MOSES, E., and A. KOHN: Polykaryocytosis induced by Rous sarcoma virus in chick fibroblasts. Exp. Cell Res. **32**, 182—186 (1963).

Moses, H. L., P. R. Glade, J. A. Kasel, A. S. Rosenthal, Y. Hirshaut, and L. N. Chessin: Infectious mononucleosis: Detection of herpes-like virus and reticular aggregates of small cytoplasmic particles in continuous lymphoid cell lines derived from peripheral blood. Proc. nat. Acad. Sci. (Wash.) 60, 489—496 (1968).

Moulton, J. E., and S. P. Garg: Effects of actinomycin D on tissue culture cells transformed by bovine papilloma virus in tissue culture. Amer. J. vet. Res. 27, 1475—1477 (1966).

Moyer, A., R. Wallace, and H. R. Cox: Limited growth period of human lung cell lines transformed by simian virus 40. J. nat. Cancer Inst. 33, 227—236 (1964).

Munk, K., H. Fischer, W. Th. Goedemans und G. Sauer: Vorgänge im Nuklein-säurestoffwechsel der mit SV 40-Virus infizierten Zelle. Z. Krebsforsch. 67, 213—220 (1965).

Munube, G. M. R., and T. M. Bell: Isolation of ECHO virus type 11 from two cases of Burkitt's tumour and three cases of other tumours. Int. J. Cancer 2, 613—618 (1967).

Munroe, S. J., F. Shipkey, R. A. Erlandson, and W. F. Windle: Tumors induced in juvenile and adult primates by chicken sarcoma virus. Nat. Cancer Inst., Monogr. 17, 365—390 (1964).

Murakami, W. T., R. Fine, M. R. Harrington, and Z. Ben Sassan: Properties and amino acid composition of polyoma virus purified by zonal ultracentrifugation. J. molec. Biol. 36, 153—166 (1968).

Murphy, J. B., and P. Rous: The behaviour of a chicken sarcoma implanted in the developing embryo. J. exp. Med. 15, p. 119 (1912).

Naegele, R. F., and F. Rapp: Enhancement of the replication of human adenoviruses in simian cells by simian adenovirus SV 15. J. Virol. 1, 838—840 (1967).

Nakata, Y., and J. P. Bader: Transformation by murine sarcoma virus: Fixation (deoxyribonucleic acid synthesis) and development. J. Virol. 2, 1255—1261 (1968).

Nazerian, K., J. J. Solomon, R. L. Witter, and B. R. Burmester: Studies on the etiology of Marek's disease. II. Finding of a herpesvirus in cell culture. Proc. Soc. exp. Biol. (N.Y.) 127, 177—182 (1968).

Nelson-Rees, W. A., A. J. Kniazeff, and N. B. Darby, Jr.: Preservation of bulk chromatin with decrease in number of chromosomes in cells of an established bovine kidney like. J. nat. Cancer Inst. 33, 347—361 (1964).

Nettleship, A., and W. R. Earle: Production of malignancy in vitro. VI. Pathology of tumors produced. J. nat. Cancer Inst. 4, 229—248 (1943).

Nichols, W. W., A. Levan, L. L. Coriell, H. Goldner, and C. G. Ahlström: Chromosome abnormalities in vitro in human leukocytes associated with Schmidt-Ruppin Rous sarcoma virus. Science 146, 248—250 (1964).

Nichols, W. W., A. Levan, W. Heneen, and M. Peluse: Synergism of the Schmidt-Ruppin strain of the Rous sarcoma virus and cytidine triphosphate in the induction of chromosome breaks in human cultured leukocytes. Hereditas 54, 213—236 (1965).

Nichols, W. W., M. Peluse, C. Goodheart, R. McAllister, and C. Bradt: Auto-radiographic studies on nuclei and chromosomes of cultured leukocytes after infection with tritium-labeled adenovirus type 12. Virology 34, 303—311 (1968).

Nilausen, K., and H. Green: Reversible arrest of growth in G1 of an established fibroblast line (3T3). Exp. Cell Res. 40, 166—168 (1965).

Nilsson, G., and L. Philipson: Cell growth inhibition of human cell lines by human tissue extracts. Exp. Cell Res. 51, 275—290 (1968).

Nilsson, K., J. Pontén, and L. Philipson: Development of immunocytes and immunoglobulin production in long-term cultures from normal and malignant human lymph nodes. Int. J. Cancer 3, 183—190 (1968).

Nordenskjöld, B. O. A., E. Klein, T. Tachibana, and E. M. Fenyö: Tissue culture assay for Moloney leukemia virus. J. nat. Cancer Inst. 44, 403—412 (1970).

NORRBY, K.: Population kinetics of normal, transforming and neoplastic cell lines. Acta path. microbiol. scand., Suppl. No. 214, **78,** 1—49 (1970).

NOVIKOFF, A. B., G. DE THÉ, D. BEARD, and J. W. BEARD: Electron microscopic study of the ATPase activity of the BAI strain A (myeloblastosis) avian tumor virus. J. Cell Biol. **15,** 451—462 (1962).

NOWELL, P.: Phytohemagglutinin: An initiator of mitosis in cultures of normal human leukocytes. Cancer Res. **20,** 462—466 (1960).

NOYES, W. F.: Development of Rous sarcoma virus antigens in cultured chick embryo cells. Virology **12,** 488—492 (1960).

NOYES, W. F.: Studies on the human wart virus. II. Changes in primary human cell cultures. Virology **25,** 358—363 (1965).

OBOSHI, S.: Continuous suspension culture of human neoplastic lymph nodes: Cell identification and detection of herpes-type EB virus. Gann, Monograph No. **7,** 191—199 (1969).

O'CONOR, G. T., and A. S. RABSON: Herpes-like particles in an American lymphoma: preliminary note. J. nat. Cancer Inst. **35,** 899—903 (1965).

O'CONOR, G. T., A. S. RABSON, I. K. BEREZESKY, and F. J. PAUL: Mixed infection with simian virus 40 and adenovirus 12. J. nat. Cancer Inst. **31,** 903—917 (1963).

O'CONNOR, T., and P. J. FISCHINGER: Physical properties of competent and defective states of a murine sarcoma (Moloney) virus. J. nat. Cancer Inst. **43,** 487—497 (1969).

O'CONNOR, T. E., and P. J. FISCHINGER: Titration patterns of a murine sarcoma-leukemia virus complex: Evidence for existence of competent sarcoma virions. Science **159,** 325—329 (1968).

O'CONNOR, T. E., F. J. RAUSCHER, and R. F. ZEIGEL: Density gradient centrifugation of a murine leukemia virus. Science **144,** 1144—1147 (1964).

ODA, K., and R. DULBECCO: Regulation of transcription of the SV 40 DNA in productively infected and in the transformed cells. Proc. nat. Acad. Sci. (Wash.) **60,** 525—532 (1968a).

ODA, K., and R. DULBECCO: Induction of cellular mRNA synthesis in BSC-1 cells infected by SV 40. Virology **35,** 439—444 (1968b).

ODAKA, T.: Inheritance of susceptibility to Friend mouse leukemia virus. V. Introduction of a gene responsible for susceptibility in the genetic complement of resistant mice. J. Virol. **3,** 543—548 (1969).

OGAWA, K. A. TSUTSUMI, K. IWATA, Y. FUJII, M. OHMORI, K. TAGUCHI, and Y. YABE: Histogenesis of malignant neoplasm induced by adenovirus type 12. Gann **57,** 43—52 (1966).

OKADA, Y.: The fusion of Ehrlich's tumor cells caused by HJV virus *in vitro*. Biken's J. **1,** 103—110 (1958).

OLD, L. J., and E. A. BOYSE: Immunology of experimental tumors. Ann. Rev. Med. **15,** 167—186 (1964).

OLD, L. J., E. A. BOYSE, and F. LILLY: Formation of cytotoxic antibody against leukemias induced by Friend virus. Cancer Res. **23,** 1063—1068 (1963).

OLD, L. J., E. A. BOYSE, H. F. OETTGEN, E. DE HARVEN, G. GEERING, B. WILLIAMSON, and P. CLIFFORD: Precipitating antibody in human serum to an antigen present in cultured Burkitt's lymphoma cells. Proc. nat. Acad. Sci. (Wash.) **56,** 1699—1704 (1966).

OPPENHEIMER, B. S., E. T. OPPENHEIMER, and A. P. STOUT: Sarcomas induced in rodents by implanting cellophane. Proc. Soc. exp. Biol. (N.Y.) **67,** 33—34 (1948).

OPPENHEIMER, B. S., E. T. OPPENHEIMER, and A. P. STOUT: Sarcomas induced in rodents by imbedding various plastic films. Proc. Soc. exp. Biol. (N.Y.) **79,** 366—369 (1952).

OROZLAN, S., L. W. JOHNS, JR., and M. A. RICH: Ultracentrifugation of a murine leukemia virus in polymer density gradients. Virology **26,** 638—645 (1965).

OROZLAN, S., and M. A. RICH: Human wart virus: *in vitro* cultivation. Science **146,** 531—533 (1964).

OSATO, T., E. A. MIRAND, and J. T. GRACE: Propagation and immunofluorescent investigations of Friend virus in tissue culture. Nature (Lond.) **201,** 52—54 (1964).

OXMAN, M. N.: Some behavioral studies of simian virus 40 (SV 40). Arch. ges. Virusforsch. **22,** 171—187 (1968).

OXMAN, M. N., S. BARON, P. H. BLACK, K. K. TAKEMOTO, K. HABEL, and W. P. ROWE: The effect of interferon on SV 40 T-antigen production in SV 40 transformed cells. Virology **32,** 122—127 (1967).

OXMAN, M. N., and P. H. BLACK: Inhibition of SV 40 T-antigen formation by interferon. Proc. nat. Acad. Sci. (Wash.) **55,** 1133—1140 (1966).

OZER, H. L., K. K. TAKEMOTO, R. L. KIRSCHSTEIN, and D. AXELROD: Immunochemical characterization of plaque mutants of simian virus 40. J. Virol. **3,** 17—24 (1969).

PARDEE, A. B.: Control of cell division: Models from microorganisms. Cancer Res. **28,** 1802—1809 (1968).

PARKER, R. C.: Cultivation of tumor cells *in vitro*. Canad. Cancer Conf. **1,** 42—54 (1955).

PARKER, R. C.: Alterations in clonal populations of monkey kidney cells. In: "Poliomyelitis: Papers and Discussions presented at the Fourth International Poliomyeliʼis Conference", Lippincott, Philadelphia, pp. 257—267 (1958).

PARKER, R. C.: Methods of Tissue Culture. Third edition. Hoeber-Harper Inc., New York, pp. 1—358 (1961).

PARKER, R. C., L. N. CASTOR, and E. A. McCULLOCH: Altered cell strains in continuous culture: A general survey. Spec. Publ. N.Y. Acad. Sci. **5,** 303—313 (1957).

PARKMAN, R., J. A. LEVY, and R. C. TING: Murine sarcoma virus: The question of defectiveness. Science **168,** 387—389 (1970).

PATULEIA, M. C., and C. FRIEND: Tissue culture studies on murine virus-induced leukemia cells: Isolation of single cells in agar-liquid medium. Cancer Res. **27,** 726—730 (1967).

PAULSON, D. F., A. S. RABSON, and E. E. FRALEY: Viral neoplastic transformation of hamster prostate tissue *in vitro*. Science **159,** 200—201 (1968).

PAYNE, F. E., J. J. SOLOMON, and H. G. PURCHASE: Immunofluorescent studies of group-specific antigen of the avian sarcoma-leukosis viruses. Proc. nat. Acad. Sci. (Wash.) **55,** 341—349 (1966).

PAYNE, L. N., and P. M. BIGGS: Differences between highly inbred lines of chicken in the response to Rous sarcoma virus of the chorioallantoic membrane and of embryonic cells in tissue culture. Virology **24,** 610—616 (1964).

PAYNE, L. N., and P. M. BIGGS: Genetic basis of cellular susceptibility to the Schmidt-Ruppin and Harris strains of Rous sarcoma virus. Virology **29,** 190—198 (1966).

PAYNE, L. N., and R. C. CHUBB: Studies on the nature and genetic control of an antigen in normal chick embryos which reacts in the COFAL test. J. gen. Virol. **3,** 379—391 (1968).

PAYNE, L. N., L. B. CRITTENDEN, and W. OKAZAKI: Influence of host genotype on responses to four strains of avian leukosis virus. J. nat. Cancer Inst. **40,** 907—916 (1968).

PEARSON, G., G. KLEIN, G. HENLE, W. HENLE, and P. CLIFFORD: Relation between Epstein-Barr viral and cell membrane immunofluorescence in Burkitt tumor cells. IV. Differentiation between antibodies responsible for membrane and viral immunofluorescence. J. exp. Med. **129,** 707—718 (1969).

PEREIRA, M. S., H. G. PEREIRA, and S. K. CLARKE: Human adenovirus type 31: a new serotype with oncogenic properties. Lancet **I,** 21—23 (1965).

PERK, K., and J. B. MOLONEY: Pathogenesis of a virus-induced rhabdomyosarcoma in mice. J. nat. Cancer Inst. **37,** 581—599 (1966).

PERK, K., J. B. MOLONEY, and E. G. JENKINS: Further studies on the relationship of a rhabdomyosarcoma virus to muscle tissue. Int. J. Cancer **2,** 43—51 (1967).

PÉTURSSON, G., D. ARMSTRONG, E. DE HARVEN, and J. FOGH: *In vitro* transformation of hamster embryonic kidney cultures exposed to human adenovirus 12. Cancer Res. **29,** 145—153 (1969).

PÉTURSSON, G., J. I. COUGHLIN, and C. MEYLAN: Long-term cultivation of diploid rat cells. Exp. Cell Res. **33**, 60—67 (1964).

PÉTURSSON, G., and R. WEIL: A study on the mechanism of polyoma-induced activation of the cellular DNA-synthesizing apparatus. Arch. ges. Virusforsch. **24**, 1—29 (1968).

PHILIPSON, L., and J. PONTÉN: Immunoglobulin synthesis in long-term cultures of lymph nodes from Hodgkin's disease. Life Sci. **6**, 2635—2641 (1967).

PIÑA, M.. and M. GREEN: Biochemical studies on adenovirus multiplication. IX. Chemical and base composition analysis of 28 human adenoviruses. Proc. nat. Acad. Sci. (Wash.) **54**, 547—551 (1965).

PIRAINO, F.: The mechanism of genetic resistance of chick embryo cells to infection by Rous sarcoma virus-Bryan strain (BS-RSV). Virology **32**, 700—707 (1967).

POGOSIANZ, H. E., E. T. BRUYAKO, and R. M. RADZIKHOVSKAYA: Susceptibility of the steppe-lemming to the oncogenic action of Rous sarcoma virus. Folia biol. (Praha) **12**, 141—144 (1966).

POLLACK, R. E., H. GREEN, and G. J. TODARO: Growth control in cultured cells: Selection of sublines with increased sensitivity to contact inhibition and decreased tumor-producing ability. Proc. nat. Acad. Sci. (Wash.) **60**, 126—133 (1968).

PONTÉN, J.: Homologous transfer of Rous sarcoma by cells. J. nat. Cancer Inst. **29**, 1147—1159 (1962a).

PONTÉN, J.: Transplantation of chicken tumor RPL12 in homologous hosts. J. nat. Cancer Inst. **29**, 1013—1021 (1962b).

PONTÉN, J.: Sex chromosomes as markers in transplanted chicken leukemic cells. Nature (Lond.) **194**, 97 (only) (1962c).

PONTÉN, J.: Chromosome analysis of three virus-associated chicken tumors: Rous sarcoma, erythroleukemia and RPL12 lymphoid tumor. J. nat. Cancer Inst. **30**, 897—921 (1963).

PONTÉN, J.: The *in vivo* growth mechanism of avian Rous sarcoma. Nat. Cancer Inst., Monogr. **17**, 131—145 (1964).

PONTÉN, J.: Spontaneous lymphoblastoid transformation of long-term cell cultures from human malignant lymphoma. Int. J. Cancer **2**, 311—325 (1967).

PONTÉN, J.: Lymphopoiesis in long-term cultures of human lymph nodes. Bibl. haemat. (Basel) **31**, 319—321 (1968).

PONTÉN, J.: The growth capacity of normal and Rous virus transformed chicken fibroblasts *in vitro*. Int. J. Cancer. **6**, 323—332 (1970).

PONTÉN, J., and B. R. BURMESTER: Transplantability of primary tumors of RPL12 virus-induced lymphoid leukosis. J. nat. Cancer Inst. **38**, 505—513 (1967).

PONTÉN, J., F. JENSEN, and H. KOPROWSKI: Morphological and virological investigation of human tissue cultures transformed with SV40. J. cell. comp. Physiol. **61**, 145—163 (1963).

PONTÉN, J., and F. LITHNER: Absence of specific chromosome alterations in bovine lung fibroblasts exposed to Rous sarcoma virus. Int. J. Cancer **1**, 589—598 (1966).

PONTÉN, J., and E. MACINTYRE: Long-term culture of normal and neoplastic human glia. Acta path. microbiol. scand. **74**, 465—486 (1968).

PONTÉN, J., and I. MACPHERSON: Interference with Rous sarcoma virus focus formation by a mycoplasma-like factor present in human cell cultures. Ann. Med. exp. Fenn. **44**, 260—264 (1966).

PONTÉN, J., and E. SAKSELA: Two established *in vitro* cell lines from human mesenchymal tumors. Int. J. Cancer **2**, 434—447 (1967).

PONTÉN, J., and B. THORELL: The histogenesis of virus-induced chicken leukemia. J. nat. Cancer Inst. **18**, 443—453 (1957).

PONTÉN, J., and B. WESTERMARK: The growth control of normal and neoplastic human glia cells *in vitro*. Abstract from the 10th Int. Cancer Congress, Houston, Texas, U.S.A., May 22—29 (1970).

PONTÉN, J., B. WESTERMARK, and R. HUGOSSON: Regulation of proliferation and movement of human glia-like cells in culture. Exp. Cell Res. **58**, 393—400 (1969).

Pope, J. H.: Establishment of cell lines from peripheral leukocytes in infectious mononucleosis. Nature (Lond.) **216**, 810—811 (1967).
Pope, J. H., B. G. Achong, and M. A. Epstein: Cultivation and fine structure of virus-
  bearing lymphoblasts from a second New Guinea Burkitt lymphoma: Establishment of sublines with unusual culture properties. Int. J. Cancer **3**, 171—182 (1968).
Pope, J. H., B. G. Achong, M. A. Epstein, and J. Biddulph: Burkitt lymphoma in
  New Guinea: Establishment of a line of lymphoblasts *in vitro* and description
  of their fine structure. J. nat. Cancer Inst. **39**, 933—945 (1967).
Pope, J. H., M. K. Horne, and W. Scott: Transformation of foetal human leukocytes
  in vitro by filtrates of a human leukaemic cell line containing herpes-like virus.
  Int. J. Cancer **3**, 847—866 (1968).
Pope, J. H., and W. P. Rowe: Detection of specific antigen in SV40 transformed cells
  by immunofluorescence. J. exp. Med. **120**, 121—128 (1964a).
Pope, J. H., and W. P. Rowe: Immunofluorescent studies of adenovirus 12 tumors
  and of cells transformed or infected by adenoviruses. J. exp. Med. **120**, 577—588
  (1964b).
Porwit-Bóbr, Z., W. Ptak, and A. Garlacz: Specific complement-fixing anti-
  bodies against polyoma tumour antigens and viral haemagglutination inhibiting
  antibodies in hamsters infected with polyoma virus. Acta virol. **11**, 357—362
  (1967).
Prage, L., U. Pettersson, and L. Philipson: Internal basic proteins in adenovirus.
  Virology **36**, 508—511 (1968).
Prince, A.: Quantitative studies on Rous sarcoma virus. II. Mechanism of resistance
  of chick embryos to chorioallantoic inoculation of Rous sarcoma virus. J. nat.
  Cancer Inst. **20**, 843—850 (1958).
Prince, A. M.: Quantitative studies on Rous sarcoma virus. V. An analysis of the
  mechanism of virulence of the Bryan "high titer" strain of RSV. Virology **11**,
  371—399 (1960a).
Prince, A. M.: Quantitative studies on Rous sarcoma virus. VI. Clonal analysis of
  in vitro infections. Virology **11**, 400—424 (1960b).
Prince, A. M.: Factors influencing the determination of cellular morphology in cells
  infected with Rous sarcoma virus. Virology **18**, 524—534 (1962).
Prince, A. M.: The biology of avian tumor viruses with especial emphasis on forms of
  virus: Cell integration and the neoplastic nature of tumors induced by these agents.
  Neoplasma **13**, 13—21 (1966).
Pruniéras, M., Y. Chardonnet et R. Sohier: Étude microcinématographique de
  l'effet cytopathogène du virus vacuolisant (SV40) sur cellules rénales de *cercopithecus aethiops*. Ann. Inst. Pasteur **106**, 1—17 (1964).
Puck, T. T.: *In vitro* studies on the radiation biology of mammalian cells. Progr.
  Biophys. **10**, 1—38 (1959).
Puck, T. T., S. J. Cieciura, and H. W. Fisher: Clonal growth *in vitro* of human
  cells with fibroblastic morphology. (Comparison of growth and genetic characteristics variety of human organs.) J. exp. Med. **106**, 145—158 (1957).
Pulvertaft, R. J. V.: Cytology of Burkitt's tumor (African lymphoma). Lancet **1**,
  238—240 (1964).
Pulvertaft, R. J. V., and J. G. Humble: Intracellular phase of existence of lymphocytes during remission of acute lymphatic leukaemia. Nature (Lond.) **194**, 194—
  195 (1962).
Purchase, H. G., and W. Okazaki: Morphology of foci produced by standard preparations of Rous sarcoma virus. J. nat. Cancer Inst. **32**, 579—589 (1964).

Rabin, E. R., M. Benyesh-Melnick, and J. P. Brunschwig: Studies on acute leukemia and infectious mononucleosis of childhood. II. Ultrastructure of cultured
  lymphoblastoid cells. Exp. molec. Path. **7**, 196—207 (1967).
Rabin, H.: Infection by Rous sarcoma virus of various chick embryo tissues cultured
  on reconstituted collagen membranes. Brit. J. exp. Path. **48**, 323—334 (1967).

RABINOWITZ, Z., and L. SACHS: Reversion of properties in cells transformed by polyoma virus. Nature (Lond.) **220**, 1203—1206 (1968).

RABINOWITZ, Z., and L. SACHS: The formation of variants with a reversion of properties of transformed cells. I. Variants from polyoma-transformed cells grown *in vivo*. Virology **38**, 336—342 (1969a).

RABINOWITZ, Z., and L. SACHS: The formation of variants with a reversion of properties of transformed cells. II. *In vitro* formation of variants from polyoma-transformed cells. Virology **38**, 343—346 (1969b).

RABOTTI, G. F., E. BUCCIARELLI, and A. J. DALTON: Presence of particles with the morphology of the avian leukosis virus complex in dog meningeal tumors induced by Rous sarcoma virus. Virology **29**, 684—686 (1966).

RABOTTI, G. F., and M. K. COOK: Specific cellular antigenic and morphologic transformation induced by Rous sarcoma ribonucleic acid. Nat. Cancer Inst., Monogr. **17**, 619—631 (1964).

RABOTTI, G. F., A. S. GROVE, R. L. SELLERS, and W. R. ANDERSON: Induction of multiple brain tumors (gliomata and leptomeningeal sarcomata) in dogs by Rous sarcoma virus. Nature (Lond.) **209**, 884—886 (1966).

RABOTTI, G. F., J. C. LANDON, T. W. PRY, L. BEADLE, J. DOLL, D. P. FABRIZIO, and A. J. DALTON: Tumors in Rhesus monkeys inoculated at birth with homologous cells converted *in vitro* by Rous sarcoma virus, Schmidt-Ruppin strain. J. nat. Cancer Inst. **38**, 821—837 (1967).

RABSON, A. S., and R. KIRSCHSTEIN: Induction of malignancy *in vitro* in newborn hamster kidney tissue infected with simian vacuolating virus (SV40). Proc. Soc. exp. Biol. (N.Y.) **111**, 323—328 (1962).

RABSON, A. S., R. KIRSCHSTEIN, and F. Y. LEGALLAIS: Autologous implantation of Rhesus monkey cells "transformed" *in vitro* by simian virus 40. J. nat. Cancer Inst. **35**, 981—991 (1965).

RABSON, A. S., R. L. KIRSCHSTEIN, and F. J. PAUL: Tumors produced by adenovirus 12 in mastomys and mice. J. nat. Cancer Inst. **32**, 77—87 (1964).

RABSON, A. S., and L. W. LAW: Studies of variation in oncogenicity of polyoma virus related to differences in cell culture media. J. nat. Cancer Inst. **30**, 367—374 (1963).

RABSON, A. S., R. A. MALMGREN, and R. L. KIRSCHSTEIN: Induction of neoplasia *in vitro* in hamster kidney tissue by adenovirus 7-SV40 "hybrid" strain (LLE46). Proc. Soc. exp. Biol. (N.Y.) **121**, 486—489 (1966).

RABSON, A. S., R. A. MALMGREN, G. T. O'CONOR, and R. KIRSCHSTEIN: Simian vacuolating virus (SV40) infection in cell cultures derived from adult human thyroid tissue. J. nat. Cancer Inst. **29**, 1123—1145 (1962).

RABSON, A. S., G. T. O'CONOR, S. BARON, J. J. WHANG, and F. Y. LEGALLAIS: Morphologic, cytogenetic and virologic studies *in vitro* of a malignant lymphoma from an African child. Int. J. Cancer **1**, 89—106 (1966).

RABSON, A. S., G. T. O'CONOR, I. K. BEREZESKY, and F. J. PAUL: Enhancement of adenovirus growth in African green monkey kidney cell cultures by SV40. Proc. Soc. exp. Biol. (N.Y.) **116**, 187—190 (1964).

RAFAJKO, R. R.: Routine establishment of serial lines of hamster embryo cells transformed by adenovirus type 12. J. nat. Cancer Inst. **38**, 581—591 (1967).

RAO, P. R., and J. W. BEARD: Lipide composition of avian BAI strain A (myeloblastosis) virus and virus-associated myeloblasts. Nat. Cancer Inst., Monogr. **17**, 673—675 (1964).

RAPP, F.: Complementation between defective oncogenic viruses. In: Malignant Transformation by Viruses (W. H. KIRSTEN, ed.) Spfinger-Verlag, New York, pp. 77—94 (1966a).

RAPP, F.: Complementation of deoxyribonucleic acid-containing viruses. In: Subviral Carcinogenesis (YOHEI ITO, ed.) "Nissha" printing Company, Mibu, Nakagyo-ku, Kyoto, pp. 311—325 (1966b).

RAPP, F., J. S. BUTEL, A. FELDMAN, T. KITAHARA, and J. L. MELNICK: Differential effects of inhibitors on the steps leading to the formation of SV40 tumor and virus antigen. J. exp. Med. 121, 935—944 (1965).

RAPP, F., J. S. BUTEL, and J. L. MELNICK: Virus-induced intranuclear antigen in cells transformed by SV40. Proc. Soc. exp. Biol. (N.Y.) 116, 1131—1135 (1964).

RAPP, F., J. S. BUTEL, and J. L. MELNICK: SV40-adenovirus "hybrid" populations: Transfer of SV40 determinants from one type of adenovirus to another. Proc. nat. Acad. Sci. (Wash.) 54, 717—724 (1965).

RAPP, F., J. S. BUTEL, S. S. TEVETHIA, and J. L. MELNICK: Comparison of ability of defective foreign genomes (para and mac) carried by human adenoviruses to induce SV40 transplantation immunity. J. Immunol. 99, 386—391 (1967).

RAPP, F., and M. JERKOFSKY: Replication of PARA (defective SV40)-adenoviruses in simian cells. J. gen. Virol. 1, 311—321 (1967).

RAPP, F., M. JERKOFSKY, and D. VANDERSLICE: Characterization of defectiveness of human adenoviruses in green monkey kidney cells. Proc. Soc. exp. Biol. 126, 782—786 (1967).

RAPP, F., K. S. KHERA, and J. S. MELNICK: Resistance of BHK21 hamster cells to SV40 papovavirus. Nature (Lond.) 201, 1349—1350 (1964).

RAPP, F., and J. L. MELNICK: Papovavirus SV40, adenovirus and their hybrids: Transformation, complementation and transcapsidation. Progr. med. Virol. 8, 349—399 (1966).

RAPP, F., J. L. MELNICK, and T. KITAHARA: Tumor and virus antigens of simian virus 40: Differential inhibition of synthesis by cytosine arabinoside. Science 147, 625—627 (1965).

RAPP, F., J. L. MELNICK, J. S. BUTEL, and T. KITAHARA: The incorporation of SV40 genetic material into adenovirus 7 as measured by intranuclear synthesis of SV40 tumor antigen. Proc. nat. Acad. Sci. (Wash.) 52, 1348—1352 (1964).

RAPP, F., S. S. TEVETHIA, and J. L. MELNICK: Papopavirus SV40 transplantation immunity conferred by an adenovirus-SV40 hybrid. J. nat. Cancer Inst. 36, 703—708 (1966).

RAPP, F. L., L. A. FELDMAN, and M. MANDEL: Synthesis of virus deoxyribonucleic acid during abortive infection of simian cells by human adenoviruses. J. Bact. 92, 931—936 (1966).

RAUSCHER, F.: A virus-induced disease of mice characterized by erythrocytopoiesis and lymphoid leukemia. J. nat. Cancer Inst. 29, 515—543 (1962).

RAUSCHER, F. J., J. A. REYNIERS, and M. R. SACKSTEDER: Response or lack of response of apparently leukosis-free Japanese quail to avian tumor viruses. Nat. Cancer Inst., Monogr. 17, 211—229 (1964).

REICH, E., R. M. FRANKLIN, A. J. SHATKIN, and E. L. TATUM: The effect of actinomycin D on cellular nucleic acid synthesis and virus production. Science 134, 556—557 (1961).

REICH, P. R., S. G. BAUM, J. A. ROSE, W. P. ROWE, and S. M. WEISSMAN: Nucleic acid homology studies of adenovirus type 7-SV40 interaction. Proc. nat. Acad. Sci. (Wash.) 55, 336—341 (1966).

REIN, A., and H. RUBIN: Effects of local cell concentrations upon the growth of chick embryo cells in tissue culture. Exp. Cell Res. 49, 666—678 (1968).

REYNIERS, J. A., and M. R. SACKSTEDER: Raising Japanese quail under germfree and conventional conditions and their use in cancer research. J. nat. Cancer Inst. 24, 1405—1421 (1960).

RHIM, J. S., R. J. HUEBNER, and R. C. TING: Transformation of hamster embryo cells in vitro by Rauscher leukemia virus. J. nat. Cancer Inst. 42, 1053—1060 (1969).

RHODE, S. L., and K. A. O. ELLEM: Control of nucleic acid synthesis in human diploid cells undergoing contact inhibition. Exp. Cell Res. 53, 184—204 (1968).

RICKARD, C. G., J. E. POST, F. NORONHA, and L. M. BARR: A transmissible virus-induced lymphocytic leukemia of the cat. J. nat. Cancer Inst. 42, 987—1014 (1969).

RIGGS, J. L., and E. H. LENNETTE: Simian virus 40: Isolation of two plaque types. Science 147, 408—409 (1965).

RIGGS, J. L., and E. H. LENNETTE: *In vitro* transformation of newborn-hamster kidney cells by simian adenoviruses. Proc. Soc. exp. Biol. (N.Y.) 126, 802—806 (1967).

RIGLER, R., and D. KILLANDER: Activation of deoxyribonucleoprotein in human leukocytes stimulated by phytohemagglutinin. II. Structural changes of deoxyribonucleoprotein and synthesis of RNA. Exp. Cell Res. 54, 171—180 (1969).

ROBBINS, F. C., J. F. ENDERS, and T. H. WELLER: Cytopathogenic effect of poliomyelitis viruses *in vitro* on human embryonic tissues. Proc. Soc. exp. Biol. (N.Y.) 75, 370—374 (1950).

ROBINSON, H.: Isolation of non-infectious particles containing Rous sarcoma virus RNA from the medium of Rous sarcoma virus-transformed non-producer cells. Proc. nat. Acad. Sci. (Wash.) 57, 1655—1662 (1967).

ROBINSON, W. S., and M. A. BALUDA: The nucleic acid from avian myeloblastosis virus compared with the RNA from the Bryan strain of Rous sarcoma virus. Proc. nat. Acad. Sci. (Wash.) 54, 1686—1692 (1965).

ROBINSON, W. S., and P. H. DUESBERG: Tumor virus RNA. In: Subviral Carcinogenesis (YOHEI ITO, ed.) 1st Int. Symp. Tumor Viruses, Nagoya, "Nisshia" Printing Comp., Mibu, Nakagyo-ku, Kyoto, pp. 3—17 (1966).

ROBINSON, W. S., and P. H. DUESBERG: The chemistry of the RNA tumor viruses. In: "Molecular Basis of Virology" (FRAENKEL-CONRAT, ed.) Reinhold, New York, pp. 306—331 (1968).

ROBINSON, W. S., A. PITKÄNEN, and H. RUBIN: The nucleic acid of the Bryan-strain of Rous sarcoma virus: Purification of the virus and isolation of the nucleic acid. Proc. nat. Acad. Sci. (Wash.) 54, 137—144 (1965).

ROBINSON, W. S., H. L. ROBINSON, and P. H. DUESBERG: Tumor virus RNA's. Proc. nat. Acad. Sci. (Wash.) 58, 825—834 (1967).

ROBL, M. G., and C. OLSON: Oncogenic action of bovine papilloma virus in hamsters. Cancer Res. 28, 1596—1604 (1968).

ROIZMAN, B., and P. R. ROANE, JR.: Studies of polyoma virus. I. Hemagglutination as a measure of virus mass and antibody to the virus. J. Immunol. 85, 418—428 (1960).

ROSKIN, G.: Versuche mit heteroplastischer Überpflanzung der bösartigen Geschwülste. Z. Krebsforsch. 24, 122—125 (1927).

ROSSI, G. B., and C. FRIEND: Erythrocytic maturation of (Friend) virus-induced leukemic cells in spleen clones. Proc. nat. Acad. Sci. (Wash.) 58, 1373—1380 (1967).

ROTH, K. F., and R. M. DOUGHERTY: Multiple antigenic components of the group-specific antigen of the avian leukosis-sarcoma viruses. Virology 38, 278—284 (1969).

ROTHFELS, K. H., A. A. AXELRAD, L. SIMINOVITCH, E. A. McCULLOCH, and R. C. PARKER: The origin of altered cell lines from mouse, monkey and man as indicated by chromosome and transplantation studies. Proc. Canad. Cancer Conf. 3, 189—214 (1959).

ROTHFELS, K. H., E. B. KUPELWIESER, and R. C. PARKER: Effects of X-irradiated feeder layers on mitotic activity and development of aneuploidy in mouse embryo cells *in vitro*. Proc. Canad. Cancer Conf. 5, 191—223 (1963).

ROTHFELS, K. H., and R. C. PARKER: The karyotypes of cell lines recently established from normal mouse tissues. J. exp. Zool. 142, 507—520 (1959).

ROUS, P.: A transmissible avian neoplasm (sarcoma of the common fowl). J. exp. Med. 12, 696—705 (1910).

ROUS, P.: A sarcoma of the fowl transmissible by an agent separable from the tumor cells. J. exp. Med. 13, 397—411 (1911).

ROUS, P.: Comment. Proc. nat. Acad. Sci. (Wash.) 58, 843—845 (1967).

Rous, P., O. H. Robertson, and J. Oliver: Experiments on the production of specific antisera for infections of unknown cause. II. The production of a serum effective against the agent causing a chicken sarcoma. J. exp. Med. **29**, 305—320 (1919).

Rowe, W. P.: Studies of adenovirus-SV40 hybrid viruses. III. Transfer of SV40 gene between adenovirus types. Proc. nat. Acad. Sci. (Wash.) **54**, 711—717 (1965).

Rowe, W. P.: Some interactions of defective animal viruses. Perspect. Virol. **5**, 123—132 (1967).

Rowe, W. P., and S. G. Baum: Evidence for a possible genetic hybrid between adenovirus type 7 and SV40 viruses. Proc. nat. Acad. Sci. (Wash.) **52**, 1340—1347 (1964).

Rowe, W. P., and S. G. Baum: Studies of adenovirus-SV40 hybrid viruses. II. Defectiveness of the hybrid particles. J. exp. Med. **122**, 955—966 (1965).

Rowe, W. P., S. G. Baum, W. E. Pugh, and D. Hoggan: Studies of adenovirus-SV40 hybrid viruses. I. Assay system and further evidence for hybridization. J. exp. Med. **122**, 943—954 (1965).

Rowe, W. P., and J. W. Hartley: A general review of the adenoviruses. Ann. N.Y. Acad. Sci. **101**, 466—474 (1962).

Rowe, W. P., J. W. Hartley, J. D. Estes, and R. J. Huebner: Studies of mouse polyoma virus infection. I. Procedures for quantitation and detection of virus. J. exp. Med. **109**, 379—391 (1959).

Rowe, W., and W. E. Pugh: Studies of adenovirus-SV40 hybrid viruses. V. Evidence for linkage between adenovirus and SV40 genetic materials. Proc. nat. Acad. Sci. (Wash.) **55**, 1126—1132 (1966).

Rowson, K. E. K., and B. W. J. Maky: Human papova (wart) virus. Bact. Rev. **31**, 110—131 (1967).

Rubin, H.: Quantitative relations between causative virus and cell in the Rous No. 1 chicken sarcoma. Virology **1**, 445—473 (1955).

Rubin, H.: An analysis of the apparent neutralization of Rous sarcoma virus with antiserum to normal chick tissues. Virology **2**, 545—558 (1956).

Rubin, H.: A virus in chick embryos which induced resistance *in vitro* to infection with Rous sarcoma virus. Proc. nat. Acad. Sci. (Wash.) **46**, 1105—1119 (1960a).

Rubin, H.: An analysis of the assay of Rous sarcoma cells *in vitro* by the infective center technique. Virology **10**, 29—49 (1960b).

Rubin, H.: The suppression of morphological alterations in cells infected with Rous sarcoma virus. Virology **12**, 14—31 (1960c).

Rubin, H.: The nature of a virus-induced cellular resistance to Rous sarcoma virus. Virology **13**, 200—206 (1961).

Rubin, H.: Conditions for establishing immunological tolerance to a tumor virus. Nature (Lond.) **195**, 342—345 (1962).

Rubin, H.: Virus defectiveness and cell transformation in the Rous sarcoma. J. cell. comp. Physiol. **64**, 173—180 (1964).

Rubin, H.: Genetic control of cellular susceptibility to pseudotypes of Rous sarcoma virus. Virology **26**, 270—276 (1965).

Rubin, H.: A substance in conditioned medium which enhances the growth of small numbers of chick embryo cells. Exp. Cell Res. **41**, 138—148 (1966a).

Rubin, H.: The inhibition of chicken embryo cell growth by medium obtained from cultures of Rous sarcoma cells. Exp. Cell Res. **41**, 149—161 (1966b).

Rubin, H.: Fact and theory about the cell surface in carcinogenesis. In: Major Problems in Developmental Biology (Michael Locke, ed.), Academic Press, N.Y., London, pp. 315—337 (1966c).

Rubin, H., M. Baluda, and J. E. Hotchin: The maturation of Western equine encephalomyelitis virus and its release from chick embryo cells in suspension. J. exp. Med. **101**, 205—212 (1955).

Rubin, H., and C. Colby: Early release of growth inhibition in cells infected with Rous sarcoma virus. Proc. nat. Acad. Sci. (Wash.) **60**, 482—488 (1968).

RUBIN, H., A. CORNELIUS, and L. FANSHIER: The pattern of congenital transmission of an avian leukosis virus. Proc. nat. Acad. Sci. (Wash.) **47,** 1058—1069 (1961).

RUBIN, H., L. FANSHIER, A. CORNELIUS, and W. F. HUGHES: Tolerance and immunity in chickens after congenital and contact infection with an avian leukosis virus. Virology **17,** 143—156 (1962).

RUBIN, H., and A. REIN: Proximity effects in the growth of animal cells. Growth regulating substances for animal cells in culture. Wistar Inst. Symp. Monogr. No. 7 (V. DEFENDI and M. STOKER, eds.), pp. 51—66 (1967).

RUBIN, H., and P. K. VOGT: An avian leukosis virus associated with stocks of Rous sarcoma virus. Virology **17,** 184—194 (1962).

RUDDLE, F. H.: Chromosome variation in cell populations derived from pig kidney. Cancer Res. **21,** 885—894 (1961).

RUFFILLI, D.: Azione del plasma di pollo eritroleucemico su tessuti coltivati *in vitro.* Boll. Lega ital. Tumori **12,** 1—6 (1938).

RUSSEL, W. C., J. S. F. NIVEN, and L. D. BERMAN: Studies on the biology of the mycoplasma-induced "stimulation" of BHK-C13 cells. Int. J. Cancer **3,** 191—202 (1968).

RYCHLIKOVA, M., and J. SVOBODA: Effects of anti-Rous sarcoma virus sera on a hemorrhagic disease of rats. Virology **10,** 545—547 (1960).

SABIN, A. B.: Manifestations of latent viral genetic material in experimentally produced cancers and search for such manifestations in human cancers. Israel J. med. Sci. **1,** 93—103 (1965).

SABIN, A. B., and M. A. KOCH: Evidence of continuous transmission of non-infectious SV40 viral genome in most or all SV40 hamster tumor cells. Proc. nat. Acad. Sci. (Wash.) **49,** 304—311 (1963a).

SABIN, A. B., and M. A. KOCH: Behavior of non-infectious SV40 viral genome in hamster cells: Induction of synthesis of infectious virus. Proc. nat. Acad. Sci. (Wash.) **50,** 407—417 (1963b).

SABIN, A. B., H. M. SHEIN, M. A. KOCH, and J. F. ENDERS: Specific complement-fixing tumor antigens in human cells morphologically transformed by SV40 virus. Proc. nat. Acad. Sci. (Wash.) **52,** 1316—1318 (1964).

SACHS, L., M. FOGEL, and E. WINOCOUR: *In vitro* analysis of a mammalian tumour virus. Nature (Lond.) **183,** 663—664 (1959).

SACHS, L., M. FOGEL, E. WINOCOUR, E. HELLER, D. MEDINA, and M. KRIM: The *in vitro* and *in vivo* analysis of mammalian tumour viruses. Brit. J. Cancer **13,** 251—265 (1959).

SACHS, L., and D. MEDINA: *In vitro* transformation of normal cells by polyoma virus. Nature (Lond.) **189,** 457—458 (1961).

SACHS, L., D. MEDINA, and Y. BERWALD: Cell transformation by polyoma virus in clones of hamster and mouse cells. Virology **17,** 491—493 (1962).

SAKSELA, E., and P. MOORHEAD: Aneuploidy in the degenerative phase of serial cultivation of human cell strains. Proc. nat. Acad. Sci. (Wash.) **50,** 390—395 (1963).

SAKSELA, E., and J. PONTÉN: Chromosomal changes of immunoglobulin producing cell lines from human lymph nodes with and without lymphoma. J. nat. Cancer Inst. **41,** 359—372 (1968).

SALAMAN, M. H., K. E. K. ROWSON, and J. J. HARVEY: The possible role of a subcellular leukaemogenic agent in homologous transplantation of mouse leukaemic tissue and the attempted passage of such an agent through tissue cultures. In: Ciba Found. Symp. Tumor Viruses of Murine Origin, pp. 214—232, Churchill Ltd., London, 1962.

SALK, J. E., and E. N. WARD: Some characteristics of a continuously propagating cell derived from monkey heart tissue. Science **126,** 1338—1339 (1957).

SAMBROOK, J., H. WESTPHAL, P. R. SRINIVASAN, and R. DULBECCO: The integrated state of viral DNA in SV40-transformed cells. Proc. nat. Acad. Sci. (Wash.) **60,** 1288—1295 (1968).

SANFORD, K. K.: Clonal studies on normal cells and on their neoplastic transformation *in vitro*. Cancer Res. **18,** 747—752 (1958).

SANFORD, K. K.: Malignant transformation of cells *in vitro*. Int. Rev. Cytol. **18,** 249—311 (1965).

SANFORD, K. K., B. E. BARKER, M. W. WOODS, P. PARSHAD, and L. W. LAW: Search for "indicators" of neoplastic conversion *in vitro*. J. nat. Cancer Inst. **39,** 705—733 (1967).

SANFORD, K. K., T. DUNN, L. COVALESKY, T. DUPREE, and W. EARLE: Polyoma virus and production of malignancy *in vitro*. J. nat. Cancer Inst. **26,** 331—357 (1961).

SANFORD, K. K., W. R. EARLE, E. SHELTON, E. L. SCHILLING, E. M. DUCHESNE, G. D. LIKELY, and M. M. BECKER: Production of malignancy *in vitro*. XII. Further transformations of mouse fibroblasts to sarcomatous cells. J. nat. Cancer Inst. **11,** 351—375 (1950).

SANFORD, K. K., and R. E. HOEMANN: Neoplastic transformation of mouse and hamster cells *in vitro* with and without polyoma virus. J. nat. Cancer Inst. **39,** 691—703 (1967).

SANFORD, K. K., G. D. LIKELY, and W. R. EARLE: The development of variations in transplantability and morphology within a clone of mouse fibroblasts transformed to sarcoma-producing cells *in vitro*. J. nat. Cancer Inst. **15,** 215—237 (1954).

SANFORD, K. K., R. M. MERWIN, G. L. HOBBS, and W. R. EARLE: Influence of animal passage on a line of tissue culture cells. J. nat. Cancer Inst. **23,** 1061—1077 (1959).

SANFORD, K. K., R. M. MERVIN, G. L. HOBBS, M. C. FIORAMONTI, and W. R. EARLE: Studies on the difference in sarcoma-producing capacity of two lines of mouse cells derived *in vitro* from one cell. J. nat. Cancer Inst. **20,** 121—145 (1958).

SANFORD, K. K., R. M. MERWIN, G. L. HOBBS, J. M. YOUNG, and W. R. EARLE: Clonal analysis of variant cell lines transformed to malignant cells in tissue culture. J. nat. Cancer Inst. **23,** 1035—1059 (1959).

SANFORD, K. K., and R. PARSHAD: Oxygen supply and neoplastic conversion of mouse embryo cells *in vitro*. J. nat. Cancer Inst. **41,** 1389—1394 (1968).

SANFORD, K. K., B. B. WESTFALL, M. C. FIORAMONTI, W. T. MCQUILIKEN, J. C. BRYANT, E. V. PEPPERS, V. J. EVANS, and W. E. EARLE: The effect of serum fractions on the proliferation of strain L mouse cells *in vitro*. J. nat. Cancer Inst. **16,** 789—802 (1956).

SARMA, P.: Discussion page 318. Nat. Cancer Inst. Monogr. No. 17 (1964).

SARMA, P. S., M. P. CHEONG, J. W. HARTLEY, and R. J. HUEBNER: A viral interference test for mouse leukemia viruses. Virology **33,** 180—184 (1967).

SARMA, P. S., R. J. HUEBNER, and W. T. LANE: Induction of tumors in hamsters with an avian adenovirus (CELO). Science **149,** 1108 (only) (1965).

SARMA, P. S., T. S. LOG, R. J. HUEBNER, and H. C. TURNER: Studies of avian leukosis group-specific complement-fixing serum antibodies in pigeons. Virology **37,** 480—483 (1969).

SARMA, P. S., H. C. TURNER, and R. J. HUEBNER: An avian leukosis group-specific complement fixation reaction. Application for the detection and assay of non-cytopathogenic leukosis viruses. Virology **23,** 313—321 (1964a).

SARMA, P. S., H. C. TURNER, and R. J. HUEBNER: Complement fixation test for the detection and assay of avian leukosis viruses. Nat. Cancer Inst. Monogr. No. 17, 481—493 (1964b).

SARMA, P. S., W. VASS, and R. J. HUEBNER: Evidence for the *in vitro* transfer of defective Rous sarcoma virus genome from hamster tumor cells to chick cells. Proc. nat. Acad. Sci. (Wash.) **55,** 1435—1442 (1966).

SARMA, P. S., W. VASS, R. J. HUEBNER, H. IGEL, W. T. LANE, and H. C. TURNER: Induction of tumours in hamsters with infectious canine hepatitis virus. Nature (Lond.) **215,** 293—294 (1967).

SATO, J., M. NAMBA, K. USUI, and D. NAGANO: Carcinogenesis in tissue culture. VIII. Spontaneous malignant transformation of rat liver cells in long-term culture. Jap. J. exp. Med. **38,** 105—118 (1968).

SAUER, G., and V. DEFENDI: Stimulation of DNA synthesis and complement-fixing antigen production by SV40 in human diploid cell cultures: Evidence for "abortive" infection. Proc. nat. Acad. Sci. (Wash.) 56, 452—457 (1966).

SAUER, G., H. FISCHER, and K. MUNK: The effect of SV40 infection on DNA synthesis in cercopithecus kidney cells. Virology 28, 765—767 (1966).

SAUER, G., and J. R. KIDWAI: The transcription of the SV40 genome in productively infected and transformed cells. Proc. nat. Acad. Sci. (Wash.) 61, 1256—1263 (1968).

SAUER, G., H. KOPROWSKI, and V. DEFENDI: The genetic heterogeneity of simian virus 40. Proc. nat. Acad. Sci. (Wash.) 58, 599—606 (1967).

SCHELL, K.: Persistence of SV40 virus in transformed rabbit kidney cells. Proc. Soc. exp. Biol. (N.Y.) 128, 1145—1148 (1968).

SCHELL, K., W. T. LANE, M. J. CASEY, and R. J. HUEBNER: Potentiation of oncogenecity of adenovirus type 12 grown in African green monkey kidney cell cultures preinfected with SV40 virus. Persistence of both T antigens in the tumors and evidence for possible hybridization. Proc. nat. Acad. Sci. (Wash.) 55, 81—88 (1966).

SCHELL, K., and J. MARYAK: Susceptibility of rabbit kidney cells of various passage levels to infection with SV40 virus. Arch. ges. Virusforsch. 19, 403—414 (1966).

SCHELL, K., and J. MARYAK: SV40 virus growth and cytopathogenicity in a serial rabbit kidney cell line. Proc. Soc. exp. Biol. (N.Y.) 124, 1099—1102 (1967).

SCHELL, K., J. MARYAK, and M. SCHMIDT: Adenovirus transformation of hamster embryo cells. III. Maintenance conditions. Arch. ges. Virusforsch. 24, 352—360 (1968).

SCHELL, K., J. MARYAK, J. YOUNG, and M. SCHMIDT: Adenovirus transformation of hamster embryo cells. II. Inoculation conditions. Arch. ges. Virusforsch. 24, 342—351 (1968).

SCHELL, K., and M. SCHMIDT: Adenovirus transformation of hamster embryo cells. I. Assay conditions. Arch. ges. Virusforsch. 24, 332—341 (1968).

SCHENDLER, S., and R. J. C. HARRIS: The effect of Rous sarcoma virus (Schmidt-Ruppin strain) on the chromosomes of human leukocytes in vitro. Int. J. Cancer 2, 109—115 (1967).

SCHLESINGER, R. W.: Adenoviruses: The nature of the virion and of controlling factors in productive or abortive infection and tumorigenesis. Advanc. Virus Res. 14, 1—61 (1969).

SCHLUMBERGER, H. D., F. A. ANDERER, and M. A. KOCH: Structure of simian virus 40. IV. The polypeptide chains of the virus particle. Virology 36, 42—47 (1968).

SCHMIDT-RUPPIN, K. N.: Versuche zur heterologen Transplantation mit frischem und gefriergetrocknetem Material des Rous sarcoma. Diskussion zum Vortrag Oberling. Krebsforschung und Krebsbekämpfung 3, 26—27 (1959).

SCHUTZ, L., and P. MORA: The need for direct cell contact in "contact" inhibition of cell division in culture. J. cell. comp. Physiol. 71, 1—6 (1968a).

SCHUTZ, L., and P. MORA: Growth of mouse fibroblasts in the presence of irradiated and lyophilized monolayers of the same type of cells. J. cell. comp. Physiol. 71, 7—16 (1968b).

SCHÄFER, W., F. A. ANDERER, H. BAUER, and L. PISTER: Studies on mouse leukemia viruses. I. Isolation and characterization of a group-specific antigen. Virology 38, 387—394 (1969).

SCHÄFER, W., FRANK H. and PISTER L.: Vergleich des latend in L-Zellen enthaltenen „Virus" mit dem Rauscher-Virus der Mäuse-Leukämie. Z. Naturforsch., 23 b, 1275-1276 (1968).

SCHÄFER, W., and E. SEIFERT: Production of a potent complement-fixing murine leukemia virus-antiserum from the rabbit and its reactions with various types of tissue culture cells. Virology 35, 323—328 (1968).

SEEMAYER, N.: Untersuchungen über die onkogene Transformation von Hamsternierenzellen in vitro durch das Simian Virus SV40. Z. Krebsforsch. 71, 186—197 (1968).

SHAH, K. V., and D. M. HESS: Presence of antibodies to simian virus 40 (SV 40) T antigen in Rhesus monkeys infected experimentally or naturally with SV 40. Proc. Soc. exp. Biol. (N.Y.) 128, 480—485 (1968).

SHAH, K. V., S. WILLARD, R. E. MYERS, D. M. HESS, and R. DiGIACOMO: Experimental infection of Rhesus with simian virus 40. Proc. Soc. exp. Biol. (N.Y.) 130, 196—203 (1969).

SHARON, N., and M. POLLARD: Spontaneous neoplastic transformation of germ-free rat embryo cell culture. Cancer Res. 29, 1523—1526 (1969).

SHARP, D. G., and J. W. BEARD: Counts of virus particles by sedimentation on agar and electron micrography. Proc. Soc. exp. Biol. (N.Y.) 81, 75—77 (1952).

SHARP, D. G., D. BEARD, and J. W. BEARD: Morphology of characteristic particles associated with avian erythroblastosis. Proc. Soc. exp. Biol. (N.Y.) 90, 168—173 (1955).

SHARP, D. G., E. A. ECKERT, D. BEARD, and J. W. BEARD: Morphology of the virus of avian erythromyeloblastic leukosis and a comparison with the agent of Newcastle disease. J. Bact. 63, 151—161 (1952).

SHARP, J. A., and R. G. BURWELL: Interaction (peripolesis) of macrophages and lymphocytes after skin homografting or challenge with soluble antigens. Nature (Lond.) 188, 474—475 (1960).

SHEIN, H. M.: Propagation of human fetal spongioblasts and astrocytes in dispersed cell cultures. Exp. Cell Res. 40, 554—569 (1965).

SHEIN, H. M.: Neoplastic transformation induced by simian virus 40 in Syrian hamster neuroglial and meningeal cell cultures. Arch. ges. Virusforsch. 22, 122—142 (1967a).

SHEIN, H. M.: Transformation of astrocytes and destruction of spongioblasts induced by a simian tumor virus (SV 40) in cultures of human fetal neuroglia. J. Neuropath. exp. Neurol. 26, 60—75 (1967b).

SHEIN, H. M.: Neoplastic transformation of hamster astrocytes in vitro by simian virus 40 and polyoma virus. Science 159, 1476—1477 (1968).

SHEIN, H. M., and J. F. ENDERS: Multiplication and cytopathogenicity of simian vacuolating virus 40 in cultures of human tissues. Proc. Soc. exp. Biol. (N.Y.) 109, 495—500 (1962a).

SHEIN, H. M., and J. F. ENDERS: Transformation induced by simian virus 40 in human renal cell cultures. I. Morphology and growth characteristics. Proc. nat. Acad. Sci. (Wash.) 48, 1164—1172 (1962b).

SHEIN, H. M., J. F. ENDERS, and J. D. LEVINTHAL: Transformation induced by simian virus 40 in human renal cell cultures. II. Cell-virus relationships. Proc. nat. Acad. Sci. (Wash.) 48, 1350—1357 (1962).

SHEIN, H. M., J. F. ENDERS, J. D. LEVINTHAL, and A. E. BURKET: Transformation induced by simian virus 40 in newborn syrian hamster renal cell cultures. Proc. nat. Acad. Sci. (Wash.) 49, 28—34 (1963).

SHEIN, H. M., J. F. ENDERS, L. PALMER, and E. GROGAN: Further studies on SV 40-induced transformation in human renal cell cultures. I. Eventual failure of sub-cultivation despite a continuing high rate of cell division. Proc. Soc. exp. Biol. (N.Y.) 115, 618—621 (1964).

SHEININ, R.: A rapid plaque assay for polyoma virus. Virology 15, 85—87 (1961).

SHEININ, R.: Procedures for the purification of polyoma T virus. Virology 17, 426—440 (1962).

SHEININ, R.: Studies on the thymidine kinase activity of mouse embryo cells infected with polyoma virus. Virology 28, 47—55 (1966a).

SHEININ, R.: Deoxyribonucleic acid synthesis in cells replicating polyoma virus. Virology 28, 621—632 (1966b).

SHEININ, R.: DNA synthesis in rat embryo cells infected with polyoma virus. Virology 29, 167—170 (1966c).

SHEININ, R.: Deoxyribonucleic acid synthesis in cells infected with polyoma virus. In: The Molecular Biology of Viruses. (COLTER and PARANCHYCH, eds.), Academic Press, New York, London, pp. 627—643 (1967).

SHEININ, R., and P. A. QUINN: Effect of polyoma virus on the replicative mechanism of mouse embryo cells. Virology **26**, 73—84 (1965).

SHELTON, E., V. J. EVANS, and G. A. PARKER: Malignant transformation of mouse connective tissue grown in diffusion chambers. J. nat. Cancer Inst. **30**, 377—391 (1963).

SHIMOJO, R., and T. YAMASHITA: Induction of DNA synthesis by adenoviruses in contact-inhibited hamster cells. Virology **36**, 422—433 (1968).

SHIMONO, H., and A. S. KAPLAN: Correlation between the synthesis of DNA and histones in polyoma virus-infected mouse embryo cells. Virology **37**, 690—694 (1969).

SHRIGLEY, E. W., H. S. N. GREENE, and F. DURAN-REYNALS: Studies on the variation of the Rous sarcoma virus following growth of the tumour in the anterior chamber of the guinea-pig eye. Cancer Res. **5**, 356—364 (1945).

SIEGEL, B. V., W. J. WEAVER, and R. D. KOLER: Mouse erythroleukaemia of viral aetiology. Nature (Lond.) **201**, 1042—1043 (1964).

SIEGLER, R., and M. A. RICH: Pathogenesis of murine leukemia. Nat. Cancer Inst. Monogr. No. **22**, 525—547 (1966).

SIGEL, M. M., TH. M. SCOTTI, V. B. SCHULZ, and TH. BURNSTEIN: Heterologous transplantation for demonstration of the oncogenic effect of Rous sarcoma virus. Cancer Res. **20**, 1338—1340 (1960).

SIMINOVITCH, L., and A. A. AXELRAD: Cell-cell interactions *in vitro*: their relation to differentiation and carcinogenesis. Fifth Canad. Cancer Conf., Honey Harbor Academic Press, pp. 149—165 (1963).

ŠIMKOVIČ, D., M. POPOVIČ, J. SVEČ, M. GRÓFOVÁ, and N. VALENTOVÁ: Continuous production of avian sarcoma virus B 77 by rat tumour cells in tissue culture. Int. J. Cancer **4**, 80—85 (1969).

ŠIMKOVIČ, D., J. SVOBODA, and N. VALENTOVÁ: Clonal analysis of line XC-te rat tumour cells (derived from tumour XC) grown *in vitro*. Folia biol. (Praha) **9**, 82—91 (1963).

ŠIMKOVIČ, D., N. VALENTOVÁ, and V. THURZO: An *in vitro* system for the detection of the Rous sarcoma virus in the cells of the rat tumour XC. Neoplasma **9**, 104—106 (1962).

SIMONS, P. J., R. H. BASSIN, and J. J. HARVEY: Transformation of hamster embryo cells *in vitro* by murine sarcoma virus (Harvey). Proc. Soc. exp. Biol. (N.Y.) **125**, 1242—1246 (1967).

SIMONS, P. J., R. R. DOURMASHKIN, A. TURANO, D. E. H. PHILLIPS, and F. C. CHESTERMAN: Morphological transformation of mouse embryo cells *in vitro* by murine sarcoma virus (Harvey). Nature (Lond.) **214**, 897—898 (1967).

SIMONS, P. J., S. S. PEPPER, and R. S. U. BAKER: Different cell culture characteristics of two strains of murine sarcoma virus. Proc. Soc. exp. Biol. (N.Y.) **131**, 454—456 (1969).

SIMONS, P. J., and M. G. R. ROSS: The isolation of herpes virus from Burkitt tumours. Europ. J. Cancer **1**, 135—136 (1965).

SINKOVICS, J. G.: Viral leukemia in mice. Ann. Rev. Microbiol. **16**, 75—100 (1962).

SINKOVICS, J. G.: Lymphoid cells in long-term cultures. Med. Rec. Ann. **61**, 50—56 (1968).

SINKOVICS, J. G., B. A. BERTIN, and C. D. HOWE: Occurrence of low leukemogenic but immunizing mouse leukemia virus in tissue culture. Nat. Cancer Inst. Monogr. No. **22**, 349—367 (1966).

SINKOVICS, J. G., C. C. SHULLENBERGER, and C. D. HOWE: Lymphoblastic reconversion in tissue cultures from human leukemia. Clin. Res. **15**, 46 (only) (1967).

SINKOVICS, J. G., J. A. SYKES, C. C. SHULLENBERGER, and C. D. HOWE: Patterns of growth in cultures deriving from human leukemic sources. Tex. Rep. Biol. Med. **25**, 446—467 (1967).

SJÖGREN, H. O.: Studies on specific transplantation resistance to polyoma virus-induced tumors. I. Transplantation resistance induced by polyoma virus infection. J. nat. Cancer Inst. **32**, 361—374 (1964a).

SJÖGREN, H. O.: II. Mechanism of resistance induced by polyoma virus infection. J. nat. Cancer Inst. 32, 375—394 (1964b).

SJÖGREN, H. O.: III. Transplantation resistance to genetically compatible polyoma tumors induced by polyoma tumor homografts. J. nat. Cancer Inst. 32, 645—649 (1964c).

SJÖGREN, H. O.: IV. Stability of the polyoma cell antigen. J. nat. Cancer Inst. 32, 661—666 (1964d).

SJÖGREN, H. O.: Studies on immunologically determined host resistance against virus-induced neoplasms. Tryckeri Balder AB, Stockholm, pp. 1—27 (1964e).

SJÖGREN, H. O.: Transplantation methods as a tool for detection of tumor-specific antigens. Progr. exp. Tumor Res. 6, 289—322 (1965).

SJÖGREN, H. O.: Specific transplantation antigens common for viral neoplasms of different animal species. Advance in Transplantation (DAUSET, HAMBURGER and MATHÉ, eds.), Munksgaard, pp. 481—482 (1968).

SJÖGREN, H. O., and I. HELLSTRÖM: Induction of the polyoma specific transplantation antigen in Moloney leukemia cells. Exp. Cell Res. 40, 208—211 (1965).

SJÖGREN, H. O., and I. HELLSTRÖM: In vivo and in vitro demonstration of the polyoma specific transplantation antigen induced in polyoma infected Moloney lymphoma cells. In: Specific Tumour Antigens (HARRIS ed.), Munksgaard, Copenhagen, pp. 162—170 (1967).

SJÖGREN, H. O., I. HELLSTRÖM, and G. KLEIN: Resistance of polyoma virus immunized mice against transplantation of established polyoma tumors. Exp. Cell Res. 23, 204—208 (1961a).

SJÖGREN, H. O., I. HELLSTRÖM, and G. KLEIN: Transplantation of polyoma virus-induced tumors in mice. Cancer Res. 21, 329—337 (1961b).

SJÖGREN, H. O., J. MINOWADA, and J. ANKERST: Specific transplantation antigens of mouse sarcomas induced by adenovirus type 12. J. exp. Med. 125, 689—701 (1967).

SLETTENMARK, B., and E. KLEIN: Cytotoxic and neutralization tests with serum and lymph node cells of isologous mice with induced resistance against Gross lymphomas. Cancer Res. 22, 947—954 (1962).

SMIDA, J., and V. SMIDOVÁ: Propagation of B77 virus and three strains of Rous sarcoma virus in Japanese quail. Neoplasma 15, 439—441 (1968).

SMIDOVÁ, V., N. VALENTOVÁ, J. SMIDA, and J. MEDZIHRADSKÝ: Neoplastic transformation of rat embryo fibroblasts by fowl sarcoma virus B77. Neoplasma 12, 453—458 (1965).

SMITH, J. D., G. FREEMAN, M. VOGT, and R. DULBECCO: The nucleic acid of polyoma virus. Virology 12, 185—186 (1960).

SOLOMON, J. J., R. L. WITTER, K. NAZERIAN, and B. R. BURMESTER: Studies on the etiology of Marek's disease. I. Propagation of the agent in cell culture. Proc. Soc. exp. Biol. (N.Y.) 127, 173—177 (1968).

SPJUT, H. J., G. L. VAN HOOSIER, and J. J. TRENTIN: Neoplasms in hamsters induced by adenovirus type 12. Arch. Path. 83, 199—203 (1967).

STANNERS, C. P., J. E. TILL, and L. SIMINOVITCH: Studies on the transformation of hamster embryo cells in culture by polyoma virus. I. Properties of transformed and normal cells. Virology 21, 448—463 (1963).

STANTON, M. F., L. W. LAW, and R. C. TING: Some biologic, immunogenic and morphologic effects in mice after infection with a murine sarcoma virus. II. Morphologic studies. J. nat. Cancer Inst. 40, 1113—1129 (1968).

STECK, F. T., and H. RUBIN: The mechanism of interference between an avian leukosis virus and Rous sarcoma virus. I. Establishment of interference. Virology 29, 628—641 (1966a).

STECK, F. T., and H. RUBIN: The mechanism of interference between an avian leukosis and Rous sarcoma virus. II. Early steps of infection by RSV of cells under conditions of interference. Virology 29, 642—653 (1966b).

STECK, T. L., S. KAUFMAN, and J. P. BADER: Glycolysis in chick embryo cell cultures transformed by Rous sarcoma virus. Cancer Res. 28, 1611—1619 (1968).

STENKVIST, B.: Long-term cultivation of human and bovine fibroblastic cells morphologically transformed *in vitro* by Rous sarcoma virus. Acta path. microbiol.
scand. **67**, 67—82 (1966a).

STENKVIST, B.: Growth rate and terminal cell density of bovine fibroblastic cells
morphologically transformed *in vitro* by Rous sarcoma virus. Acta path. microbiol.
scand. **67**, 180—190 (1966b).

STENKVIST, B.: Clonal analysis of bovine cells morphologically transformed *in vitro*
by Rous sarcoma virus. Acta path. microbiol. scand. **67**, 201—219 (1966c).

STENKVIST, B.: Attempts to demonstrate infectious virus in human and bovine
fibroblastic cells morphologically transformed *in vitro* by Rous sarcoma virus. Acta
Soc. Med. upsalien. **71**, 130—140 (1966d).

STENKVIST, B., and J. PONTÉN: Morphological changes in bovine and human fibroblasts exposed to two strains of Rous sarcoma virus *in vitro*. Acta path. microbiol.
scand. **62**, 315—330 (1964).

STEPLEWSKI, Z., B. KNOWLES, and H. KOPROWSKI: The mechanism of internuclear
transmission of SV40 induced complement fixation antigen in heterokaryocytes.
Proc. nat. Acad. Sci. (Wash.) **59**, 769—776 (1968).

STEWART, S. E., and B. E. EDDY: Properties of a tumor-inducing virus recovered
from mouse neoplasms. In: Perspect. Virol., Wiley & Sons, Inc. N.Y., Chapman &
Hall, Ltd., London, pp. 245—255 (1959).

STEWART, S. E., B. E. EDDY, and N. BORGESE: Neoplasms in mice inoculated with
a tumor agent carried in tissue culture. J. nat. Cancer Inst. **20**, 1223—1243 (1958).

STEWART, S. E., B. E. EDDY, A. M. GOCHENOUR, N. G. BORGESE, and G. E. GRUBBS:
The induction of neoplasms with a substance released from mouse tumors by
tissue culture. Virology **3**, 380—400 (1957).

STEWART, S. E., B. E. EDDY, V. M. HAAS, and N. G. BORGESE: Lymphocytic chorio-
meningitis virus as related to chemotherapy studies and to tumor induction in
mice. Ann. N.Y. Acad. Sci. **68**, 419—429 (1957).

STEWART, S. E., D. GLAZER, TH. BEN, and B. J. LLOYD, JR.: Studies on the hamster-
brain passage virus recovered from human lymphoma cultures. J. nat. Cancer
Inst. **40**, 423—428 (1968).

STEWART, S. E., J. LANDON, E. LOVELACE, and G. PARKER: Burkitt tumor: Brain
lesions in hamsters induced with an extract from the SL-1 cell line. Methodological
approaches to the study of leukemias. (V. Defendi, ed.), Wistar Inst. Symp.
Monogr. No. 4, 93—101 (1965).

STEWART, S. E., E. LOVELACE, J. J. WHANG, and A. NGU: Burkitt tumor: Tissue
culture, cytogenetic and virus studies. J. nat. Cancer Inst. **34**, 319—327 (1965).

STICH, H. F., G. L. VAN HOOSIER, and S. S. TRENTIN: Viruses and mammalian chromosomes. Chromosome aberrations by human adenovirus type 12. Exp. Cell Res. **34**,
400—403 (1964).

STOKER, M.: Characteristics of normal and transformed clones arising from BHK21
cells exposed to polyoma virus. Virology **18**, 649—651 (1962).

STOKER, M.: Delayed transformation by polyoma virus. Virology **20**, 366—371 (1963a).

STOKER, M.: Regulation of growth and orientation in hamster cells transformed by
polyoma virus. Virology **24**, 164—174 (1964).

STOKER, M.: Transfer of growth inhibition between normal and virus-transformed
cells: Autoradiographic studies using marked cells. J. Cell Sci. **2**, 293—304 (1967).

STOKER, M.: Abortive transformation by polyoma virus. Nature (Lond.) **218**, 234—
238 (1968).

STOKER, M., and P. ABEL: Conditions affecting transformation by polyoma virus.
Cold Spr. Harb. Symp. quant., Biol. **27**, 375—386 (1962).

STOKER, M., and I. MACPHERSON: Studies on transformation of hamster cells by
polyoma virus *in vitro*. Virology **14**, 359—370 (1961).

STOKER, M., and I. MACPHERSON: The Syrian hamster fibroblast cell line BHK 21
and its derivates. Nature (Lond.) **203**, 1355—1357 (1964).

STOKER, M., and H. RUBIN: Density dependent inhibition of cell growth in culture.
Nature (Lond.) **215**, 171—172 (1967).

STOKER, M., M. SHEARER, and C. O'NEILL: Growth inhibition of polyoma-transformed cells by contact with static normal fibroblasts. J. Cell Sci. 1, 297—310 (1966).

STOKER, M., and A. SMITH: Characteristics of normal cells in mixed clones arising after delayed transformation by polyoma virus. Virology 24, 175—178 (1964).

STOKER, M., and M. SUSSMAN: Studies on the action of feeder layers in cell culture. Exp. Cell Res. 38, 645—653 (1965).

STROHL, W. A.: The response of BHK 21 cells to infection with type 12 adenovirus. I. Cell killing and T antigen synthesis as correlated viral genome functions. Virology 39, 642—652 (1969a).

STROHL, W. A.: The response of BHK 21 cells to infection with type 12 adenovirus. II. Relationship of virus-stimulated DNA synthesis to other viral functions. Virology 39, 653—665 (1969b).

STROHL, W. A., A. S. RABSON, and H. ROUSE: Adenovirus tumorigenesis: Role of the viral genome in determining tumor morphology. Science 156, 1631—1633 (1967).

STROHL, W. A., H. C. ROUSE, and W. R. SCHLESINGER: Properties of cells derived from adenovirus-induced hamster tumors by long-term in vitro cultivation. II. Nature of the restricted response to type 2 adenovirus. Virology 28, 645—658 (1966).

STROHL, W. A., and R. W. SCHLESINGER: Quantitative studies of natural and experimental adenovirus infections of human cells. I. Characteristics of viral multiplication in fibroblasts derived by long-term culture from tonsils. Virology 26, 199—207 (1965a).

STROHL, W. A., and W. R. SCHLESINGER: Quantitative studies of natural and experimental adenovirus infections of human cells. II. Primary cultures and the possible role of asynchronous viral multiplication in the maintenance of infection. Virology 26, 208—220 (1965b).

SUBAK-SHARPE, H., R. R. BÜRK, and J. D. PITTS: Metabolic co-operation by cell to cell transfer between genetically different mammalian cells in tissue culture. Heredity 21, 342—343 (1966).

SULTANIAN, I. V., and G. FREEMAN: Enhanced growth of human embryonic cells infected with adenovirus 12. Science 154, 665—667 (1966).

SUNDELIN, P.: Microphotometric determination of DNA and RNA in single chick embryo fibroblasts during morphological transformation induced by Rous sarcoma virus. Exp. Cell Res. 46, 581—592 (1967).

SUNDELIN, P.: DNA and RNA content at successive postmitotic intervals in chick embryo fibroblasts infected with Rous sarcoma virus. Exp. Cell Res. 50, 233—238 (1968).

SUNDELIN, P., and R. ADAMS: DNA, RNA and hemoglobin cytophotometry of maturing erythroid cells from liver nodules in Rauscher virus infected mice. Int. J. Cancer 2, 544—550 (1967).

SUSKIND, R. G., T. W. PRY, and G. F. RABOTTI: Effect of Rous sarcoma virus infection on recovery of nucleolar RNA, ribonucleoprotein, and DNA synthesis from inhibition by actinomycin D. Cancer Res. 29, 1598—1605 (1969).

SVET-MOLDAVSKY, G. J.: Development of multiple cysts and of haemorrhagic affection of internal organs in albino rats treated during the embryonic or newborn period with Rous sarcoma virus. Nature (Lond.) 180, 1299—1300 (1957).

SVET-MOLDAVSKY, G. J.: Sarcoma in albino rats treated during the embryonic stage with Rous virus. Nature (Lond.) 182, 1452—1453 (1958).

SVET-MOLDAVSKY, G. J., and A. SKORIKOVA: The development of multiple cysts in rats after inoculating them with Rous sarcoma virus. Vop. Oncol. 6, 673—677 (1957).

SVET-MOLDAVSKY, G. J., L. TRUBCHENINOVA, and L. I. RAVKINA: Pathogenicity of the chicken sarcoma virus (Schmidt-Ruppin) for amphibians and reptiles. Nature (Lond.) 214, 300—302 (1967).

SVOBODA, J.: Presence of chicken tumour virus in the sarcoma of the adult rat inoculated after birth with Rous sarcoma virus. Nature (Lond.) 186, 980—981 (1960).

SVOBODA, J.: The tumorigenic action of Rous sarcoma in rats and the permanent production of Rous virus by the induced rat sarcoma XC. Folia biol. (Praha) 7, 46—60 (1961).

SVOBODA, J.: Further findings on the induction of tumours by Rous sarcoma in rats and on the Rous-virus producing capacity of one of the induced tumours (XC) in chicks. Folia biol. (Praha) 8, 215—219 (1962).

SVOBODA, J.: Malignant interaction of Rous virus with mammalian cells *in vivo* and *in vitro*. Nat. Cancer Inst. Monogr. No. 17, 277—292 (1964).

SVOBODA, J., and P. CHÝLE: Malignization of rat embryonic cells by Rous sarcoma virus *in vitro*. Folia biol. (Praha) 9, 329—342 (1963).

SVOBODA, J., P. CHÝLE, D. SIMKOVIC, and I. HILGERT: Demonstration of the absence of infectious Rous virus in rat tumour XC, whose structurally intact cells produce Rous sarcoma when transferred to chicks. Folia biol. (Praha) 9, 77—81 (1963).

SVOBODA, J., T. HLOZÁNEK, and O. MACHALA: Rescue of Rous sarcoma virus in mixed cultures of virogenic mammalian and chicken cells, treated and untreated with Sendai virus and detected by focus assay. J. gen. Virol. 2, 461—464 (1968).

SVOBODA, J., and V. KLEMENT: Formation of delayed tumours in hamsters inoculated with Rous virus after birth and finding of infectious Rous virus in induced tumour P₁. Folia biol. (Praha) 9, 403—411 (1963).

SVOBODA, J., O. MACHALA, and I. HLOZÁNEK: Influence of Sendai virus on RSV formation in mixed culture of virogenic mammalian cells and chicken fibroblasts. Folia biol. (Praha) 13, 155—157 (1967).

SWEET, B. H., and M. R. HILLEMAN: Detection of a "non-detectable" simian virus (vacuolating agent) present in rhesus and cynomologous monkey-kidney cell culture material. Proc. Soc. exp. Biol. (N.Y.) 105, 420—427 (1960).

SWIFT, M. R., and G. J. TODARO: Membrane potentials of human fibroblast strains in culture. J. cell. comp. Physiol. 71, 61—63 (1968).

SWIM, H. E., and R. F. PARKER: Culture characteristics of human fibroblasts propagated serially. Amer. J. Hyg. 66, 235—243 (1957).

TAI, H. T., and R. L. O'BRIEN: Multiplicity of viral genomes in an SV40 transformed hamster cell line. Virology 38, 698—701 (1969).

TAKAHASHI, M., T. OGINO, K. BABA, and M. ONAKA: Synthesis of deoxyribonucleic acid in human and hamster kidney cells infected with human adenovirus types 5 and 12. Virology 37, 513—520 (1969).

TAKEMOTO, K. K., and K. HABEL: Hamster tumor cells doubly transformed by SV40 and polyoma viruses. Virology 30, 20—28 (1966).

TAKEMOTO, K. K., R. L. KIRSCHSTEIN, and K. HABEL: Mutants of simian virus 40 differing in plaque size, oncogenicity and heat sensitivity. J. Bact. 92, 990—994 (1966).

TAKEMOTO, K. K., R. A. MALMGREN, and K. HABEL: Immunofluorescent demonstration of polyoma tumor antigen in lytic infection of mouse embryo cells. Virology 28, 485—488 (1966a).

TAKEMOTO, K. K., R. A. MALMGREN, and K. HABEL: Heat-labile serum factor required for immunofluorescence at polyoma tumor antigens. Science 153, 1122—1123 (1966b).

TAKEMOTO, K. K., R. C. Y. TING, H. L. OZER, and P. FABISCH: Establishment of a cell line from an inbred mouse strain for viral transformation studies: Simian virus 40 transformation and tumor production. J. nat. Cancer Inst. 41, 1401—1409 (1968).

TAKEMOTO, K. K., G. J. TODARO, and K. HABEL: Recovery of SV40 virus with genetic markers of original inducing virus from SV40-transformed mouse cells. Virology 35, 1—8 (1968).

TANIGAKI, N., Y. YAGI, G. E. MOORE, and D. PRESSMAN: Immunoglobulin production in human leukemia cell lines. J. Immunol. 97, 634—646 (1966).

TEMIN, H. M.: The control of cellular morphology in embryonic cells infected with Rous sarcoma virus *in vitro*. Virology 10, 182—197 (1960).

TEMIN, H. M.: Mixed infection with two types of Rous sarcoma virus. Virology 13, 159—163 (1961).

TEMIN, H. M.: The effects of actinomycin D on growth of Rous sarcoma virus in vitro. Virology 20, 577—582 (1963).

TEMIN, H. M.: The participation of DNA in Rous sarcoma virus production. Virology 23, 486—494 (1964a).

TEMIN, H. M.: Nature of the provirus of Rous sarcoma. Nat. Cancer Inst. Monogr. No. 17, 557—570 (1964b).

TEMIN, H. M.: Homology between RNA from Rous sarcoma virus and DNA from Rous sarcoma virus-infected cells. Proc. nat. Acad. Sci. (Wash.) 52, 323—329 (1964c).

TEMIN, H. M.: On the mechanism of carcinogenesis by avian sarcoma viruses. I. On cell multiplication and differentiation. J. nat. Cancer Inst. 35, 679—693 (1965).

TEMIN, H. M.: Studies on carcinogenesis by avian sarcoma viruses. III. The differential effect of serum and polyanions on multiplication of uninfected and converted cells. J. nat. Cancer Inst. 37, 167—175 (1966).

TEMIN, H. M.: Studies on carcinogenesis by avian sarcoma viruses. IV. In: Molecular Biology of Viruses (J. COLTER and W. PARANCHYCH, ed.), pp. 709—715, Academic Press, New York (1967a).

TEMIN, H. M.: Studies on carcinogenesis by avian sarcoma viruses. V. Requirement for new DNA synthesis and for cell division. J. cell. comp. Physiol. 69, 53—64 (1967b).

TEMIN, H. M.: Control by factors in serum of multiplication of uninfected cells and cells infected and converted by avian sarcoma viruses. Growth Reg. Sub. Anim. Cells Cult., The Wistar Symp. Monogr. No. 7, 103—116 (1967c).

TEMIN, H. M.: Studies on carcinogenesis by avian sarcoma viruses: VIII. Glycolysis and cell multiplication. Int. J. Cancer 3, 273—282 (1968a).

TEMIN, H. M.: Altered properties of Fujinami virus-infected duck cells and murine sarcoma virus-infected rat cells. Int. J. Cancer 3, 491—503 (1968b).

TEMIN, H. M.: Carcinogenesis by avian sarcoma viruses. X. The decreased requirement for insulin-replaceable activity in serum for cell multiplication. Int. J. Cancer 3, 771—787 (1968c).

TEMIN, H. M.: Control of cell multiplication in uninfected chicken cells and chicken cells converted by avian sarcoma viruses. J. cell. comp. Physiol. 74, 9—16 (1969).

TEMIN, H. M., and S. MIZUTANI: RNA-dependent DNA polymerase in virions of Rous sarcoma virus. Nature (Lond.) 226, 1211—1213 (1970).

TEMIN, H. M., and H. RUBIN: Characteristics of an assay of Rous sarcoma virus and Rous sarcoma cells in tissue culture. Virology 6, 669—688 (1958).

TEMIN, H. M., and H. RUBIN: A kinetic study of infection of chick embryo cells in vitro by Rous sarcoma virus. Virology 8, 209—222 (1959).

TENENBAUM, E., and L. DOLJANSKI: Studies on Rous sarcoma cells cultivated in vitro. Morphological properties of Rous sarcoma cells. Cancer Res. 3, 585—603 (1943).

TEUTSCHLAENDER, O.: Über die Biologie eines übertragbaren Hühnersarkoms. Z. Krebsforsch. 20, 79—110 (1923).

TEVETHIA, S., L. A. COUVILLION, and F. RAPP: Development in hamsters of antibodies against surface antigens present in cells transformed by papovavirus SV 40. J. Immunol. 100, 358—362 (1968).

TEVETHIA, S., G. TH. DIAMANDOPOULOS, F. RAPP, and J. F. ENDERS: Lack of relationship between virus-specific surface and transplantation antigens in hamster cells transformed by simian papovavirus SV 40. J. Immunol. 101, 1192—1198 (1968).

TEVETHIA, S. S., M. KATZ, and F. RAPP: New surface antigen in cells transformed by simian papovavirus SV 40. Proc. Soc. exp. Biol. (N.Y.) 119, 896—901 (1965).

TEVETHIA, S. S., and F. RAPP: Demonstration of new surface antigens in cells transformed by papovavirus SV 40 by cytotoxic tests. Proc. Soc. exp. Biol. (N.Y.) 120, 455—458 (1965).

THOMAS, M., M. BOIRON, Y. STOYTCHKOW, and J. LASNERET: *In vitro* replication of mouse sarcoma virus (Moloney strain) in bovine embryo skin cells. Virology 36, 514—518 (1968).

THOMAS, M., M. BOIRON, J. TANZER, J. P. LÉVY, and J. BERNARD: *In vitro* transformation of mice cells by bovine papilloma virus. Nature (Lond.) 202, 709—710 (1964).

THOMAS, M., and G. LE BOUVIER: Transformation of bovine cell cultures by preparations of polyoma virus. J. gen. Virol. 1, 125—130 (1967).

THOMAS, M., J.-P. LÉVY, J. TANZER, M. BOIRON et J. BERNARD: Transformation *in vitro* de cellules de peau de veau embryonnaire sous l'action d'extraits acellulaires de papillomes bovins. C. R. Acad. Sci. (Paris) 257, 2155—2158 (1963).

THORELL, B.: Induktion von Nierentumoren durch Leukämievirus. Zbl. allg. Path. path. Anat. 98, 314—315 (1958).

THORELL, B.: Changes in the cytotropism of tumor virus. Acta path. microbiol. scand., Suppl. 144, 51, 215—217 (1961).

THORELL, B.: "Tumorvirology", Yearbook Swedish Cancer Society, 8, 258—262 (1962).

THORNE, H. V., W. HOUSE, and A. L. KISCH: Electrophoretic properties and purification of large and small plaque-forming strains of polyoma virus. Virology 27, 37—43 (1965).

TING, R. C.: *In vitro* transformation of rat embryo cells by a murine sarcoma virus. Virology 28, 783—785 (1966).

TING, R. C.: Tumor induction of thymectomized rats by murine sarcoma virus (Moloney) and properties of the induced virus-free tumor cells. Proc. Soc. exp. Biol. (N.Y.) 126, 778—781 (1967).

TJIO, J. H., and T. T. PUCK: Genetics of somatic mammalian cells. II. Delineation of chromosomal constitution in tissue culture. J. exp. Med. 108, 259—268 (1958).

TOCKSTEIN, G., H. POLASA, M. PIÑA, and M. GREEN: A simple purification procedure for adenovirus type 12 T and tumor antigens and some of their properties. Virology 36, 377—386 (1968).

TODARO, G., and R. J. HUEBNER: Oncogenes of RNA tumor viruses as determinants of cancer. Proc. nat. Acad. Sci. (Wash.) 64, 1087—1094 (1969).

TODARO, G., Y. MATSUYA, S. BLOOM, A. ROBBINS, and H. GREEN: Stimulation of RNA synthesis and cell division in resting cells by a factor present in serum. Growth Regulating Substances for Animal Cells in Culture, Wistar Inst. Symp. Monogr. No. 7 (DEFENDI and STOKER, eds.), pp. 87—101 (1967).

TODARO, G. J., and S. A. AARONSON: Properties of clonal lines of murine sarcoma virus transformed Balb/3T3 cells. Virology 38, 174—179 (1969).

TODARO, G. J., and S. BARON: The role of interferon in the inhibition of SV 40 transformation cell line 3T3. Proc. nat. Acad. Sci. (Wash.) 54, 752—756 (1965).

TODARO, G. J., and H. GREEN: Quantitative studies of the growth of mouse embryo cells in culture and their development into established lines. J. Cell Biol. 17, 299—313 (1963).

TODARO, G. J., and H. GREEN: An assay for cellular transformation by SV 40. Virology 23, 117—119 (1964a).

TODARO, G. J., and H. GREEN: Enhancement by thymidine analogs of susceptibility of cells to transformation by SV 40. Virology 24, 393—400 (1964b).

TODARO, G. J., and H. GREEN: Serum albumin supplemented medium for long-term cultivation of mammalian fibroblast strains. Proc. Soc. exp. Biol. (N.Y.) 116, 688—692 (1964c).

TODARO, G. J., and H. GREEN: Successive transformations of an established cell line by polyoma virus and SV 40. Science 147, 513—514 (1965).

TODARO, G. J., and H. GREEN: Cell growth and the initiation of transformation by SV 40. Proc. nat. Acad. Sci. (Wash.) 55, 302—308 (1966a).

Todaro, G. J., and H. Green: High frequency of SV 40 transformation of mouse cell line 3T3. Virology 28, 756—759 (1966b).

Todaro, G. J., H. Green, and B. D. Goldberg: Transformation of properties of an established cell line by SV 40 and polyoma virus. Proc. nat. Acad. Sci. (Wash.) 51, 66—73 (1964).

Todaro, G. J., H. Green, and M. R. Swift: Susceptibility of human diploid fibroblast strains to transformation by SV 40 virus. Science 153, 1252—1254 (1966).

Todaro, G. J., G. K. Lazar, and H. Green: The initiation of cell division in a contact inhibited mammalian cell line. J. cell. comp. Physiol. 66, 325—333 (1965).

Todaro, G. J., and G. M. Martin: Increased susceptibility of Down's syndrome fibroblasts to transformation by SV 40. Proc. Soc. exp. Biol. (N.Y.) 124, 1232—1236 (1967).

Todaro, G. J., K. Nilausen, and H. Green: Growth properties of polyoma virus-induced hamster tumor cells. Cancer Res. 23, 825—832 (1963).

Todaro, G. J., and K. K. Takemoto: "Rescued" SV40: Increased transforming efficiency in mouse and human cells. Proc. nat. Acad. Sci. (Wash.) 62, 1031—1037 (1969).

Todaro, G. J., S. R. Wolman, and H. Green: Rapid transformation of human fibroblasts with low growth potential into established cell lines by SV 40. J. cell. comp. Physiol. 62, 257—265 (1963).

Todorov, T. G., and M. Yakimov: Changes in the cultures of chicken bone marrow cells infected with the virus of chicken myelocytomatosis (Strains Mc-29 and Mc-31). Neoplasma 14, 613—617 (1967).

Toplin, I., D. Riccardo, and E. M. Jensen: Large-scale production of Moloney murine leukemia virus in tissue culture. Cancer 18, 1377—1383 (1965).

Toplin, I., and G. Schidlovsky: Partial purification and electron microscopy of virus in the EB-3 cell line derived from a Burkitt lymphoma. Science 152, 1084—1085 (1966).

Toshima, S., N. Takagi, J. Minowada, G. E. Moore, and A. A. Sandberg: Electron microscopic and cytogenetic studies of cells derived from Burkitt's lymphoma. Cancer Res. 27, 753—771 (1967).

Tournier, P., R. Cassigena, R. Wicker, J. Coppey et H. Suarez: Étude du méchanisme de l'induction chez des cellules de hamster syrien transformées par le virus SV 40. Propriétés d'une lignée cellulaire clonale. Int. J. Cancer 2, 117—132 (1967).

Trager, G. W., and H. Rubin: Quantitative studies on cell transformation following infection with Rous sarcoma virus. Nat. Cancer Inst. Monogr. No. 17, 575—585 (1964).

Trager, G. W., and H. Rubin: Rous sarcoma virus production from clones of non-transformed chick embryo fibroblasts. Virology 30, 266—274 (1966a).

Trager, G. W., and H. Rubin: Mixed clones produced following infection of chick embryo cell cultures with Rous sarcoma virus. Virology 30, 275—285 (1966b).

Trávníček, M., L. Bukič, J. Říman, and F. Šorm: The nucleotide composition of the RNA of the avian myeloblastosis virus (BAI strain A) and of the nucleic acids of leukaemic myeloblasts. Neoplasma 11, 571—592 (1964).

Trentin, J. J., and E. Bryan: Virus-induced transplantation immunity to human adenovirus type 12 tumors of the hamster and mouse. Proc. Soc. exp. Biol. (N.Y.) 121, 1216—1219 (1966).

Trentin, J. J., G. L. Van Hoosier, and L. Samper: The oncogenicity of human adenoviruses in hamsters. Proc. Soc. exp. Biol. (N.Y.) 127, 683—689 (1968).

Trentin, J. J., Y. Yabe, and G. Taylor: The quest for human cancer viruses. Science 137, 835—841 (1962).

Trowell, O. A.: A modified technique for organ culture in vitro. Exp. Cell Res. 9, 246—248 (1954).

Trowell, O. A.: The culture of mature organs in a synthetic medium. Exp. Cell Res. 16, 118—147 (1959).

Trowell, O. A.: Radiosensitivity of the cortical and medullary lymphocytes in the thymus. Int. J. Radiat. Biol. 4, 163—173 (1961).

TROWELL, O. A.: Lymphocytes. In: Cells and Tissues in Culture. Vol. 2, 96—172 (1965) Academic Press, London.

TRUJILLO, J. M., J. J. BUTLER, M. J. AHEARN, C. C. SHULLENBERGER, B. LIST-YOUNG, C. GOTT, H. ANSTALL, and J. A. SHIVELY: Long-term culture of lymph node tissue from a patient with lymphocytic lymphoma. Cancer 20, 215—224 (1967).

TRUJILLO, J. M., B. LIST-YOUNG, J. J. BUTLER, C. C. SHULLENBERGER, and C. GOTT: Long-term culture of lymph node tissue from a patient with lymphocytic lymphoma. Nature (Lond.) 209, 310—311 (1966).

TSUDA, T.: Establishment of two cell lines from Syrian hamster embryonic tissue in vitro. Tohoku J. exp. Med. 86, 380—393 (1965).

TYNDALL, R. L., E. TEETER, J. A. OTTEN, N. D. BOWLES, J. G. VIDRINE, A. C. UPTON, and H. E. WALBURG, JR.: Further observations on in vitro cytopathic effects associated with murine leukemia virus infection. Int. J. Cancer 1, 565—572 (1966).

TYNDALL, R. L., J. G. VIDRINE, E. TEETER, A. C. UPTON, W. W. HARRIS, and M. A. FINK: Cytopathogenic effects in a cell culture infected with a murine leukemia virus. Proc. Soc. exp. Biol. (N.Y.) 119, 186—189 (1965).

UCHIDA, S., and S. WATANABE: Tumorigenicity of the antigen-forming defective virions of simian virus 40. Virology 35, 166—169 (1968).

UCHIDA, S., K. YOSHIIKE, S. WATANABE, and A. FURUNO: Antigen-forming defective viruses of simian virus 40. Virology 34, 1—8 (1968).

UCHIDA, S., S. WATANABE, and M. KATO: Incomplete growth of simian virus 40 in African green monkey kidney culture induced by serial undiluted passages. Virology 28, 135—141 (1966).

USHIJIMA, R. N., C. E. GARDNER, and E. CATE: Transformation of renal cells from a prosimian by simian virus 40 (SV 40). Proc. Soc. exp. Biol. (N.Y.) 122, 676—679 (1966).

VAINIO, T., L. SAXÉN, and S. TOIVONEN: Viral susceptibility and embryonic differentiation. III. Correlation between an inductive tissue interaction and the onset of viral resistance. J. nat. Cancer Inst. 31, 1533—1547 (1963).

VALENTI, C., and E. A. FRIEDMAN: Long-term cultivation of diploid rabbit skin cells. Tex. Rep. Biol. Med. 26, 363—380 (1968).

VALENTINE, A. F., and J. P. BADER: Production of virus by mammalian cells transformed by Rous sarcoma and murine sarcoma viruses. J. Virol. 2, 224—237 (1968).

VANDEPUTTE, M., and P. DE SOMER: Transplantation of polyoma tumours in rats. Nature (Lond.) 199, 391—392 (1963).

VAN DER NORDAA, J.: Transformation of rat cells by adenovirus types 1, 2 and 3. J. gen. Virol. 3, 303—304 (1968a).

VAN DER NORDAA, J.: Transformation of rat kidney cells by adenovirus type 12. J. gen. Virol. 2, 269—272 (1968b).

VARON, S. S., C. W. RAIBORN, JR., T. SETO, and C. M. POMERAT: A cell line from trypsinized adult rabbit brain tissue. Z. Zellforsch. 59, 35—46 (1963).

VASILIEV, J. M., I. M. GELFAND, L. V. DOMNINA, and R. I. RAPPOPORT: Wound healing processes in cell cultures. Exp. Cell Res. 54, 83—93 (1969).

VAUGHAN, R. B., and J. P. TRINKAUS: Movements of epithelia cell sheets in vitro. J. Cell Sci. 1, 407—413 (1966).

VENDRELY, C., F. KASTEN, P. TOURNIER et R. WICKER: Étude cytophotométrique du contenu en acide déoxyribonucléique de cellules en culture infectées par le virus SV 40. C. R. Acad. Sci. (Paris) 255, 797—799 (1962).

VESELÝ, P., L. DONNER, and M. KUČEROVÁ: Spontaneous malignant transformation of embryonic rat fibroblasts from an inbred Lewis strain in vitro. Folia biol. (Praha) 14, 409—410 (1968).

VESELÝ, P., and J. SVOBODA: Malignant transformation of Syrian hamster embryonic cells with Rous virus of the Schmidt-Ruppin strain in vitro. Folia biol. (Praha) 11, 78—80 (1965).

Veselý, P., J. Svoboda, and L. Donner: Line of hamster cells transformed by the Schmidt-Ruppin strain of Rous virus *in vitro*. Folia biol. (Praha) **12**, 81—87 (1966).

Vigier, P.: Reliability of titration of Rous sarcoma virus by the count of the pocks produced on the egg chorioallantoic membrane. Virology **8**, 41—59 (1959).

Vigier, P.: Persistance du génome du virus de Rous dans des cellules de hamster converties *in vitro* par un virus non-défectif et un virus défectif. C. R. Acad. Sci. (Paris) **262**, 2554—2557 (1966).

Vigier, P.: Persistance du génome du virus de Rous dans des cellules du hamster converties *in vitro*, et action du virus Sendai inactivé sur la transmission aux cellules de poule. C. R. Acad. Sci. (Paris) **264**, 422—425 (1967).

Vigier, P., and A. Goldé: Growth curve of Rous sarcoma virus on chick embryo cells *in vitro*. Virology **8**, 60—79 (1959).

Vigier, P., et A. Goldé: Culture *in vitro* du virus de Friend. Bull. Cancer **49**, 374—381 (1962).

Vigier, P., et A. Goldé: Étude du système virus de Rous-cellules d'embryon de poule à l'aide d'analogues et d'antibiotiques. Bull. Cancer **51**, 73—81 (1964a).

Vigier, P., and A. Goldé: Effects of actinomycin D and of mitomycin C on the development of Rous sarcoma virus. Virology **23**, 511—519 (1964b).

Vigier, P., et J. Svoboda: Étude, en culture, de la production du virus de Rous par contact entre les cellules du sarcome XC du rat et les cellules d'embryon de poule. C. R. Acad. Sci. (Paris) **261**, 4278—4281 (1965).

Vinograd, J., J. Lebovitz, R. Radloff, R. Watson, and P. Laipis: The twisted circular form of polyoma viral DNA. Proc. nat. Acad. Sci. (Wash.) **53**, 1104—1111 (1965).

Virolainen, M., and V. Defendi: Dependence of macrophage growth *in vitro* upon interaction with other cell types. In: Growth Regulating Substances for Animal Cells in Culture. (Defendi and Stoker, eds.), Wistar Inst. Symp. Monogr. No. **7**, 67—85 (1967).

Vogt, M., and R. Dulbecco: Virus-cell interaction with a tumor-producing virus. Proc. nat. Acad. Sci. (Wash.) **46**, 365—370 (1960).

Vogt, M., and R. Dulbecco: Properties of cells transformed by polyoma virus. Cold Spr. Harb. Symp. quant. Biol. **27**, 367—373 (1962).

Vogt, M., and R. Dulbecco: Steps in the neoplastic transformation of hamster embryo cells by polyoma virus. Proc. nat. Acad. Sci. (Wash.) **49**, 171—179 (1963).

Vogt, M., R. Dulbecco, and B. Smith: Induction of cellular DNA synthesis by polyoma virus. III. Induction in productively infected cells. Proc. nat. Acad. Sci. (Wash.) **55**, 956—960 (1966).

Vogt, P. K.: The cell surface in tumor virus infection. Cancer Res. **23**, 1519—1527 (1963a).

Vogt, P. K.: The infection of chicken fibroblast cultures by avian myeloblastosis virus. In: Viruses, Nucleic Acids, and Cancer. Williams and Wilkins Comp., pp. 374—386 (1963b).

Vogt, P. K.: Fluorescence microscopic observations on the defectiveness of Rous sarcoma virus. Nat. Cancer Inst. Monogr. No. **17**, 523—535 (1964).

Vogt, P. K.: A heterogeneity of Rous sarcoma virus revealed by selectively resistant chick embryo cells. Virology **25**, 237—247 (1965).

Vogt, P. K.: Virus-directed host responses in the avian leukosis and sarcoma complex. Perspect. Virol. **5**, 199—228 (1967a).

Vogt, P. K.: Phenotypic mixing in the avian tumor virus group. Virology **32**, 708—717 (1967b).

Vogt, P. K.: A virus released by "nonproducing" Rous sarcoma cells. Proc. nat. Acad. Sci. (Wash.) **58**, 801—808 (1967c).

Vogt, P. K.: Nonproducing state of Rous sarcoma cells: Its contagiousness in chicken cell cultures. J. Virol. **1**, 729—737 (1967d).

Vogt, P. K., and R. Ishizaki: Reciprocal patterns of genetic resistance to avian tumor viruses in two lines of chickens. Virology **26**, 664—672 (1965).

VOGT, P. K., and R. ISHIZAKI: Patterns of virus interference in the avian leukosis and sarcoma complex. Virology **30**, 368—374 (1966).

VOGT, P. K., R. ISHIZAKI, and R. DUFF: Studies on the relationships among avian tumor viruses. In: Subviral Carcinogenesis, (YOHEI ITO, ed.), 1st Int. Symp. Tumor Viruses, Nagoya, pp. 297—310 (1966a).

VOGT P. K. and R. ISHIZAKI: Criteria for the classification of avian tumor viruses. In: Viruses inducing Cancer: Implications for Therapy. (W. J. BURDETTE, ed.) Univ. of Utah Press, Salt Lake City, 71-90 (1966b)

VOGT, P. K., and N. LUYKX: Observations on the surface of cells infected with Rous sarcoma virus. Virology **20**, 75—87 (1963).

VOGT, P. K., and H. RUBIN: Localization of infectious virus and viral antigen in chick fibroblasts during successive stages of infection with Rous sarcoma virus. Virology **13**, 528—544 (1961).

VOGT, P. K., and H. RUBIN: Studies on the assay and multiplication of avian myeloblastosis virus. Virology **19**, 92—104 (1963).

VOGT, P. K., P. S. SARMA, and R. J. HUEBNER: Presence of avian tumor virus group-specific antigen in nonproducing Rous sarcoma cells of the chicken. Virology **27**, 233—236 (1965).

VONKA, V., L. KUTINOVÁ, and H. ZÁVADOVÁ: Transformation of dog kidney cells infected with SV 40 and SV 40-adeno 7 hybrid virus. Acta virol. **13**, 1—7 (1969).

VONKA, V., H. ZAVADOVÁ, L. KUTONOVÁ, and D. ŘEZÁČOVÁ: Development of antibodies against viral and tumor antigens of papovavirus SV 40 in monkeys. Proc. Soc. exp. Biol. (N.Y.) **125**, 790—794 (1967).

WAHREN, B.: Immunologic aspects of virus-induced mouse leukemias. Tryckeri Balder, Stockholm (1966).

WAKEFIELD, J. D., G. J. THORBECKE, L. J. OLD, and E. A. BOYSE: Production of immunoglobulins and their subunits by human tissue culture cell lines. J. Immunol. **99**, 308—319 (1967).

WALLBANK, A. M., F. G. SPERLING, E. L. STUBBS, and K. HUBBEN: Studies of avian sarcoma and erythroblastosis (strain 13). II. Virus susceptibility to ether and chloroform. Proc. Soc. exp. Biol. (N.Y.) **110**, 809—811 (1962).

WARREN, G. H., E. C. WILLIAMS, H. E. ALBURN, and J. SEIFTER: Rous chicken sarcoma as source for hyaluronic acid. Arch. Biochem. **20**, 300—304 (1949).

WATERS, N. F., and B. R. BURMESTER: Mode of inheritance of resistance to Rous sarcoma virus in chickens. J. nat. Cancer Inst. **27**, 655—662 (1961).

WATERS, N. F., and A. K. FONTES: Genetic response of inbred lines of chickens to Rous sarcoma virus. J. nat. Cancer Inst. **25**, 351—358 (1960).

WATKINS, J. F., and R. DULBECCO: Production of SV 40 virus in heterokaryons of transformed and susceptible cells. Proc. nat. Acad. Sci. (Wash.) **58**, 1396—1403 (1967).

WEIL, R.: A quantitative assay for a subviral infective agent related to polyoma virus. Virology **14**, 46—53 (1961).

WEIL, R.: Some aspects of polyoma virus and its oncogenic action. In: Chemotherapy of Cancer Proc. int. Symp. Elsevier Publish. Co., pp. 263—277 (1964).

WEIL, R., M. R. MICHEL, and G. K. RUSCHMANN: Induction of cellular DNA synthesis by polyoma virus. Proc. nat. Acad. Sci. (Wash.) **53**, 1468—1476 (1965).

WEIL, R., G. PÉTURSSON, J. KÁRA, and H. DIGGELMANN: On the interaction of polyoma virus with the genetic apparatus of host cells. In: The Molecular Biology of Viruses (COLTER and PARANCHYCH, eds.), Academic Press, New York-London, pp. 593—626 (1967).

WEIL, R., and J. VINOGRAD: The cyclic helix and cyclic coil forms of polyoma viral DNA. Proc. nat. Acad. Sci. (Wash.) **50**, 730—738 (1963).

WEINSTEIN, D., and P. S. MOORHEAD: The relation of karyotypic change to loss of contact inhibition of division in human diploid cells after SV 40 infection. J. cell. comp. Physiol. **69**, 367—376 (1967).

WEISBERG, R. A.: Virus multiplication and cell killing in polyoma-infected mouse embryo cultures. Virology **21**, 658—661 (1963).

WEISBERG, R. A.: The morphology and transplantability of mouse embryo cells transformed *in vitro* by polyoma virus. Virology **23**, 553—564 (1964).

WEISS, L.: The cell periphery metastasis and other contact phenomena. North Holland, P. Amsterdam, 1967.

WEISS, L.: Studies on cellular adhesion in tissue culture. Electrophoretic mobility and contact phenomena. Exp. Cell Res. **51**, 609—625 (1968).

WEISS, M. C.: Further studies on loss of T-antigen from somatic hybrids between mouse cells and SV40-transformed human cells. Proc. nat. Acad. Sci. (Wash.) **66**, 79—86(1970).

WEISS, M. C., B. EPHRUSSI, and L. J. SCALETTA: Loss of T-antigen from somatic hybrids between mouse cells and SV40-transformed human cells. Proc. nat. Acad. Sci. (Wash.) **59**, 1132—1135 (1968).

WEISS, P.: *In vitro* experiments on the factors determining the course of the outgrowing nerve fiber. J. exp. Zool. **68**, 393—448 (1934).

WEISS, P.: Principles of Development. New York: Holt, Rinehart and Winston, p. 601 (1939).

WEISS, P.: Experiments on cell and axon orientation *in vitro:* the role of colloidal exudates in tissue organization. J. exp. Zool. **100**, 353—386 (1945).

WEISS, P.: Cell contact. Int. Rev. Cytol. **7**, 391—423 (1958).

WEISS, R.: Spontaneous virus production from "non-virus producing" Rous sarcoma cells. Virology **32**, 719—723 (1967).

WEISS, R.: The host range of Bryan strain Rous sarcoma virus synthesized in the absence of helper virus. J. gen. Virol. **5**, 511 (1969a).

WEISS, R.: Interference and neutralization studies with Bryan strain Rous sarcoma virus synthesized in the absence of helper virus. J. gen. Virol. **5**, 529 (1969b).

WEISS, R.: Studies on the growth and behaviour of cells infected with avian tumour viruses. Manus 1—192 (1969c).

WELLS, S. A., A. S. RABSON, R. A. MALMGREN, and A. S. KETCHAM: *In vitro* neoplastic transformation of newborn hamster salivary-gland tissue by oncogenic DNA viruses. Cancer **19**, 1411—1415 (1966).

WELLS, S. A., R. J. WURTMAN, and A. S. RABSON: Viral neoplastic transformation of hamster pineal cells *in vitro:* Retention of enzymatic function. Science **154**, 278—279 (1966).

WERCHAU, H., H. WESTPHAL, G. MAASS und R. HAAS: Untersuchungen über den Nucleinsäurestoffwechsel von Affennierengewebe-Kulturzellen nach Infektion mit SV40. II. Einfluss der Infektion auf den DNS-Stoffwechsel der Zelle. Arch. ges. Virusforsch. **19**, 351—360 (1966).

WERCHAU, E. E. KAUKEL, G. MAASS, G. BRANDNER und R. HAAS: Untersuchungen über den Nucleinsäurestoffwechsel von Affennierengewebekulturen nach Infektion mit SV40. V. Virusvermehrung unter Hemmung der zellulären DNS- Sythese. Arch. ges. Virusforsch. **25**, 115—125 (1968).

WESTERMARK, B.: Density dependent cell cycle inhibition — a universal growth control among normal animal cells. Acta path. microbiol. scand., in press (1970).

WESTPHAL, H., and R. DULBECCO: Viral DNA in polyoma- and SV40-transformed cell lines. Proc. nat. Acad. Sci. (Wash.) **59**, 1158—1165 (1968).

WESTWOOD, J. C. N., I. A. MACPHERSON, and D. H. J. TITMUSS: Transformation of normal cells in tissue culture: its significance relative to malignancy and virus vaccine production. Brit. J. exp. Path. **38**, 138—154 (1957).

WEVER, G. H., S. KIT, and D. R. DUBBS: Initial site of synthesis of virus during rescue of simian virus 40 from heterokaryons of simian virus 40-transformed and susceptible cells. J. Virol. **5**, 578—585 (1970).

WHITCUTT, J. M., and H. S. GEAR: Transformation of newborn hamster cells with simian adenovirus SA 7. Int. J. Cancer, **3**, 566—571 (1968).

WILDY, P., M. G. P. STOKER, I. A. MACPHERSON, and R. W. HORNE: The fine structure of polyoma virus. Virology **11**, 444—457 (1960).

WILLIAMS, J. F., and J. E. TILL: Transformation of rat embryo cells in culture by polyoma virus. Virology 24, 505—508 (1964).

WILLIAMS, M. G., and R. SHEININ: Cytological studies of mouse embryo cells infected with polyoma virus, using acridine orange and fluorescent antibody. Virology 13, 368—370 (1961).

WILLIS, R. A.: Pathology of Tumours. Butterwords & Co. Printed by Love and Malcomson Ltd. (1967).

WINOCOUR, E.: Purification of polyoma virus. Virology 19, 158—168 (1963).

WINOCOUR, E.: Attempts to detect an integrated polyoma genome by nucleic acid hybridization. I. "Reconstruction" experiments and complementarity tests between synthetic polyoma RNA and polyoma tumor DNA. Virology 25, 276—288 (1965).

WINOCOUR, E.: Studies on the basis for the observed homology between DNA from polyoma virus and DNA from normal mouse cells. In: The Molecular Biology of Viruses (COLTER and PARANCHYCH, eds.), Academic Press, New York—London, pp. 577—591 (1967a).

WINOCOUR, E.: On the apparent homology between DNA from polyoma virus and normal mouse synthetic RNA. Virology 31, 15—28 (1967b).

WINOCOUR, E.: Further studies on the incorporation of cell DNA into polyoma-related particles. Virology 34, 571—582 (1968).

WINOCOUR, E., A. M. KAYE, and V. STOLLAR: Synthesis and transmethylations of DNA in polyoma-infected cultures. Virology 27, 156—169 (1965).

WINOCOUR, E., and L. SACHS: Cell-virus interactions with the polyoma virus. I. Studies on the lytic interaction in the mouse embryo system. Virology 11, 699—721 (1960).

WINOCOUR, E., and L. SACHS: Cell-virus interactions with the polyoma virus. II. Studies on the nature of the interaction in tumor cells. Virology 13, 207—226 (1961).

WINOCOUR, E., and L. SACHS: Challenge infection of polyoma parotid tumor cells. Virology 16, 496—498 (1962).

WOODALL, J. P., M. C. WILLIAMS, D. I. H. SIMPSON, and A. J. HADDOW: The isolation in mice of strains of herpes virus from Burkitt tumours. Europ. J. Cancer 1, 137—140 (1965).

WRIGHT, B. S., and J. C. LASFARGUES: Long-term propagation of the Rauscher murine leukemia virus in tissue culture. J. nat. Cancer Inst. 33, 319—327 (1965).

WRIGHT, B. S., and J. C. LASFARGUES: Attenuation of the Rauscher murine leukemia virus through serial passages in tissue culture. Nat. Cancer Inst. Monogr. No. 22, 685—700 (1966).

YABE, Y., L. SAMPER, E. BRYAN, G. TAYLOR, and J. TRENTIN: Oncogenic effect of human adenovirus type 12, in mice. Science 143, 46—47 (1964).

YABE, Y., J. J. TRENTIN, and G. TAYLOR: Cancer induction in hamsters by human type 12 adenovirus. Effect of age and of virus dose. Proc. Soc. exp. Biol. (N.Y.) 111, 343—344 (1962).

YAFFE, D., and D. GERSHON: The effect of polyoma virus on differentiating multinucleated muscle fibers. Israel J. med. Sci. 3, 329—330 (1967).

YAMAGUCHI, J.: Localization of RNA, protein and lipid in Rous sarcoma virus by electron microcytochemistry. In: Electron Microscopy, Fifth Int. Congr. Electron Microscopy, Philadelphia (S. S. BREESE, JR., ed.), Academic Press, New York, Inc. 2, p. 1 (1962).

YAMAGUCHI, N., M. TAKEUCHI, and T. YAMAMOTO: Rous sarcoma virus production in mixed culture of mouse tumour cells and chicken embryo fibroblasts by the addition of UV-irradiated HVJ. Jap. J. exp. Med. 37, 83—86 (1967).

YAMANE, I., and T. KUSANO: In vitro transformation of hamster brain cells by human adenovirus type 12. Nature (Lond.) 213, 187 (only) (1967).

YAMANE, I., and T. TSUDA: Malignant transformation and chromosome change of two cell lines from Syrian hamster embryonic tissue in vitro. Tohoku J. exp. Med. 88, 171—180 (1966).

YEH, J., and H. W. FISHER: A diffusible factor which sustains contact inhibition of replication. J. Cell Biol. **40**, 382—388 (1969).

YERGANIAN, G.: Relative growth and oncogenic potential of normal, malignant and virus-transformed euploid cells of dwarf species of hamster. In: Recent Results in Cancer Research. Malignant Transformation by Viruses (W. H. KIRSTEN, ed.), Springer-Verlag, Berlin-Heidelberg-New York, pp. 112—122 (1966).

YERGANIAN, G., and M. J. LEONARD: Maintenance of normal *in situ* chromosomal features in long-term tissue cultures. Science **133**, 1600—1601 (1961).

YERGANIAN, G., M. T. LEONARD, and H. J. GAGNON: Chromosomes of the Chinese hamster *"cricetulus griseus"*. II. Onset of malignant transformation *"in vitro"* and the appearance of the X_1 chromosome. Extrait de Pathologie-Biologie, Xe Congrès International de Biologie Cellulaire **9**, 533—541 (1961).

YERGANIAN, G., H. M. SHEIN, and J. F. ENDERS: Chromosomal disturbances observed in human fetal renal cells transformed *in vitro* by simian virus 40 and carried in culture. Cytogenesis **1**, 314—324 (1962).

YOSHIDA, K., K. L. SMITH, and D. PINKEL: Studies of murine leukemia viruses. I. Detection of Rauscher and Moloney leukemia viruses by indirect immunofluorescence. Proc. Soc. exp. Biol. (N.Y.) **121**, 72—81 (1966).

YOSHIDA, M. C., and S. MAKINO: A chromosome study of non-treated and an irradiated human *in vitro* cell line. Jap. J. hum. Genet. **5**, 39—45 (1963).

YOSHIKAWA-FUKADA, M., and J. D. EBERT: Hybridization of RNA from Rous sarcoma virus with cellular and viral DNA's. Proc. nat. Acad. Sci. (Wash.) **64**, 870—877 (1969).

YOSHIIKE, K.: Studies on DNA from low-density particles of SV40. I. Heterogeneous defective virions produced by successive undiluted passages. Virology **34**, 391—401 (1968a).

YOSHIIKE, K.: Studies on DNA from low-density particles of SV40. II. Non-infectious virions associated with a large-plaque variant. Virology **34**, 402—409 (1968b).

YOSHIKURA, H., and Y. HIROKAWA: Induction of cell replication. Exp. Cell Res. **52**, 439—444 (1968).

YOSHIKURA, H., Y. HIROKAWA, Y. IKAWA, and H. SUGANO: Transformation of a mouse cell line by murine sarcoma virus (Moloney). Int. J. Cancer **3**, 743—750 (1968).

YOSHIKURA, H., Y. HIROKAWA, M. YAMADA, and H. SUGANO: Production of Friend leukemia virus in a mouse lung cell line. Jap. J. med. Sci. Biol. **20**, 225—236 (1967).

ZEIGEL, R. F.: Morphological evidence for the association of virus particles with the pancreatic acinar cells of the chick. J. nat. Cancer Inst. **26**, 1011—1039 (1961).

ZETTERBERG, A., and G. AUER: Proliferative activity and cytochemical properties of nuclear chromatin related to local cell density of epithelial cells *in vitro*. Exp. Cell Res. **62**, 262—270 (1970).

ZETTERBERG, S., and G. AUER: Early changes in the binding between DNA and histone in human leucocytes exposed to phytohemagglutinin. Exp. Cell Res. **56**, 122—126 (1968).

ZEVE, V. H., L. S. LUCAS, and R. A. MANAKER: Continuous cell culture from a patient with chronic myelogenous leukemia. II. Detection of a herpes-like virus by electron microscopy. J. nat. Cancer Inst. **37**, 761—773 (1966).

ZILBER, L. A.: Pathogenicity and oncogenicity of Rous sarcoma virus for mammals. Progr. exp. Tumor Res. **7**, 1—48 (1965).

ZILBER, L. A., B. A. LAPIN, and F. I. ADGIGHYTOV: Pathogenicity of Rous sarcoma virus for monkeys. Nature (Lond.) **205**, 1123—1124 (1965).

ZILBER, L. A., and I. N. KRYUKOVA: Haemorrhagic disease of rats due to the virus of chick sarcoma. Acta virol. **1**, 150—160 (1957a).

ZILBER, L. A., and I. N. KRYUKOVA: Haemorrhagic disease in white rats caused by Rous sarcoma virus. Vop. Virus **4**, 239—243 (1957b).

ZILBER, L. A., and I. N. KRYUKOVA: Fibromatosis in rabbits caused by Rous sarcoma virus. Vop. Virus. **3**, 166—170 (1958).

Zur Hausen, H.: Chromosomal changes of similar nature in 7 established cell lines derived from the peripheral blood of patients with leukemia. J. nat. Cancer Inst. **38**, 683—696 (1967).

Zur Hausen, H.: Association of adenovirus type 12 deoxyribonucleic acid with host cell chromosomes. J. Virol. **2**, 218—223 (1968a).

Zur Hausen, H.: Chromosomal aberrations and cloning efficiency in adenovirus type 12-infected hamster cells. J. Virol. **2**, 915—917 (1968b).

Zur Hausen, H.: Persistence of the viral genome in adenovirus type 12-infected hamster cells. J. Virol. **2**, 918—924 (1968c).

Zur Hausen, H., and F. Sokol: Fate of adenovirus type 12 genomes in non-permissive cells. J. Virol. **4**, 256—263 (1969).

Zylberbaum, S., et H. Febvre: Hétérogreffe de cellules humaines diploides transformées *in vitro* par le virus du sarcome de Rous du poulet (souche de Bryan). C. R. Acad. Sci. (Paris) **261**, 5735—5738 (1965).

VIROLOGY
MONOGRAPHS
DIE VIRUSFORSCHUNG IN EINZELDARSTELLUNGEN

SPRINGER-VERLAG
WIEN · NEW YORK